Lecture Notes in Physics

Springer
Berlin
Heidelberg
New York
Barcelona
Hong Kong
London
Milan
Paris
Tokyo

The Editorial Policy for Edited Volumes

The series *Lecture Notes in Physics* (LNP), founded in 1969, reports new developments in physics research and teaching - quickly, informally but with a high degree of quality. Manuscripts to be considered for publication are topical volumes consisting of a limited number of contributions, carefully edited and closely related to each other. Each contribution should contain at least partly original and previously unpublished material, be written in a clear, pedagogical style and aimed at a broader readership, especially graduate students and nonspecialist researchers wishing to familiarize themselves with the topic concerned. For this reason, traditional proceedings cannot be considered for this series though volumes to appear in this series are often based on material presented at conferences, workshops and schools.

Acceptance

A project can only be accepted tentatively for publication, by both the editorial board and the publisher, following thorough examination of the material submitted. The book proposal sent to the publisher should consist at least of a preliminary table of contents outlining the structure of the book together with abstracts of all contributions to be included. Final acceptance is issued by the series editor in charge, in consultation with the publisher, only after receiving the complete manuscript. Final acceptance, possibly requiring minor corrections, usually follows the tentative acceptance unless the final manuscript differs significantly from expectations (project outline). In particular, the series editors are entitled to reject individual contributions if they do not meet the high quality standards of this series. The final manuscript must be ready to print, and should include both an informative introduction and a sufficiently detailed subject index.

Contractual Aspects

Publication in LNP is free of charge. There is no formal contract, no royalties are paid, and no bulk orders are required, although special discounts are offered in this case. The volume editors receive jointly 30 free copies for their personal use and are entitled, as are the contributing authors, to purchase Springer books at a reduced rate. The publisher secures the copyright for each volume. As a rule, no reprints of individual contributions can be supplied.

Manuscript Submission

The manuscript in its final and approved version must be submitted in ready to print form. The corresponding electronic source files are also required for the production process, in particular the online version. Technical assistance in compiling the final manuscript can be provided by the publisher's production editor(s), especially with regard to the publisher's own LaTeX macro package which has been specially designed for this series.

LNP Homepage (http://www.springerlink.com/series/lnp/)

On the LNP homepage you will find:
—The LNP online archive. It contains the full texts (PDF) of all volumes published since 2000. Abstracts, table of contents and prefaces are accessible free of charge to everyone. Information about the availability of printed volumes can be obtained.
—The subscription information. The online archive is free of charge to all subscribers of the printed volumes.
—The editorial contacts, with respect to both scientific and technical matters.
—The author's / editor's instructions.

H. M. Antia   A. Bhatnagar   P. Ulmschneider   (Eds.)

Lectures on Solar Physics

Editors

H. M. Antia
Tata Institute of Fundamental Research
Homi Bhabha Road
Mumbai 400005, India

Arvind Bhatnagar
Udaipur Solar Observatory
Physical Research Laboratory
P. O. Box No. 198
Udaipur 313001, India

Peter Ulmschneider
Institut für Theoretische Astrophysik
Univ. Heidelberg
Tiergartenstr. 15
69121 Heidelberg, Germany

Cataloging-in-Publication Data applied for

A catalog record for this book is available from the Library of Congress.

Bibliographic information published by Die Deutsche Bibliothek

Die Deutsche Bibliothek lists this publication in the Deutsche Nationalbibliografie; detailed bibliographic data is available in the Internet at http://dnb.ddb.de

ISSN 0075-8450
ISBN 3-540-01528-0 Springer-Verlag Berlin Heidelberg New York

Springer-Verlag Berlin Heidelberg New York
a member of BertelsmannSpringer Science+Business Media GmbH

http://www.springer.de

Printed in Germany

Typesetting: Camera-ready by the authors/editor
Camera-data conversion by Steingraeber Satztechnik GmbH Heidelberg
Cover design: *design & production*, Heidelberg

Printed on acid-free paper
54/3141/du - 5 4 3 2 1 0

Preface

The idea for these lecture series arose at a Workshop on solar physics which was held at the Inter University Centre for Astronomy and Astrophysics (IUCAA), Pune/India in December 2000. This Workshop aimed to present a comprehensive and up-to-date overview of solar physics for interested students and faculty in other branches of astrophysics. It was intended to show that this field, concentrating on our closest star, is a vital and exciting field of research. For this purpose a number of comprehensive reviews were organised which assumed that the audience would have only a basic physics background but had no prior knowledge about solar physics. The set of lectures covered topics ranging from the solar core to the convection zone, the photosphere, chromosphere, and corona and extending to the solar wind in the interplanetary medium

During and after the Workshop there was much enthusiasm for this form of presentation and it was felt that these lectures, augmented by including the latest research findings in the field, would be beneficial to a much larger audience. Thus the plan for this book originated which could then be realised thanks to the publishers, Springer-Verlag.

There are 9 articles based on the lectures given at the Workshop. The article by Chitre on "Overview of Solar Physics" gives an introduction to the whole variety of phenomena of solar physics, the problems and their solutions and salient results. The article on "Instrumentation and Observational techniques related to Solar Physics" by Bhatnagar describes in detail the principles of solar instrumentation normally used to take simple white light, monochromatic and spectroscopic observations. Practical methods to measure important basic parameters, like area, position and the classification of sunspots are described in detail. Antia's article on "Solar Interior and Seismology" describes the solar interior, the technique of helioseismology and how this new technique allows a determination of the internal structure and dynamics of the Sun and constrains theories of stellar structure, evolution and angular momentum transport. Ambastha's article on "The Active and Explosive Sun" gives an overview of highly time-dependent phenomena in the photosphere, chromosphere and corona of the Sun and provides some theoretical models of the solar flares. Hasan's article on "Magnetic Flux Tubes and Activity on the Sun" discusses the generation, storage and emergence of magnetic fields in the form of small-scale flux tubes and examines their role in heating of the chromosphere. Ventakrishan's article on "Solar Magnetic Fields" gives a theoretical overview of the generation of magnetic fields by the dynamo mechanism, the general magnetic field topology and how the magnetic fields are measured. Ulmschneider's contribution on "The Physics of Chromospheres and Coronae" discusses why all stars like the Sun have hot outer chromospheric and coronal layers. It identifies the heating mechanisms and dynamical processes which take place both in the presence and absence of magnetic fields. The article by Dwivedi on "The Solar Corona" gives a general overview of the solar corona, how it is observed and what the physical processes leading to its formation are. Finally Manoharan's contribution on "The Solar Wind" describes the generation and measurement of the solar wind derived from in situ observations by spacecraft and interplanetary scintillation studies.

We hope that by reading these lectures, interested people, amateurs, graduate and postgraduate students will be motivated to take up solar physics as an area of research, and share our excitement about the wonders of our nearest star – the Sun.

We are thankful to T. Padmanabhan and the Inter University Centre for Astronomy and Astrophysics, Pune for organising and hosting this Workshop on Solar Physics.

Mumbai, Udaipur, Heidelberg
February 2003

H. M. Antia
A. Bhatnagar
P. Ulmschneider

Table of Contents

List of Contributors

Ashok Ambastha
Udaipur Solar Observatory
Physical Research Laboratory
P. O. Box No. 198
Udaipur 313001, India
ambastha@prl.ernet.in

H.M. Antia
Tata Institute
of Fundamental Research
Homi Bhabha Road
Mumbai 400005, India
antia@tifr.res.in

Arvind Bhatnagar
Udaipur Solar Observatory
Physical Research Laboratory
P. O. Box No. 198
Udaipur 313001, India
arvind@prl.ernet.in

S.M. Chitre
Department of Physics
University of Mumbai
Mumbai 400098, India
kumarchitre@hotmail.com

Bhola N. Dwivedi
Department of Applied Physics
Institute of Technology
Banaras Hindu University
Varanasi 221005, India
dwivedi@banaras.ernet.in

S.S. Hasan
Indian Institute of Astrophysics
Bangalore 560034, India
hasan@iiap.ernet.in

P.K. Manoharan
Radio Astronomy Centre
Tata Institute
of Fundamental Research
P. O. Box 8
Udhagamandalam (Ooty) 643001, India
mano@racooty.ernet.in

P. Ulmschneider
Institut für Theoretische Astrophysik
Univ. Heidelberg
Tiergartenstr. 15
69121 Heidelberg, Germany
ulm@ita.uni-heidelberg.de

P. Venkatakrishnan
Udaipur Solar Observatory
Physical Research Laboratory
P. O. Box No. 198
Udaipur 313001, India
pvk@prl.ernet.in

Overview of Solar Physics

S.M. Chitre

Department of Physics, University of Mumbai, Mumbai 400098, India
email: kumarchitre@hotmail.com

Abstract. The Sun has been aptly described as the Rosetta Stone of astronomy. Even though the interior of the Sun is not directly accessible to observations, it has been possible to unravel its internal structure with the help of equations governing the mechanical and thermal equilibrium along with the boundary conditions provided by observations of its mass, radius, luminosity and surface chemical composition. The external layers of the Sun display astonishingly rich dynamics with a host of energetic phenomena occurring at the surface and in the outer atmosphere. An interaction between solar differential rotation, turbulent convection and magnetic field seems to provide an effective mechanism that maintains the solar dynamo and drives the cyclic activity seen at the surface in the form of sunspots. The magnetic field appears to be the guiding force that can effectively supply the energy required for heating the chromospheric and coronal regions. The Sun thus turns out to be an ideal cosmic laboratory for testing atomic and nuclear physics, high-temperature plasma physics and magnetohydrodynamics, neutrino physics and general relativity.

1 Introduction

The Sun has played a major role in the development of physics and mathematics for the past several centuries. Thus, Kepler's laws provided the framework for describing motions of planets under the influence of the Sun's gravitational field. The Newtonian theory of gravitation explained the planetary and lunar motions with a remarkable degree of precision. The Newtonian theory has, in fact, successfully expounded the mechanics of planetary motions and the precession of their elliptical orbits. The measurements were refined by the end of the nineteenth century to the extent that the unaccounted precession of the orbit of planet Mercury was observed to be close to 43 seconds of arc per century. The excellent agreement between the prediction of general theory of relativity and the observed precession of the perihelion of Mercury was a great triumph for Einstein's geometrised formulation of gravitation. Another prediction of general relativity, namely, the gravitational deflection of light rays from a background star grazing the solar limb was measured during the total solar eclipse expedition of 1919, and found to be approximately the same as the predicted value of 1.75 arc seconds (precisely twice the Newtonian value). A longer transit time for radio waves propagating close to the solar body, across its deep gravitational potential well was also verified. It is clear our Sun has played a vital role in verification of general relativity (e.g., Weinberg 1972).

The Sun has been widely regarded as the Rosetta Stone of astronomy. This is a very apt description since our star has provided a readymade laboratory for studying a variety of processes and phenomena operating both within and outside this object. Solar studies have also served as a valuable guide for the development of the theory of structure and evolution of stars in general and pulsating stars in particular. The proximity of our Sun has enabled a close enough scrutiny of its atmospheric layers and provided data of high spatial resolution of its surface features which is clearly not possible for other stars.

More than a century ago all that was known about the Sun was from the study of its face and the visible layers. Indeed, the early astronomers had noticed that the solar disk is dotted with dark blotches. These sunspots were, in fact, known to the Chinese and Greek astronomers, but it was Galileo who first made scientific observations of the march of these dark spots across the solar disk. The appearance of sunspots first in mid-latitudes ($\sim 30^\circ$) and then their migration towards the equator following a cycle with a period of approximately 11 years, has been systematically observed and encapsulated in the well-known "butterfly diagram" due to Maunder (e.g., Ambastha, Venkatakrishnan, this volume). Astronomers keep track of the spots appearing and disappearing on the visible disk of the Sun, hoping to gain insight into the processes that drive the solar cycle as well as to link solar activity with terrestrial climatic changes. Observational techniques and instruments used for solar observations are described by Bhatnagar (this volume).

There is a well-defined hierarchy of magnetic elements at the solar surface: magnetic flux tubes or fibrils, faculae, pores, plages and sunspots. Sunspots were the first significant markings observed on the face of the Sun, in the vicinity of which were also noticed the bright, irregular patches called faculae. Later observations made in chromospheric spectral lines revealed the presence of bright areas known as plages overlying the regions of enhanced magnetic fields. There are also widely separated concentrations of magnetic elements or fibrils with field strength of 1000–2000 G, over scales of the general order of 100 km.

The solar atmosphere displays a rich variety of features and complex phenomena which can be witnessed in their awesome splendour during the occurrence of a total solar eclipse. The chromosphere appears fleetingly just before and after totality as a fiery red ring around the disk and lingers for several seconds before disappearing. At totality the pearly white solar corona comes into view which changes its shape synchronously with the activity cycle, forming a jagged ring around the Sun at the peak of the activity cycle and transforming into trailing plumes and streamers by the end of the cycle. The corona is an extremely hot, tenuous and inhomogeneous region of the solar atmosphere consisting of complex loop structures with radiation emitted mainly in the UV and X-ray wavelengths (e.g., Ulmschneider, Dwivedi this volume).

Contrary to thermodynamic expectations, the outer atmosphere of the Sun is hotter than the visible photospheric layers from which much of the solar radiation is emitted. The temperature at the surface of the Sun where the particle density is about $10^{17}\,\mathrm{cm}^{-3}$ is approximately 5700 K, which decreases to a value of 4200 K at about 500 km above the photosphere and then rises up to a value of several tens of thousand degrees in the chromospheric layers made up of the network and active regions, at heights of around a few thousand km above the visible surface. The overlying coronal regions have temperatures approaching $(1\text{–}2)\times10^6$ K and are composed mainly of protons and electrons with number densities typically of order $10^8\,\mathrm{cm}^{-3}$ with an admixture of small amounts of heavier ions. Both the chromosphere and corona are observed to be highly structured and show clear evidence of association with the solar magnetic fields (e.g., Narain & Ulmschneider 1996). High resolution images from the Transition Region and Coronal Explorer (TRACE) spacecraft show that a large part of the corona has a fine-structure down to sub-arcsecond scale.

In the interior, it is the solar material that controls the magnetic field lines, while outside the solar body it is the magnetic field that dictates the behaviour of the plasma causing a variety of dynamic and transient phenomena. In fact, the magnetic field serves as an effective agent and provides a conduit to transport energy of the sub-photospheric motions and waves to the chromospheric/coronal regions and at other times acts as a detonator displaying spectacular events in the form of flares. Prominences of various kinds (e.g., quiescent, loop, hedgerow, eruptive, etc.) are seen to rise above active regions on the solar surface, providing striking evidence for the presence of magnetic fields in the outer atmosphere capturing and controlling the motion of the plasma along the field lines. It is evident that if the Sun were not to possess any magnetism, its external layers would not have presented such a spectacularly dazzling and explosive picture.

2 Composition and Structure of the Sun

More than a century ago all that was known about the Sun was by studying its visible layers and its surface markings. The early investigations in solar physics were largely devoted to an extensive collection of spectroscopic data for studying the surface temperature, density and chemical composition. Spectroscopy of the photospheric layers showed a spectrum dominated by the lines of elements such as carbon, silicon, sodium, iron, magnesium etc. Helium, even though relatively inconspicuous in the solar spectrum, was first discovered on the Sun before it was known in the laboratory. It was the spectroscopy of the chromosphere (with its somewhat higher temperature than that at the photosphere) which established, during a total solar eclipse, that hydrogen is the most abundant element in the Sun with helium being about one in ten atoms and heavier elements being present at the level of approximately 0.1 percent.

With a handle on the surface chemical composition, the attention of solar physicists turned to working out the internal structure of the Sun. For several centuries astronomers believed that the interior of the Sun and stars, shielded by the material beneath the visible surface, will never be accessible. This prompted the nineteenth century French philosopher, Auguste Comte to proclaim: "We can never learn their internal constitution". It is, therefore, a triumph of the theory of stellar structure that one has been able to construct a reasonable picture of the Sun's inside with the help of a set of mathematical equations governing its mechanical as well as thermal equilibrium and the nuclear energy generation, together with the boundary conditions supplied by observations. The earlier analytical efforts were mainly concentrated on the study of polytropic models for inferring the physical conditions inside the Sun. With the advent of high-speed computers, the structure equations were numerically integrated with the auxiliary input of physics, supplemented by appropriate boundary conditions. For this purpose, the Standard Solar Model (SSM) based on a minimum number of assumptions and physical processes was developed (e.g., Christensen-Dalsgaard et al. 1996; Bahcall, Pinsonneault & Basu 2001). In the SSM the Sun is taken to be a spherically symmetric object with negligible effects of rotation, magnetic fields, mass loss and tidal forces on its global structure. It is supposed to be in a quasi-stationary state maintaining hydrostatic and thermal equilibrium. The energy generation takes place in the central regions by thermonuclear reactions which convert hydrogen into helium mainly, by the proton-proton chain. The energy is transported outward from the core principally by radiative processes, but in the outer third of the solar radius it is carried largely by convection. There is supposed to be no mixing of nuclear reaction products outside the convection zone, except for the slow gravitational settling of helium and heavy elements by diffusion beneath the convection zone into the radiative interior. There is no energy transport by wave motion and the standard nuclear and neutrino physics is adopted for constructing theoretical solar models to obtain the present luminosity and radius by adjusting the initial helium abundance and the mixing-length parameter which controls the convective energy transport.

2.1 Equations of Stellar Structure

The central problem of solar structure is to determine the march of thermodynamic quantities with depth with the help of equations governing mechanical and thermal equilibrium. The mechanical equilibrium ensures that the pressure gradient balances the gravitational forces (e.g., Cox & Giuli 1968) and may be expressed as

$$\frac{\mathrm{d}P(r)}{\mathrm{d}r} = -\frac{Gm(r)}{r^2}\rho(r) \,, \tag{1}$$

$$\frac{\mathrm{d}m(r)}{\mathrm{d}r} = 4\pi r^2 \rho(r) \,. \tag{2}$$

Here $P(r)$ is the pressure, $\rho(r)$, the density and $m(r)$, the mass interior to the radius r, for a spherically symmetric Sun.

For maintaining thermal equilibrium, the energy radiated by the Sun, as measured by its luminosity, must be balanced by the nuclear energy generated throughout the solar interior,

$$\frac{\mathrm{d}L(r)}{\mathrm{d}r} = 4\pi r^2 \rho(r)\epsilon \,, \tag{3}$$

where ϵ is the energy generation rate per unit mass and $L(r) = 4\pi r^2(F_{\mathrm{rad}}+F_{\mathrm{conv}})$ is the luminosity. F_{rad} and F_{conv} are respectively, the radiative and convective energy flux (energy per cm^2 per s). The energy generation takes place in the central regions by thermonuclear reactions converting hydrogen into helium mainly by the proton-proton chain outlined in Table 1, where the numbers in the parentheses represent the energy of the neutrinos.

Table 1. pp Chain

$$
\begin{array}{lrl}
 & p + p \rightarrow d + e^{+} + \nu_e & (\leq 0.42\,\mathrm{MeV}) \\
 & p + e^{-} + p \rightarrow d + \nu_e & (1.44\,\mathrm{MeV}) \\
 & p + d \rightarrow {}^{3}\mathrm{He} + \gamma & \\
\text{pp-I:} & {}^{3}\mathrm{He} +{}^{3}\mathrm{He} \rightarrow {}^{4}\mathrm{He} + 2p & \\
 & {}^{3}\mathrm{He} + p \rightarrow {}^{4}\mathrm{He} + e^{+} + \nu_e & (\leq 18.8\,\mathrm{MeV}) \\
\text{pp-II:} & {}^{3}\mathrm{He} +{}^{4}\mathrm{He} \rightarrow {}^{7}\mathrm{Be} + \gamma & \\
 & {}^{7}\mathrm{Be} + e^{-} \rightarrow {}^{7}\mathrm{Li} + \nu_e & (0.38,\ 0.86\,\mathrm{MeV}) \\
 & {}^{7}\mathrm{Li} + p \rightarrow {}^{8}\mathrm{Be} + \gamma & \\
 & {}^{8}\mathrm{Be} \rightarrow 2\,{}^{4}\mathrm{He} & \\
\text{pp-III:} & {}^{3}\mathrm{He} +{}^{4}\mathrm{He} \rightarrow {}^{7}\mathrm{Be} + \gamma & \\
 & {}^{7}\mathrm{Be} + p \rightarrow {}^{8}\mathrm{B} + \gamma & \\
 & {}^{8}\mathrm{B} \rightarrow {}^{8}\mathrm{Be} + e^{+} + \nu_e & (\leq 14.6\,\mathrm{MeV}) \\
 & {}^{8}\mathrm{Be} \rightarrow 2\,{}^{4}\mathrm{He} &
\end{array}
$$

The Sun derives more than 98% of its energy from the proton-proton chain; there is an additional contribution of less than 2% from the CNO cycle reactions outlined in Table 2:

Table 2. CNO Cycle

$$
\begin{array}{lll}
{}^{12}\mathrm{C} + p & \rightarrow {}^{13}\mathrm{N} + \gamma & \\
{}^{13}\mathrm{N} & \rightarrow {}^{13}\mathrm{C} + e^{+} + \nu_e & (\leq 1.2\,\mathrm{MeV}) \\
{}^{13}\mathrm{C} + p & \rightarrow {}^{14}\mathrm{N} + \gamma & \\
{}^{14}\mathrm{N} + p & \rightarrow {}^{15}\mathrm{O} + \gamma & \\
{}^{15}\mathrm{O} & \rightarrow {}^{15}\mathrm{N} + e^{+} + \nu_e & (\leq 1.7\,\mathrm{MeV}) \\
{}^{15}\mathrm{N} + p & \rightarrow {}^{12}\mathrm{C} + {}^{4}\mathrm{He} & \\
 & \text{or} & \\
{}^{15}\mathrm{N} + p & \rightarrow {}^{16}\mathrm{O} + \gamma & \\
{}^{16}\mathrm{O} + p & \rightarrow {}^{17}\mathrm{F} + \gamma & \\
{}^{17}\mathrm{F} & \rightarrow {}^{17}\mathrm{O} + e^{+} + \nu_e & (\leq 1.7\,\mathrm{MeV}) \\
{}^{17}\mathrm{O} + p & \rightarrow {}^{14}\mathrm{N} + {}^{4}\mathrm{He} &
\end{array}
$$

The energy generated by these reaction networks is transported from the centre to the surface of the Sun from where it is radiated into the outside space. In about two-thirds of the solar interior the energy flux is carried by radiative processes and the radiative flux, $F_{\rm rad}$ is related to the temperature gradient by,

$$F_{\rm rad} = -\frac{4acT^3}{3\kappa\rho}\frac{{\rm d}T}{{\rm d}r} . \tag{4}$$

Here a is the Stefan-Boltzmann constant, c the speed of light and κ the opacity of solar material caused by a host of atomic processes involving many elements and several stages of ionisation (e.g., Rogers & Iglesias 1992; Iglesias & Rogers 1996). In the zone extending approximately one third of the solar radius below the surface, the radiative temperature gradient steepens because of the sharp rise in opacity while the adiabatic gradient drops in the ionisation zones. In such a situation the Schwarzschild instability criterion is readily satisfied and the energy flux is carried largely by convection and modelled in the framework of a local mixing-length formulation (Böhm-Vitense 1958) expressed as

$$F_{\rm conv} = -\kappa_t \rho T\frac{{\rm d}S}{{\rm d}r} . \tag{5}$$

Here κ_t is the turbulent diffusivity given by $\kappa_t \propto wl$, w being the mean vertical velocity, l the local mixing-length ($= \alpha H_P$, where H_P is the local pressure scale-height), S the entropy and α is a parameter of order unity. The mean convective velocity is given by

$$w = \left(\beta\frac{g}{H_P}Ql^2(\nabla - \nabla_{\rm ad})\right)^{1/2} . \tag{6}$$

In this expression β is supposed to represent the effect of viscous breaking of the convective elements and the factor $Q = -\frac{T}{\rho}\left(\frac{\partial\rho}{\partial T}\right)_P$ takes into account variation of the degree of ionisation in the moving elements. A value of $\alpha \approx 2$ seems to be indicated by time-dependent hydrodynamical simulation of stellar convection (Steffen 1992; Trampedach et al. 1997) as well as by a careful fitting of evolutionary tracks of the Sun with its present luminosity, radius and age (Schröder & Eggleton 1996; Hünsch & Schröder 1996).

An additional requirement is the knowledge of the thermodynamic state of matter throughout the solar body. For most parts except for the outermost layers, the material inside the Sun is essentially completely ionised and the perfect gas law is an adequate description of the equation of state which expresses the gas pressure as

$$P_g = \frac{k_B}{m_H\mu}\rho T , \tag{7}$$

where k_B is the Boltzman constant, m_H the mass of hydrogen atom and μ the mean molecular weight which is given by (Schwarzschild 1958)

$$\mu = \frac{1}{2X + \frac{3}{4}Y + \frac{1}{2}Z} . \tag{8}$$

Here X, Y, Z refer to the fractional abundance by mass of hydrogen, helium and heavy elements respectively. The perfect gas law description of the state of matter is clearly an idealisation. There are corrections, of course, amounting to a few per cent, to this ideal gas law arising from effects due to electron degeneracy, plasma screening, pressure ionisation and Coulomb free energy between charged particles (Eggleton, Faulkner & Flannery 1973; Mihalas, Däppen & Hummer 1988; Christensen-Dalsgaard & Däppen 1992; Rogers, Swenson & Iglesias 1996). In the sub-surface layers of the Sun, both hydrogen and helium undergo various stages of ionisation until temperatures upwards of 2×10^5 K are reached. The partial ionisation leads to a local decrease both in the adiabatic index, $\Gamma_1 = (\partial \ln P / \partial \ln \rho)_S$ and the logarithmic adiabatic temperature gradient, $\nabla_{\mathrm{ad}} \equiv (\partial \ln T / \partial \ln P)_S$. Note that Γ_1 dips to a value of 1.21 in the ionisation zone of hydrogen and singly ionised helium and to a value of about 1.58 in the second helium ionisation zone, thus showing departures from the ideal gas value of 5/3. Moreover, the superadiabatic gradient $(\nabla - \nabla_{\mathrm{ad}})$ (where $\nabla = \frac{\mathrm{d} \ln T}{\mathrm{d} \ln P}$ is the dimensionless temperatute gradient) has a pronounced peak in the ionisation zone near the surface.

Assuming all atoms to be in the ground state, the fraction of atoms ionised in the solar interior may be determined by using Saha's ionisation equation which relates the number densities of electrons, n_e and the number densities, n_i and n_{ii} of atoms in two successive stages of ionisation by the relation:

$$\frac{n_e n_{ii}}{n_i} = 2 \frac{u_{ii}}{u_i} \frac{(2\pi m_e k_B T)^{3/2}}{h^3} \mathrm{e}^{-I/k_B T} \,. \tag{9}$$

Here, u_i and u_{ii} are the partition functions of the two ionisation levels, m_e the electron mass, h the Planck constant and I is the ionisation potential of state i. This equation can be written for each stage of ionisation and all these equations can be solved to get the fractional abundance in each ionisation stage as well as the number density of electrons which are contributed by these ions.

In the standard solar model there is supposed to be no mixing of material outside the convection zone. But because of the momentum exchange between heavier and lighter elements, there is a slow gravitational diffusion of helium and heavy elements relative to hydrogen beneath the base of the convection zone into the radiative interior (e.g., Guzik & Cox 1993). In addition, the presence of a temperature gradient can cause thermal diffusion and so also can the radiation pressure acting on partially ionised or neutral atoms. It turns out for the solar conditions, the gravitational settling of helium and heavy elements relative to hydrogen is a more important process.

2.2 The Standard Solar Model

The structure equations supplemented by auxiliary input physics describing the thermodynamic state of the matter, the opacity and the nuclear energy generation rate are then numerically integrated to construct theoretical solar models which satisfy constraints, namely, the observed mass, radius, luminosity and ratio of chemical abundances by mass, Z/X. The resultant model profiles of temperature T, density ρ, pressure P, sound speed, adiabatic index Γ_1, hydrogen abundance X and helium abundance Y profiles through the solar interior are displayed in Fig. 1. The interior model is matched to the atmospheric model of Vernazza, Avrett & Loeser (1981) at the photosphere, above which the model profiles are calculated using this atmospheric model. The pressure decreases monotonically with increasing radius and the scale height of its variation becomes progressively small closer to the surface giving rise to a steep fall as we approach the surface. As we had noted earlier, the temperature reaches a minimum at about 500 km above the surface and then starts increasing towards the chromospheric and coronal regions. The density falls off monotonically with radial distance except in a very thin region just below the surface, where it increases as a result of a very strong superadiabatic temperature gradient prevailing in a narrow region. The occurrence of this feature which is referred to as the density inversion, depends on the treatment of convection and may be absent in some solar models. The sound speed also has a minimum at the temperature minimum and starts increasing as we move up to the chromosphere. The second dip in sound speed profile beyond the temperature minimum is due to steep fall in Γ_1 due to ionisation. The adiabatic index, Γ_1 has a value close to 5/3 when the solar material is either fully ionised as in the interior, or when there is no ionisation as in the region just above the surface.

It is customary in the theory of solar structure to assume the Sun has a homogeneous initial chemical composition, say, $X = 73\%$, $Y = 25\%$ with a small admixture of heavy elements, $Z = 2\%$, and its total mass, $M_\odot = 1.989 \times 10^{33}$ gm. The Sun is then evolved with a few adjustable parameters, to yield the present luminosity, $L_\odot = 3.846 \times 10^{33}$ erg s^{-1}, a radius $R_\odot = 6.9599 \times 10^{10}$ cm and a composition ratio $Z/X = 0.0245$ at the surface (Grevesse, Noels & Sauval 1996) after 4.6 billion years which is the estimated age of the Sun inferred from meteoritic data; for example, the Allende meteorite is dated to be 4.566 billion years old (Allégre, Manhès & Göpel 1995). These boundary conditions are generally satisfied by varying the initial composition and a parameter in the mixing-length formulation to calculate the convective flux in the convection zone. Thus, effectively there are no free parameters in the SSM, as all the unknown parameters are adjusted to satisfy the boundary conditions. Nevertheless, by varying input physics, like the opacities, the equation of state, the nuclear reaction rates or the diffusion coefficient it is possible to get different solar models. Further, it should be noted that the solar mass is not directly measured, but rather it is the product $GM_\odot$ which is accurately known from the study of planetary orbits. The solar mass is then determined from the knowledge of G, which is not known to very high accuracy. Thus the values of G and $M_\odot$ should be chosen to yield the correct value for the product $GM_\odot$.

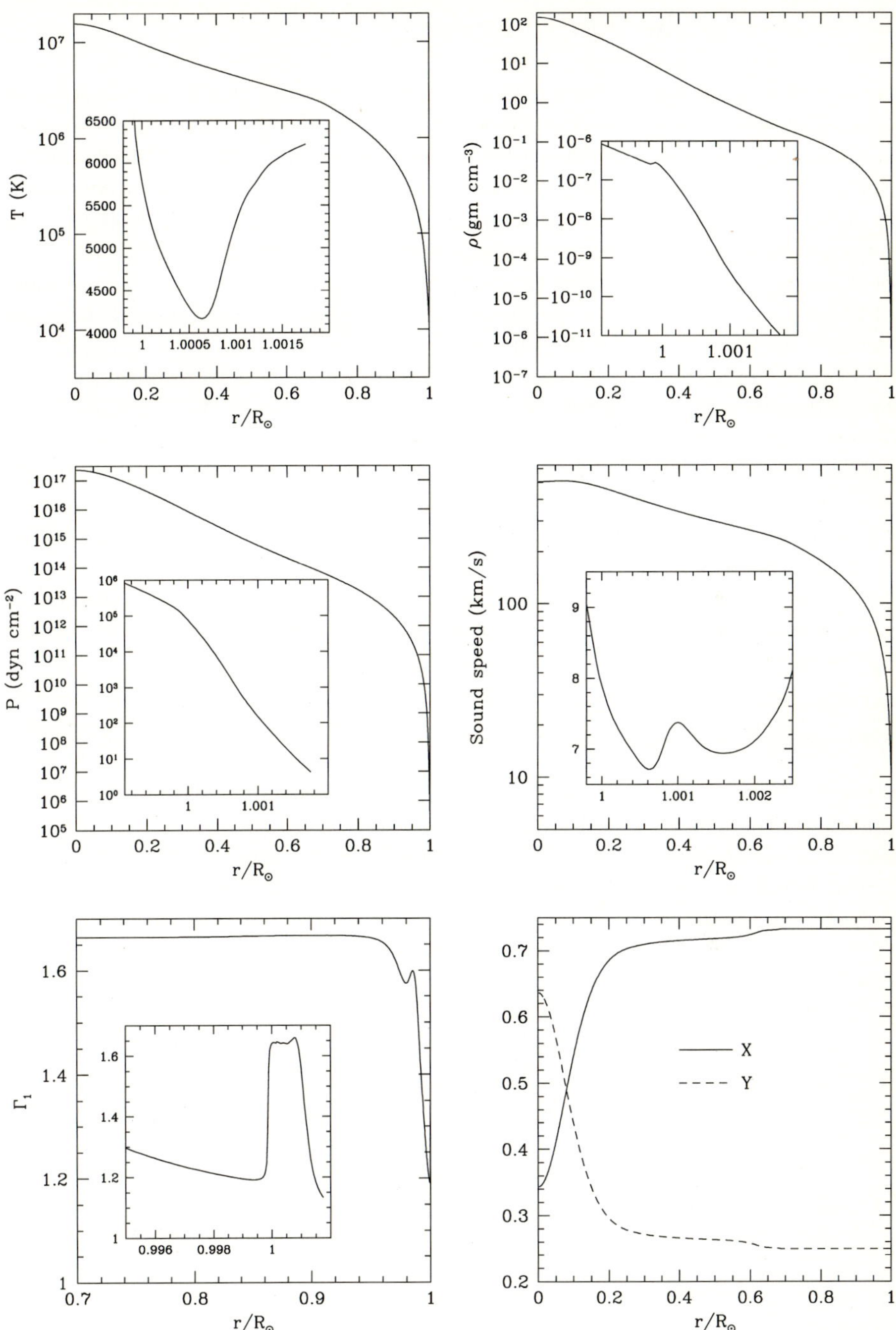

Fig. 1. The temperature, density, pressure, sound speed, adiabatic index, hydrogen and helium abundance profiles as a function of radial distance inside the Sun in a standard solar model of Brun, Turck-Chièze & Zahn (1999). The inset shows a blowup of the region close to the surface

3 Probes of the Sun's Interior

It turns out from the theoretical calculations that there is a large variation of temperature from about 5700 K at the surface to upwards of 15 million degrees at the centre; likewise, the density varies from about 10^{-7} gm cm^{-3} to some 150 gm cm^{-3} between the surface and the core of the Sun. There is also a very steep variation of density and temperature through the overlying atmosphere. The principal questions concerning the structure of the Sun are: Is there any way of checking the correctness of these theoretical models? Are there any means of measuring the central temperature and finding out if the chemical make-up inside is the same as that at the surface? "What appliance can pierce through the outer layers of a star and test the conditions within?", asked Eddington (1926), in his classic book, The Internal Constitution of the Stars. As it happens, the Sun is, indeed, transparent to neutrinos released in the nuclear reaction network operating in the energy-generating core and also to waves generated through bulk of the solar body. These valuable probes complement each other and enable us "to see" inside the Sun. The deduced thermal and chemical composition profiles as well as rotation and magnetic fields prevailing in the solar interior can then be related to the phenomena occurring in the solar atmosphere. The internal and external layers of the Sun, it turns out, furnish an ideal cosmic laboratory for testing various branches of physics.

3.1 Solar Neutrino Problem

Historically, the measurement of neutrinos produced in the reaction network operating in the solar core was the first probe conceived to surmise the physical conditions in the deep interior. The neutrino fluxes are sensitive to the temperature and composition profiles in the central regions of the Sun. It was, therefore, hoped that the steep temperature dependence of some of the nuclear reaction rates involved in the production of neutrinos would enable a determination of the Sun's central temperature to better than a few per cent. "The use of a radically different observational probe may reveal wholly unexpected phenomena; perhaps, there is some great surprise in store for us when the first experiment in neutrino astronomy is completed", said Bahcall in 1967. There have been valiant efforts undertaken since the 1960s to set up experiments designed for the exceedingly difficult measurement of neutrinos from the Sun. Ray Davis's Chlorine experiment (Davis 1964) has been operating for well over 35 years and is sensitive to intermediate and high energy neutrinos released in the thermonuclear network. It has a tank containing 615 tons of liquid perchloroethylene, located some 1480 m underground in the Homestake gold mine in South Dakota, in which the Chlorine nuclei are the solar neutrino absorbers according to the reaction

$$^{37}\mathrm{Cl} + \nu_\odot \rightarrow {}^{37}\mathrm{Ar} + e^- \qquad (\text{threshold} = 0.814\,\mathrm{MeV}) \,. \tag{10}$$

The capture rate is dominated by the ^{8}B neutrinos contributing 5.9 SNU, with the ^{7}Be neutrinos making a contribution of 1.1 SNU. The sole motivation of the

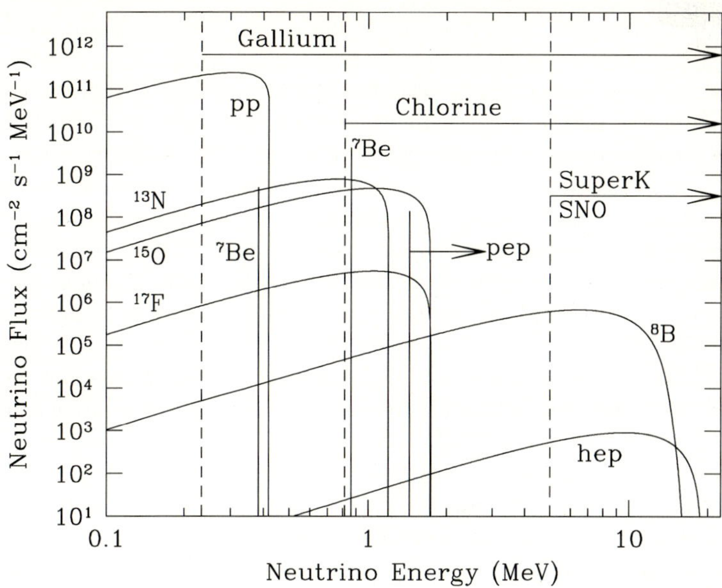

Fig. 2. The energy spectrum of neutrinos emitted by each of the 8 nuclear reactions that generate neutrinos in the solar core. For each curve the source of neutrinos is marked in the figure. The dashed vertical lines mark the threshold energy for each of the operating experiment as marked in the figure

Chlorine experiment was "to see into the interior of the Sun and thus verify directly the hypothesis of nuclear energy generation in stars". The Homestake solar neutrino experiment which is sensitive to intermediate and high energy neutrinos admirably fulfilled its objective. Unfortunately, over the years Davis has been reporting measurements of the solar neutrino counting rate of 2.56 ± 0.23 SNU (1 SNU $= 10^{-36}$ captures per target atom per second), which is at variance with the counting rate of 7.6 ± 1.2 SNU predicted by the standard solar model for the Chlorine experiment. This puzzling deficit in the neutrino counting rate, by nearly a factor of 3 over the SSM prediction, constitutes the solar neutrino problem which has been haunting the community for well over three decades. (e.g., Cleveland et al. 1998). Figure 2 shows the energy spectrum of neutrino fluxes from each of the 8 nuclear reactions that produce neutrinos in the standard solar model of Bahcall, Pinsonneault & Basu (2001). This figure also shows the energy threshold of all currently operating neutrino detectors. Only neutrinos above this energy are detected.

There have been a number of ingenious suggestions to account for the observed deficit in the solar neutrino flux: partial mixing in the solar interior which brings additional fuel of hydrogen and helium to the centre, thus maintaining the nuclear energy production at a slightly lower temperature; the presence of a small admixture of Weakly Interacting Massive Particles in the solar core which would effectively contribute to an increase in the thermal conductivity, in the process diminishing the temperature gradient required to transport the flux;

the rapidly rotating solar core; the centrally concentrated magnetic field; lower heavy element abundance. All such proposals lead to a slight reduction in the central temperature causing a lowering of the flux of high-energy neutrinos.

A Japanese experiment (Fukuda et al. 1996) consisting of a 680 ton water tank was located about 1 km underground in the Kamioka mine where charged particles are detected by measuring Cerenkov light through the elastic scattering reaction, $\nu_x + e^- \rightarrow \nu_x + e^-$ (threshold $= 5.5\,\mathrm{MeV}$). This and the upgraded SuperKamiokande experiment (Fukuda et al. 1999) are sensitive only to the high-energy ^{8}B neutrinos released by the pp-chain of nuclear reactions. The measured flux from the SuperKamiokande experiment is again deficient by about 50% over the total flux predicted by the standard solar model. The Homestake and SuperKamiokande experimental measurements are clearly inconsistent with the proposition of a cooler solar core being a viable solution for the solar neutrino problem. Such a reduction in the central temperature will lead to even larger suppression of the high energy ^{8}B neutrino flux to which the SuperKamiokande experiment is exclusively sensitive; this is because of the extremely high-temperature dependence of the ^{8}B neutrino reaction rate. Paradoxically, the Homestake experiment that detects the intermediate as well as high energy neutrinos shows an even larger reduction in the neutrino counting rate. Thus by reducing the core temperature it is not possible to get a solar model which simultaneously matches both the Homestake and SuperKamiokande measurements.

Besides these experiments there are three other radiochemical experiments (GALLEX, SAGE and GNO) that use a gallium detector with a relatively low threshold of 0.233 MeV and are capable of detecting the low-energy pp-neutrinos. The GALLEX, SAGE and GNO experiments (Hampel et al. 1999) report measurement of the solar neutrino counting rate of 74.7 ± 5.0 SNU, while the SSM prediction of the neutrino capture rate for the gallium experiments is 128 ± 8 SNU, again showing a deficit in the measured neutrino counting rate. Over the past three decades, experimental efforts and more refined theoretical models have only confirmed the discrepancy between the measured and calculated neutrino fluxes.

One of the primary goals of contemporary solar neutrino experiments was to understand the physics of thermonuclear reactions operating in the Sun and more importantly, to constrain the properties of neutrinos. It became clear that none of the measurements of neutrino fluxes by the Chlorine, Water and Gallium experiments were consistent with each other, provided one makes the following assumptions: neutrinos have standard physical properties, namely, no mass and hence no magnetic moment and no flavour-mixing during transit and that the Sun is in thermal equilibrium generating a constant luminosity. There are considerations based on fairly general arguments independent of any underlying solar model which can be demonstrated to lead to unphysical situations such as a negative flux of beryllium neutrinos. A possible resolution of this conundrum is to endow neutrinos with a tiny mass and permit oscillations of neutrino flavours during propagation. The electron neutrinos could get transformed into neutrinos of a different flavour along their flight path through the interior of the Sun and the Earth, or through space between the Sun and Earth. A fraction of electron neutrinos exclusively produced in the Sun's nuclear reaction network would then go undetected in some of the solar neutrino experiments. This raises the exciting possibility of nonstandard neutrino physics being responsible for the deficit in the measured neutrino fluxes and for the need to go beyond the Standard Model of Particle Physics. The first compelling evidence for such neutrino oscillations came a few years ago from the SuperKamiokande's analysis of the data on the high-energy cosmic ray neutrinos from the atmosphere. The SuperKamiokande experiment measured the difference in the up and down fluxes of neutrinos produced by cosmic ray interaction with the terrestrial atmosphere to show that neutrino oscillations, indeed, take place. This asymmetry in the up and down fluxes arises because upward moving neutrinos have to pass through the solid material of the Earth, while the downward moving neutrinos, coming from overhead and being generated afresh in the Earth's atmosphere are less likely to undergo any flavour oscillations.

The recent results from the Sudbury Neutrino Observatory (SNO) have claimed convincing evidence that the solar neutrinos, indeed, change from one flavour to another during their journey from the Sun to Earth. (Ahmad et al. 2001). The SNO experiment located at a depth of over 6000 meters of water equivalent in Sudbury uses 1000 tons of heavy water containing the deuterium isotopes of hydrogen for detecting solar neutrinos, while the SuperKamiokande detector contains ordinary water for capturing the neutrinos. In both heavy and ordinary water neutrinos can elastically scatter electrons to produce Cerenkov radiation, but such electron scattering can be caused by any of the three neutrino flavours: electron-, muon- and tau-neutrino. The Sudbury Neutrino Observatory is capable of measuring the ^{8}B neutrinos through the following reactions:

$$\nu_e + d \rightarrow p + p + e^- \qquad \text{(charged current)} , \tag{11}$$

$$\nu_x + e^- \rightarrow \nu_x' + e^- \qquad \text{(elastic scattering)} \qquad (x = e, \mu, \tau) , \tag{12}$$

$$\nu_x + d \rightarrow \nu_x' + p + n \qquad \text{(neutral current)} \qquad (x = e, \mu, \tau) . \tag{13}$$

SNO's heavy water detector is capable of isolating electron neutrinos, because that flavour alone can be absorbed by a deuterium nucleus to produce two protons and an electron. The neutral current (NC) reaction is equally sensitive to all neutrino flavours, while the elastic scattering (EC) has significantly low sensitivity to mu- and tau-neutrinos. SNO has reported the elastic scattering count rate which equals the SuperKamiokande event rate, to within experimental errors. However, SNO's count of the charged current reaction which is sensitive exclusively to the electron-neutrinos is lower than the SNO/SuperKamiokande event rate of all flavours. This difference in the ^{8}B flux deduced from the charged current and elastic scattering rates, at the level of 1.6σ, provides reasonably firm evidence that some of the Sun's electron-neutrinos are transformed into mu- or tau-neutrinos by the time they reach the experimental setup on Earth. Recently, the neutral current reaction results have been announced by SNO reporting the flux of mu- or tau-neutrino at 5.3σ level (Ahmad et al. 2002). Furthermore, the total ^{8}B neutrino flux as measured by the NC reactions is $(5.09 \pm 0.62) \times 10^6$ cm^{-2} s^{-1}, in agreement with that predicted by the standard solar model of Bahcall, Pinsonneault & Basu (2001). The neutrino oscillations have been further confirmed by the KamLAND experiment (Eguchi et al. 2003) which has detected oscillations in anti-neutrinos produced by nuclear reactors. The KamLAND results combined with results from other solar neutrino experiments have succeeded in determining the parameters governing mixing of neutrino flavours in favour of the large mixing angle (Bahcall, Gonzalez-Garcia & Pena-Garay 2003; Bandyopadhyay et al. 2002), thus effectively solving the solar neutrino problem. These results reassure solar physicists that the theoretical models of the Sun's internal structure are essentially correct and that the resolution of the solar neutrino puzzle should be sought in the realm of particle physics. It has also prompted the community to explore an independent, complementary tool to probe the physical conditions inside the Sun and this was provided by helioseismic studies.

3.2 Helioseismology

The surface of the Sun undergoes a series of mechanical vibrations which manifest themselves as Doppler shifts oscillating with a period centred around 5 minutes (e.g., Leighton, Noyes & Simon 1962; Antia, this volume). These have now been identified as acoustic modes of pulsation of the entire Sun representing a superposition of millions of standing waves with amplitude of an individual mode of the order of a few cm s^{-1} (Ulrich 1970; Leibacher & Stein 1971; Deubner 1975). The frequencies of many of these modes have been determined to an accuracy of better than 1 part in 10^5. The accurately measured oscillation frequencies provide very stringent constraints on the admissible solar models. The determination of the mode frequencies to a high accuracy, of course, requires continuous observations extending over very long periods of time and this is achieved with the help of ground-based network observing the Sun almost continuously. The most prominent amongst these networks is the Global Oscillation Network Group (GONG) which comprises six stations located in contiguous longitudes around the world (Harvey et al. 1996). Satellite-borne instruments have also been observing the solar oscillations and particularly, the Michelson Doppler Imager (MDI) on board the Solar and Heliospheric Observatory (SOHO) with its higher spatial resolution has been able to study solar oscillations with small associated length scales (Scherrer et al. 1995).

Despite considerable progress in the field of helioseismology over the past 25 years, the basic mechanism responsible for the excitation of solar oscillations is still not adequately understood. The acoustic modes may be either intrinsically overstable, or they could be stochastically excited by nonlinear interactions with other motions. In the solar envelope, except for the top several tens of kilometres, convection is responsible for transporting a major fraction of the heat flux. The turbulent conductivity also far exceeds the corresponding radiative conductivity for most part of the convection zone. The convective turbulence and radiative exchange are, therefore, expected to control both the excitation and damping of solar p-mode oscillations.

The linear growth rates of five-minute oscillations for realistic solar models were studied by Ulrich (1970), Ando & Osaki (1975) and Antia, Chitre & Narasimha (1982) using a highly simplified description of radiative transfer and incorporating mechanical and thermal effects of convective turbulence in an approximate manner. It was demonstrated that many of the p-modes in the five-minute period range could be overstable. However, there are many uncertainties in these calculations such as the diffusion approximation for radiative transfer which breaks down near the surface layers, inadequacies of our knowledge of the atmospheric opacity and its derivatives and the lack of knowledge to treat the pulsation-convection coupling. In any case the observed amplitudes of p-modes are extremely small and if these modes were indeed overstable, there should be present some nonlinear amplitude-limiting mechanism. It is difficult to imagine any nonlinear mechanism which becomes effective at such small amplitudes. The linear stability of solar p-modes is rather sensitively dependent on the interaction of pulsation with radiation and convection and many studies have found all modes to be stable (e.g., Balmforth 1992). It turns out, the mechanism of stochastic excitation by turbulent convection, on the other hand, yields amplitudes of individual modes that are in rough agreement with observations (e.g., Goldreich & Keeley 1977; Christensen-Dalsgaard & Frandsen 1983).

The coupling of solar convection with acoustic oscillation was studied by Kumar & Goldreich (1989) by assuming the p-modes to be stable and driven by acoustic emission from turbulent convection. The outstanding question concerns the basic energy source for driving these oscillations. The reservoir of energy available in the form of radiation and convection is certainly quite adequate for the purpose of exciting the p-modes to observed levels. Unfortunately, all the proposed mechanisms for extracting energy from such a reservoir necessarily involve overstable modes which would lead to an unacceptably large build-up of mode amplitudes for the Sun.

Another source of energy for driving p-modes is provided by the mechanical energy of fully developed turbulent convective motions. The theory of acoustic emission from homogeneous turbulence was developed by Lighthill (1952) and it is well known that turbulent flow field can generate sound waves with frequency bandwidth equal to the inverse of the energy cascade time. The acoustic emission could arise from a monopole, dipole or quadrupole sources (e.g., Ulmschneider, this volume). The equipartition between mode energy and the kinetic energy of a resonant eddy for compressible turbulence was derived by Goldreich & Keeley (1977), by taking into account the quadrupole emission and absorption due to Reynolds stresses. In the solar case, the mechanism responsible for exciting turbulence can itself cause acoustic emission and absorption (e.g., Kumar & Goldreich 1989). It turns out the acoustic emission associated with the buoyancy forces is, in fact, more efficient compared to Reynolds stresses by (Mach no.)$^{-2}$, and there is a monopole emission when, near the surface of the Sun, there is a loss of heat by radiation. However, the contributions from monopole and dipole radiation can cancel each other for energy-bearing eddies, with a residue left that is comparable with the quadrupolar emission from Reynolds stresses. It would thus, appear that the forcing of p-modes through coupling with acoustic noise generated by turbulent convection is a viable mechanism for their excitation to the desired amplitude level.

The accurate helioseismic data of oscillation frequencies may be analysed in two ways: i) Forward method; ii) Inverse method (e.g., Antia, this volume). In the Forward method, an equilibrium standard solar model is perturbed in a linearised theory to obtain the eigenfrequencies of solar oscillations, and these are compared with the accurately measured mode frequencies (e.g., Elsworth et al. 1990). The fit naturally is seldom perfect, but a comparison between the observed and theoretically computed frequencies indicated the thickness of the convection zone to be close to 200 000 km and the helium abundance, Y in the solar envelope to be 0.25. It was noted that an improved treatment of convection due to Canuto & Mazzitelli (1991) led to a significantly better accord between calculated and observed p-mode frequencies (Basu & Antia 1994). The forward method has had only a limited success. A number of inversion techniques have, therefore, been developed using the equations of mechanical equilibrium to infer the acoustic structure of the Sun (Gough & Thompson 1991).

One of the major accomplishments of the inversion methods was an effective use of the accurately measured solar oscillation frequencies for a reliable inference of the internal structure of the Sun (Gough et al. 1996; Kosovichev et al. 1997). The profile of the sound speed can now be determined through the bulk of the solar interior to an accuracy of better than 0.1% and the profiles of density to a somewhat lower accuracy. The agreement between the sound speed profile deduced from helioseismic inversions and the SSM is remarkably close except for a pronounced discrepancy near the base of the convection zone and a noticeable difference in the energy-generating core. The hump at the base of the convection zone may be attributed to a sharp change in the gradient of the helium abundance profile on account of diffusion. A moderate amount of rotationally-induced mixing immediately beneath the convection zone, can smooth out this feature (Richard et al. 1996; Brun, Turck-Chièze & Zahn 1999). The dip in the relative sound speed difference in the core may be due to ill-determined composition profiles in the SSM, possibly resulting from the use of inaccurate nuclear reaction rates.

From the recently available seismic data, the helium abundance in the solar envelope is deduced to be 0.249 ± 0.003 (Basu & Antia 1995) and the depth of the convection zone is estimated to be $(0.2865 \pm 0.0005) R_{\odot}$ (Christensen-Dalsgaard, Gough & Thompson 1991; Basu 1998). It has also been possible to surmise the extent of overshoot of convective eddies beneath the base of the convection zone. The measured oscillatory signal is found to be consistent with no overshoot, with an upper limit of $0.05 H_P$ (H_P being the local pressure scale height) (Monteiro, Christensen-Dalsgaard & Thompson 1994; Basu, Antia & Narasimha 1994; Basu 1997).

The seismic structure of the Sun discussed so far is based on the equations of mechanical equilibrium. The equations of thermal equilibrium have not been used because on oscillatory time scales of several minutes, the modes are not expected to exchange significant amounts of energy. The frequencies of solar oscillations are, therefore, largely unaffected by the thermal processes in the interior. However, in order to determine the temperature and chemical composition profiles one needs to supplement the seismically inferred structure, obtained through primary inversions, by the equations of thermal equilibrium, together with the auxiliary input physics such as the opacity, equation of state and nuclear energy generation rates (Gough & Kosovichev 1990; Antia & Chitre 1998; Takata & Shibahashi 1998). It turns out that the inverted sound speed, density, temperature and composition profiles, and consequently the neutrino fluxes, come pretty close to those given by the SSM. In general, the computed total luminosity resulting from these inverted profiles would not necessarily match the observed solar luminosity. The discrepancy between the computed and observed solar luminosity, $L_{\odot}$ can, then be effectively used to provide a test of the input nuclear physics; in particular, it can be demonstrated that the cross-section for the proton-proton reaction needs to be increased slightly to $(4.15 \pm 0.25) \times 10^{-25}$ MeV barns (Antia & Chitre 1998). Note this cross-section has a crucial influence on the nuclear energy generation and neutrino fluxes, but it has not been measured in the laboratory. Indeed, it can be readily shown that the current best estimates (Adelberger et al. 1998) for the proton-proton reaction cross-section and metallicity, Z are only marginally consistent with the helioseismic constraints and probably need to be increased slightly by a few per cent (Antia & Chitre 1999a). The extent to which the proton-proton reaction cross-section needs to be increased also depends on the treatment of electron screening (e.g., Antia this volume). With the use of intermediate screening treatment due to Mitler (1977) the theoretically computed cross-section is essentially consistent with helioseismic constraints (Antia & Chitre 2002).

The seismic models enable a determination of the central temperature of the Sun which is found to be $(15.6 \pm 0.4) \times 10^{6}$ K, allowing for up to 10% uncertainty in the opacities (Antia & Chitre 1995). It turns out that it is possible to determine only one parameter specifying the chemical composition and we assume the heavy element abundance, Z to be known and attempt to surmise the helium abundance profile, Y. The inferred helium abundance profile is in fairly good agreement with that in the SSM which includes diffusion, except in the regions just beneath the convection zone where the profile is essentially flat (Antia & Chitre 1998). This is suggestive of some sort of mixing, possibly arising from a rotationally-induced instability. Interestingly, the temperature at the base of the solar convection zone is 2.2×10^{6} K, which is not high enough to burn lithium. However, if there is some amount of mixing that extends a little beyond the base of the convection zone to a radial distance of $0.68 R_{\odot}$, say, temperatures exceeding 2.5×10^{6} K will be attained for the destruction of lithium by nuclear burning, and this may explain the low lithium abundance observed at the solar surface.

The remarkable feature that emerges from these computations is that even if we allow for arbitrary variations in the input opacities and relax the requirement of thermal equilibrium, but assume standard properties for neutrinos, it turns out to be difficult to construct a seismic model that is simultaneously consistent with any two of the three existing solar neutrino experiments within 2σ of the measured fluxes (Roxburgh 1996; Antia & Chitre 1997). It has been suggested that mixing of ^{3}He in the solar core can alter the neutrino fluxes significantly (e.g., Cumming & Haxton 1996). However, such a modification of the standard solar model can be ruled out on the basis of helioseismic constraints (Bahcall et al. 1997). It is unlikely that any substantial mixing can take place in the solar core, as, otherwise, the chemical composition profile will need to be fine-tuned to reproduce the helioseismically inferred sound speed profile. This argument is applicable to any general type of mixing process. On the other hand, it is conceivable that ^{3}He abundance may not have been estimated correctly in the solar interior which will, of course, not affect the mean molecular weight and hence the sound speed because of the very low ^{3}He abundance compared to ^{4}He and H. But it can be demonstrated that the solar neutrino problem is unlikely to be solved with an arbitrary redistribution of ^{3}He or arbitrary heavy element abundance or any non-Maxwellian equilibrium energy distribution, provided the observed luminosity constraint is maintained (Antia & Chitre 1999b). This suggests that the persistent discrepancy between measured and predicted solar neutrino fluxes is likely to be due to non-standard neutrino physics. In this sense, helioseismology may be regarded to have highlighted the importance of the Sun as a cosmic laboratory for studying the novel properties of neutrinos.

3.3 Rotation Rate in the Solar Interior

Helioseismology has also made it possible to determine the rotation rate in the interior from the accurately measured rotational splittings (e.g., Antia, this volume). The first order effect of rotation yields splittings which depend on odd powers of the azimuthal order. These odd splitting coefficients can be used to deduce the rotation rate as a function of depth and latitude. It is found that the surface differential rotation persists through the solar convection zone, while in the radiative interior the rotation rate appears to be relatively uniform (Thompson et al. 1996; Schou et al. 1998). A transition region near the base of the convection zone (the tachocline) is centred at a radial distance, $r = (0.7050 \pm 0.0027) R_\odot$ (Basu 1997). The seat of the solar dynamo is widely believed to be located in this tachocline region. There is also a shear layer present just beneath the solar surface extending to $r \simeq 0.94 R_\odot$ where the rotation rate is found to increase with depth. It will be instructive to examine its role in sustaining a secondary dynamo.

It may be recalled, an important aspect of solar internal rotation is that it can provide a crucial test of Einstein's general theory of relativity. The test is based on the measurements of planetary orbits which should be elliptical under Newton's inverse square law. In practice, however, on account of the mutual gravitational interaction between planets, the orbits are somewhat different. After correcting for these perturbations, the residual orbit of planet Mercury, for example, was found to be a rotating ellipse that precesses about the Sun at 43 seconds of arc per century. The excellent agreement between the theoretical prediction of general relativity and observed precession of the perihelion of Mercury was heralded as a great triumph for the theory of relativity. The prediction of Einstein general theory is, of course, based on the crucial assumption that the Sun is a spherically symmetric body. The presence of both rotation and magnetic field in the interior is liable to cause a bulge at the equator and a flattening at the poles, in the process contributing a higher order term to its gravitational potential. Such an oblateness would modify the Sun's gravitational field in a way to induce the observed precession of Mercury's orbit from purely Newtonian effects. It turns out in order to account for full precession of 43 arc second per century the Sun would have to rotate much faster than what is inferred from helioseismic inversions. The helioseismically inferred rotation rate is, indeed, consistent with the measured solar oblateness of approximately 10^{-5} (Kuhn et al. 1998). The resulting quadrupole moment turns out to be $(2.18 \pm 0.06) \times 10^{-7}$ (Pijpers 1998), implying a precession of perihelion of the orbit of planet Mercury by about 0.03 arcsec/century, which is clearly consistent with the general theory of relativity. The even order terms in the splittings of solar oscillation frequencies reflect the Sun's effective acoustic asphericity and can provide a valuable handle to probe the presence of a large-scale magnetic field or a latitude-dependent thermal fluctuation in the solar interior.

It has now been well demonstrated that the frequencies of solar oscillations vary with time and that these variations are correlated with the solar activity (e.g., Bhatnagar, Jain & Tripathy 1999). It is expected that these frequency variations should result from structural changes in the layers close to the solar surface for explaining fluctuations over timescales of order 11 years. With accumulating GONG and MDI data over nearly seven years during the rising phase of solar cycle 23, it has, indeed, been possible to study temporal variations of the solar rotation rate and other characteristic features associated with the solar envelope. In fact, helioseismic inversions have revealed small temporal variations of the rotation rate in the subsurface layers. These alternating bands of fast and slow rotation appear to migrate towards the equator as the solar cycle progresses, reminiscent of the torsional oscillations detected at the solar surface, but extending to a depth of at least 60 Mm (e.g., Antia, this volume).

The frequencies of fundamental, or f-modes which are surface modes, are largely determined by the surface gravity and thus provide a valuable tool to probe the near-surface regions as well as an accurate measurement of the solar radius. An important application of the accurately measured f-mode frequencies is their potential use as a diagnostic of solar oblateness and of magnetic fields just beneath the solar surface, in addition to studying the solar cycle variations of these quantities.

The ongoing efforts in helioseismology will hopefully, reveal the nature and strength of magnetic fields present inside the Sun and will also help in highlighting the processes that drive the cyclical magnetic activity and also locate the seat of the solar dynamo. The accumulating seismic data during the ascending and descending phases of cycle 23 will enable us to study the temporal variations of mode frequencies and amplitudes which should be indicative of the changes in the solar structure and dynamics. In the process, we may also learn how the magnetic field of the Sun changes with the solar cycle and what causes the solar irradiance to vary synchronously with the sunspot cycle. Finally, an unambiguous detection of buoyancy driven gravity modes would furnish a powerful tracer of the energy-generating regions of our Sun!

4 Magnetically Controlled Solar Phenomena

The existence of magnetic fields on the Sun was established by Hale from the Zeeman splitting of spectral lines in sunspots, indicating magnetic fields of order 2000–3000 G in the dark central regions of the spot. The general background magnetic field in the Sun, detected with sensitive magnetographs, was shown by the Babcocks to have an average strength of about 10 G. The overall magnetic field structures are oppositely directed in the northern and southern hemispheres, and the reversals in the field polarities are observed to take place near the maximum phase of the sunspot cycle. It is now widely believed that the global magnetic field of the Sun is not uniformly spread over its surface, but rather the field is distributed in separate clusters of magnetic flux tubes (fibrils) with field strength $\sim$ 1000–2000 G and diameters of order 100 km (e.g., Hasan, this volume). The active regions with which the sunspots and large flaring events are normally associated are found to lie in the midst of extended bipolar regions of $\sim$ 100 G fields. The outstanding question that is continuing to puzzle the solar astronomers is the origin and seat of the solar dynamo and the nature of the mechanism that drives the activity cycle with such a regularity. The observed nonuniform rotation of the Sun, namely, faster rotation at the equatorial latitudes than near the polar regions continually shears the dipole magnetic field to generate a toroidal component, while the cyclonic turbulent convection interacting with the toroidal loops reinforces the dipole field configuration (e.g., Venkatakrishnan, this volume). It is fair to say that the issues relating to the formation of sunspots, their emergence at the surface, their evolution and decay and in fact, the basic underlying mechanism responsible for the origin of the solar activity cycle are not adequately understood.

The Sun has evidently a large reservoir of free magnetic energy available, but the process for its explosive release is not altogether clear. The generally accepted mechanism for sudden energy release is a process called magnetic reconnection which involves splicing and rejoining magnetic lines of force. Thus, the flare phenomena occurring in the vicinity of active regions are evidently hydromagnetic manifestations which involve a rapid conversion of magnetic free energy into fast particles and hot plasma.

The production of prominences also results from the strong, large-scale magnetic fields existing in active regions playing a major role. The solar plasma is guided along the lines of force condensing into regions of higher density and lower temperature and raining down back towards the photosphere. The solar flares are observed on various scales ranging from the largest with energy $\sim 10^{32}$ erg over dimensions of 10^4 km down to the limit of detection with energy $\sim 10^{25}$ ergs over 100–1000 kms (microflares). The basic mechanism seems to involve rapid diffusion and reconnection of magnetic field lines (e.g., Ambastha, this volume). The prominent feature associated with magnetic fields embedded in a plasma and undergoing continuous deformation would be the occurrence of current sheets with steep magnetic field gradients. These current sheets provide the sites for fast reconnection and explosive dissipation of magnetic energy. The recent results from the SUMER instrument aboard the SOHO satellite provide plausible evidence for magnetic reconnections on the Sun from the formation of bi-directional outflow jets at these sites.

The outer solar atmosphere presents a rich variety of designs and complex structures for close scrutiny in a cosmic setting. The chromosphere and corona are observed to be highly structured with a clear evidence of association with magnetic fields. Thus, the chromospheric network closely coincides with the network of locally strong and mainly vertical magnetic field with strength $\sim$10–20 G (e.g., Ambastha, this volume). An indication of magnetic activity in the solar atmosphere is the presence of plages, (incandescent bright regions of gas with a higher density than the surrounding atmosphere), which are caused by enhanced magnetic fields. It has also been observed that the tenuous gas above the strong ($\sim$ 100 G) bipolar fields of active regions is heated to temperatures of $\sim 4 \times 10^6$ K, while the broad regions of weak (5–10 G) fields are heated to temperatures of 1.5×10^6 K. The active region corona is thus appreciably hotter than that associated with the quiet regions (e.g., Dwivedi, this volume). The solar corona is a magnetically structured region consisting of X-ray bright points, coronal loops and coronal holes with open streamer structures. The coronal loops are closed magnetic structures spread over a wide range of scales with their footpoints anchored in the surface layers. The large loops interconnecting active regions are likely to be responsible for the diffuse coronal emission, while the smallest loops probably form the X-ray bright points. It appears that most of the loops are heated within about 10 000–20 000 km of the solar surface and the upper atmospheric layers of the Sun probably respond to the evolution of magnetic fields that are anchored in the photosphere. The existence of coronal holes as persistent depression in the coronal intensity has been known from the ground-based coronagraphic observations since the 1950s. Later satellite observations from the Skylab and Yohkoh further established that high-speed solar winds approaching velocities of order 800 km s^{-1} originate in coronal holes where field lines are open to interplanetary space (e.g., Manoharan, this volume). The classical solar wind model of Parker is based upon thermally driven effects, but the mechanism for the acceleration of high-speed winds in coronal holes is still not clear, as are the agents responsible for the coronal mass ejections.

The importance of magnetic fields in supplying the energy for heating the solar atmosphere is being widely recognised. The presence of kilogauss magnetic fields at the boundaries of supergranules, the detection of coronal loops and bright points in soft X-ray photographs have served to highlight the dominant role of magnetic fields in controlling the energetic phenomena in outer layers of the Sun.

The temperatures in the outer atmosphere of the Sun exceed that at the surface by about one to two orders of magnitude. But the nature of the mechanisms responsible for heating the chromospheric and the coronal layers to such high temperatures has continued to be intriguing (e.g., Ulmschneider, this volume). It is known that the sub-photospheric turbulent convection in the Sun generates waves of different kind: acoustic, gravity and hydromagnetic (Alfvén) waves. Biermann (1946) and Schwarzschild (1948) were the first to suggest a mechanism for heating the solar atmosphere by sound waves generated in the sub-surface turbulent convection zone, steepening into shock waves during their propagation outwards. It is now generally believed that the dissipation of acoustic waves is perhaps important only for the lower chromospheric regions for which the heating needed for the energy-balance is about 4×10^6 erg cm^{-2} s^{-1} (Withbroe & Noyes 1977). Alternatively, acoustic waves impinging on the chromospheric magnetic canopy can be resonantly absorbed and subsequently dissipated in narrow layers by resistive effects (e.g., Chitre & Davila 1991). In the overlying corona, however, the required heating is only about 3×10^5 erg cm^{-2} s^{-1} for the quiet regions and 5×10^6 erg cm^{-2} s^{-1} for active regions. But basically, both the chromosphere and corona of the Sun are heated by some mechanical input of energy and the underlying mechanism for heating the upper chromosphere and corona is very likely to be of magnetic origin. The observational support for the acoustic heating of the lower solar atmosphere comes from the profiles of spectral lines which are broadened by the presence of some nonthermal motions (e.g., propagating sound waves) that appear to increase in magnitude outward.

Several different mechanisms have been proposed for heating the solar corona. There are two main contenders capable of supplying the requisite amount of energy: hydromagnetic waves generated by the sub-photospheric turbulence propagating outwards and getting damped in the upper layers of the chromosphere and corona and formation of current sheets and small-scale reconnection leading to an explosive release of energy for coronal heating (e.g., Dwivedi, this volume).

A fresh insight into the nature and location of the process responsible for heating the solar corona has been provided by the recent observations from the SOHO and TRACE missions (Dwivedi & Mohan 1997). The inhomogeneously structured corona is seen to be made up of a large number of loops of different sizes down to a few hundred kilometres wide loops revealed by TRACE imagery. There is an evident relationship between such loops and the large-scale coronal arches with the photospheric magnetic fields. The earlier theoretical studies envisaged a fairly uniform heating extended over the whole length of the coronal loops (Rosner, Tucker & Vaiana 1978). The X-ray observations of the diffuse corona seem to validate such a model with the uniform distribution of energy describing the observed temperature variations along large loops (Priest et al. 1998). The recent TRACE images of active regions reveal a continual localised brightening indicating dynamic events occurring near the footpoints of the small active region loops (Aschwanden, Nightingale & Alexander 2000). This would place the source of coronal heating in the lower atmosphere of the Sun within about 10 000 km of the surface. Earlier balloon-borne measurements by Lin et al. (1984) had reported the detection of impulsive, bursts of X-ray emission (microflares). With Yohkoh data on active regions, Shimizu (1995) also found numerous small brightening events associated with active region loops. It is plausible that part of the coronal heating responsible for the presence of X-ray bright points results from the process of reconnection of magnetic loops which are driven by the motions of their footpoints by the sub-surface convection. The diffuse coronal emission is likely to arise from regions of in-situ heating that is uniformly distributed along the large-scale loops. The dissipation of long wavelength Alfvén waves by the mechanism of resonant absorption was previously thought to be a promising candidate for heating large coronal loops. Such a heating process tends to be non-uniformly distributed and is, therefore, unlikely to explain the observations.

A viable mechanism that is currently in favour for heating the coronal plasma is the Ohmic dissipation of many narrow current sheets. It appears that the energy input for the coronal holes and the associated high-speed solar wind may be supplied mainly by microflares occurring among the magnetic fibrils that are present on the surface of the Sun. The dense X-ray corona is heated to temperatures in excess of a few million degrees by even smaller flares (nanoflares) that take place in the small current sheets produced in the stronger ($\sim 100\,\mathrm{G}$) bipolar magnetic regions by continuous shuffling and buffeting of the footpoints of the field by the sub-surface convective motions. The measurements with the extreme ultraviolet imaging telescope on board the SOHO satellite have also highlighted the role of numerous tiny flaring events (nanoflares) as plausible feeders of energy into the extended loops to heat the corona to temperatures of the order of a few million degrees. However, a major theoretical problem with any coronal heating mechanism involving magnetic fields is the requirement of an efficient diffusion process followed by the reconnection of field lines, and also distribution of the heat from the small volume where the energy dissipation occurs to the larger coronal regions. The recent observations with SOHO and TRACE have provided evidence for the outward transfer of magnetic energy from the solar surface up to the coronal regions. The presence of a magnetic carpet made up of loops is probably responsible in heating the corona to its temperature of several million degrees. These magnetic concentrations are spread all over the surface with their foot points anchored in the photosphere. Each of these magnetic loops carries substantial amount of energy so that when they interact, they cause electrical and magnetic short circuits. The strong electric currents that are produced in these thin sheets can then release adequate amount of energy to heat the solar corona to high temperatures.

It used to be thought that the solar wind streamed outwards from points on the solar surface in all directions. The observations from spacecrafts have revealed the solar corona to be highly structured by magnetic fields. In some places the magnetic field lines form large loop-like structures trapping the solar plasma within them, while at other places on the Sun where the field lines are open, the unconfined coronal plasma flows out into space at high speed as solar wind. SOHO observation have shown that the plumes near the polar caps of the Sun are found within coronal holes which are sites of denser and possibly cooler streams of solar wind. The high speed solar wind (~ 800 km s^{-1}) associated with open field lines occupies most of the Sun during the phases of solar minimum, and it seems to carry the imprint of the 27 days (synodic) rotation period of the Sun. The coronal holes, in fact, appear to display rigid rotation as if they are attached to the solar body. The slow solar wind (~ 400 km s^{-1}) is limited by the closed magnetic field lines and its velocity increases polewards from ~ 400 km s^{-1} in the equatorial regions to ~ 600–700 km s^{-1} in the polar latitudes.

The polar regions may be the seats of plumes and coronal holes, but it is the great streamers and huge eruptions called coronal mass ejection (CMEs) that dominate the solar wind pattern in the equatorial latitudes. The CMEs are huge clouds of solar material lifting off from the corona and travelling out into interplanetary space like great blobs of plasma. These outbursts are occasionally seen to travel in opposite directions, nearly simultaneously, resembling ejections girdling the equatorial belt. Observations of the solar corona with Large Angle and Spectrometric Coronagraph (LASCO) and Extreme ultraviolet Imaging Telescope (EIT) instruments aboard the SOHO should provide an opportunity to study CMEs from their initiation to gain an understanding of the sources regions from which they originate and their association with active regions on the surface of the Sun.

It is no exaggeration that the internal and external layers of our Sun furnish unlimited opportunities to study various branches of physics in the cosmic environment. Equally, some of the violent events occurring in its atmosphere have profound implications for the life here on Earth.

References

Adelberger, E. C., Austin, S. M., Bahcall, J. N., et al. 1998, Rev. Mod. Phys., 70, 1265
Ahmad, Q. R., Allen, R. C., Andersen, T. C., et.al. 2001, Phys. Rev. Lett., 87, 071301
Ahmad, Q. R., Allen, R. C., Andersen, T. C., et.al. 2002, Phys. Rev. Lett., 89, 011301
Allègre, C. J., Manhès, G., & Göpel, C. 1995, Geochim. Cosmochim. Acta, 59, 1445
Ando, H., & Osaki, Y. 1975, PASJ, 27, 581
Antia, H. M., & Chitre, S. M. 1995, ApJ, 442,434
Antia, H. M., & Chitre, S. M. 1997, MNRAS, 289, L1
Antia, H. M., & Chitre, S. M. 1998, A&A, 339, 239
Antia, H. M., & Chitre, S. M. 1999a, A&A, 347, 1000
Antia, H. M., & Chitre, S. M. 1999b, Bull. Astron. Soc. India, 27, 69
Antia, H. M., & Chitre, S. M. 2002, A&A, 393, L95
Antia, H. M., Chitre, S. M., & Narasimha, D. 1982, Sol. Phys., 77, 303
Aschwanden, M. J., Nightingale, R. W., & Alexander, D. 2000, ApJ, 541, 1059
Bahcall, J. N., Pinsonneault, M. P., Basu, S., & Christensen-Dalsgaard, J. 1997, Phys. Rev. Lett., 78, 171

Bahcall, J. N., Pinsonneault, M. P., & Basu, S. 2001, ApJ, 555, 990.

Bahcall, J. N., Gonzalez-Garcia, M. C., Pena-Garay, C. 2003, J. High Ener. Phys., 02, 009 (hep-ph/0212147)

Balmforth, N. J. 1992, MNRAS, 255, 603

Bandyopadhyay, A., Choubey, S., Gandhi, R., Goswami, S., & Roy, D. P. 2002, hep-ph/0212146

Basu, S., 1997, MNRAS, 288, 572

Basu, S., 1998, MNRAS, 298, 719

Basu, S., & Antia, H. M. 1994, J. Astroph. Astron., 15,143

Basu, S., & Antia, H. M. 1995, MNRAS, 276, 1402

Basu, S., Antia, H. M., & Narasimha, D. 1994, MNRAS, 267, 209

Bhatnagar, A., Jain, K., & Tripathy, S. C. 1999, ApJ, 521, 885

Biermann, L. 1946, Naturwissenschaften, 33, 118

Böhm-Vitense, E. 1958, Z. Astrophys., 46, 108

Brun, A. S., Turck-Chièze, S., & Zahn, J. P. 1999, ApJ, 525, 1032

Canuto, V. M., & Mazzitelli, I. 1991, ApJ, 370, 295

Chitre, S. M., & Davila, J. M. 1991, ApJ, 371, 785

Christensen-Dalsgaard, J., & Däppen, W. 1992, Astron. Astroph. Rev., 4, 267

Christensen-Dalsgaard, J., & Frandsen, S. 1983, Sol. Phys., 82, 165

Christensen-Dalsgaard, J., Gough, D. O., & Thompson, M. J. 1991, ApJ, 378, 413

Christensen-Dalsgaard, J., Däppen, W., Ajukov, S. V., et al. 1996, Science, 272, 1286

Cleveland, B. T., Daily, T., Davis, R., Jr., Distel, J. R., Lande, K., Lee, C. K., Wildenhain, P. S., & Ullman, J. 1998, ApJ, 496, 505

Cox, J. P., & Giuli, R. T. 1968 Principles of Stellar Structure (Gordon & Breach, New York)

Cumming, A., & Haxton, W. C. 1996, Phys. Rev. Lett., 77, 4286

Davis, R. 1964, Phys. Rev. Lett., 12, 302

Deubner, F.-L. 1975, A&A, 44, 371

Dwivedi, B. N., & Mohan, A. 1997, Curr. Science, 72, 437

Eddington, A. S. 1926, The Internal Constitution of the Stars (Cambridge University Press, Cambridge)

Eggleton, P. P., Faulkner, J., & Flannery, B. P. 1973, A&A, 23, 325

Eguchi, K., Enomoto, S., Furuno, K., et al. 2003, Phys. Rev. Lett., 90, 021802

Elsworth, Y., Howe, R., Isaak, G. R., McLeod, C. P., & New, R. 1990, Nature, 347, 536

Fukuda, Y., Hayakawa, T., Inoue, K., et al. 1996, Phys. Rev. Lett., 77, 1683

Fukuda, Y., Hayakawa, T., Ichihara, E., et al. 1999, Phys. Rev. Lett., 82, 1810

Goldreich, P., & Keeley, D. A. 1977, ApJ, 212, 243

Gough, D. O., & Kosovichev, A. G. 1990, in Inside the Sun, Proc. IAU Colloquium No 121, eds. G. Berthomieu & M. Cribier (Kluwer, Dordrecht), 327

Gough, D. O., & Thompson, M. J., 1991, in Solar Interior and Atmosphere, eds. A. N. Cox, W. C. Livingston, M. Matthews, (University of Arizona Press, Tucson) 519

Gough, D. O., Kosovichev, A. G., Toomre, J., et al. 1996, Science, 272, 1296

Grevesse, N., Noels, A., & Sauval, A. J. 1996, in Cosmic abundances, eds. S. S. Holt & G. Sonneborn, ASP Conf. Ser., 99, 117

Guzik, J. A., & Cox, A. N. 1993, ApJ, 411, 394

Hampel, W., Handt, J., Heusser, G., et al. 1999, Phys. Lett. B, 447, 127

Harvey, J. W., Hill, F., Hubbard, R., et al. 1996, Science, 272, 1284

Hünsch, M., & Schröder, K.-P. 1996, A&A 309, L51

Iglesias, C. A., & Rogers, F. J. 1996, ApJ, 464, 943

Kosovichev, A. G., Schou, J., Scherrer, P. H., et.al. 1997, Sol. Phys., 170, 43
Kumar, P., & Goldreich, P. 1989, ApJ, 342, 558
Kuhn, J. R., Bush, R. I., Scheick, X., & Scherrer, P. 1998, Nature, 392, 155
Leibacher, J. W., & Stein, R. F. 1971, Astrophys. Lett., 7, 191
Leighton, R. B., Noyes, R. W., & Simon, G. W. 1962, ApJ, 135, 474
Lighthill, M. J. 1952, Proc. Roy. Soc. London, 211A, 564
Lin, R. P., Schwartz, R. A., Kane, S. R., Pelling, R. M., & Hurley, K. C. 1984, ApJ, 283, 421
Mihalas, D., Däppen, W., & Hummer, D. G., 1988, ApJ, 331, 815
Mitler, H. E. 1977, ApJ, 212, 513
Monteiro, M. J. P. F. G., Christensen-Dalsgaard, J., & Thompson, M. J. 1994, A&A, 283, 247
Narain, U., & Ulmschneider, P. 1996, Space Sci. Rev., 75, 453
Pijpers, F. P. 1998, MNRAS, 297, L76
Priest, E. R., Foley, C. R., Heyvaerts, J., Arber, T. D., Culhane, J. L., & Acton, L. N., 1998, Nature, 393, 545
Richard, O., Vauclair, S., Charbonnel, C., & Dziembowski, W. A. 1996, A&A, 312, 1000
Rogers, F. J., & Iglesias, C. A. 1992, ApJS, 79, 507
Rogers, F. J., Swenson, F. J., & Iglesias, C. A., 1996, ApJ, 456, 902
Rosner, R., Tucker, W. H. & Vaiana, G. S., 1978, ApJ, 220, 643
Roxburgh, I. W., 1996, Bull. Astro. Soc. India, 24, 89
Scherrer, P. H., Bogart, R. S., Bush, R. I., et al. 1995, Sol. Phys., 162, 129
Schou, J., Antia, H. M., Basu, S., et al. 1998, ApJ, 505, 390
Schröder, K.-P., & Eggleton, P. P. 1996, Rev. Mod. Astr., 9, 221
Schwarzschild, M. 1948, ApJ, 107, 1
Schwarzschild, M. 1958, Structure and Evolution of Stars (Princeton University Press, Princeton)
Shimizu, T. 1995, PASJ, 47, 251
Steffen, M. 1992, Habil. Thesis, Univ. Kiel, Germany
Takata, M., & Shibahashi, H. 1998, ApJ, 504, 1035
Trampedach, R., Christensen-Dalsgaard, J., Nordlund, A., & Stein, R. F. 1997, in Solar Convection and Oscillations and their relationship, eds. F. P. Pijpers, J. Christensen-Dalsgaard & C. S. Rosenthal (Kluwer Academic Publishers, Dordrecht), 73
Thompson, M. J., Toomre, J., Anderson, E., et al. 1996, Science, 272, 1300
Ulrich, R. K. 1970, ApJ, 162, 993
Vernazza, J. E., Avrett, E. H., & Loeser, R. 1981, ApJS, 45, 635
Weinberg, S. 1972, Gravitation and Cosmology: principles and applications of the general theory of relativity (John Wiley, New York)
Withbroe, G. L., & Noyes, R. W. 1977, ARA&A, 15, 363

Instrumentation and Observational Techniques in Solar Astronomy

A. Bhatnagar

Udaipur Solar Observatory, Physical Research Laboratory
P. O. Box no. 198, Udaipur 313001, India
email: arvind@prl.ernet.in

Abstract. Basic concepts in solar physics are described with an attempt to bring out the importance of the Sun as a "Rosetta Stone" to understand other stars. Our Sun being the nearest star, shows intricate surface details and a wide variety of dynamic phenomena. These range in size from a few kilometres to millions of kilometres and in the temporal domain from a fraction of a second to decades. In addition, the Sun displays a great variety of magnetic and velocity fields, as well as radiative energy spectra. The close connection between the solar activity and the Earth's ionosphere, atmosphere and geomagnetic field makes the Solar–Terrestrial relations a very interesting and valuable field of study for a whole variety of disciplines. The principles and details of various kinds of solar instruments are described, especially solar telescopes from the simplest to the advanced types, along with several kinds of back-end instruments, such as monochromatic filters, spectrographs, spectroheliographs, magnetographs, etc., used for photospheric, chromospheric and coronal observations. Standard techniques for making solar observations, e.g., measurements of sunspot areas, coordinates and position of solar features, etc., are described. Solar observations made even with simple equipments are of great importance for short and long term synoptic studies and can even be taken up as a hobby by amateur solar astronomers along with professional solar physicists.

1 Introduction

Our Sun is the nearest star which presents its disk, displaying gamut of phenomena, ranging in sizes from a few hundred kilometres to several thousands of kilometres, in the temporal domain from few milliseconds to several decades, and temperatures ranging from thousands of degree to several million degrees. Besides displaying a large variety of phenomena, our Sun is a source of enormous photon flux, which is a great advantage for detailed study, even with small telescopes and simple equipment. The Sun acts as a 'Rosetta stone' and a celestial dynamic laboratory to help us to understand the physics of other stars and to test theoretical models of astrophysical interest. A whole discipline of Solar–Terrestrial physics has emerged, as a result of the interaction of the solar radiation with the interplanetary medium, ionosphere, our atmosphere and terrestrial magnetic field. The study of Sun–Earth relationship is a very fascinating subject, a large number of scientists all over the world are engaged in this exciting field of research and with the advent of space missions, the monitoring of solar activity and a world wide watch of the Sun have become still more important.

In this chapter, we describe some of the basic techniques and principles for taking solar optical observations, using small telescopes and simple equipment. Methods for measuring important fundamental parameters for observations, and some practical hints about telescopes, filters, spectrographs, spectroheliographs, magnetographs are also given. In the following sections, we describe the basic techniques for observing the solar photosphere, chromosphere, corona, magnetic fields with appropriate instruments and methods for determination of heliographic coordinates of solar features, activity indices, sunspot area, classification of sunspots group etc.

For general introductions to Solar Physics, the readers may consult books by Beck et al. (1995), Mitton (1981), Phillips (1992), Taylor (1991) Taylor (1996) and Zirin (1988).

2 How to Observe the Sun?

Observing the Sun is rather simple but requires common sense and good amount of patience. The main considerations for good solar observations are:

- that the observing site should have minimum atmospheric turbulence, or in other words solar 'seeing' should remain good over long periods of time,
- that the telescope should have minimum thermal currents along the optical path within the telescope tube,
- the exposure time should be as short as possible, to 'freeze' the image otherwise disturbed by the seeing fluctuations,
- 2-dimensional and multi-waveband observations are generally required, taken preferably at a rapid rate.

The Sun is like an onion, you peel one layer after another and see each time a different view of the Sun and also to a different depth of the solar atmosphere. Hence to observe a certain layer of the Sun one has to choose the corresponding radiation emanating from that particular layer. For example, to observe the photosphere – the topmost layer of the visible disk of the Sun's surface, the integrated 'white light' is used. Which is essentially emanating as incandescent continuum radiation from an extremely hot dense plasma. To view other layers of the Sun, such as the chromosphere, one has to observe the hydrogen Hα line radiation or the ionised Calcium lines and other strong lines, or the radio continua emanating from these tenuous solar layers. To observe the solar corona, one has to use line radiation originating from highly ionised ions of Iron, Calcium, Magnesium, Nickel and Argon, or detect the continuum light scattered by free electrons in these layers. Earlier, coronal observations were made only during a total solar eclipse. But with the invention of the coronagraph by Lyot in 1930, it has become possible to observe the inner corona both in line emission and the continuum even without an eclipse. Soft X-rays are used to observe from innermost to outermost layers of the solar corona, which extends to several tens of solar radii out into the interplanetary space. These observations are made by spacecraft flying above the Earth's atmosphere. Since the photosphere emits hardly any X-rays, or in other words is 'dark in X-rays', one doesn't need to block the solar disk for observing the corona in X-rays. Thus it is possible to see the entire corona both on the disk and outside it, in X-rays.

2.1 Observing the Photosphere

To study the photosphere, or the solar surface, generally the integrated white light is used and mainly the following methods are employed:

- Projection method,
- Objective filter method,
- Eyepiece filters, or
- Special solar eyepieces.

Projection Method

Unfiltered sunlight is gathered by a telescope, preferably by a refractor of suitable aperture and focal length. The primary solar image is enlarged by an ordinary eyepiece, used as a projection optics and the enlarged image is projected onto a screen. Depending on the distance and the focal length of the eyepiece, the solar image can be enlarged to any desired size. However, image sizes bigger than 15 cm diameter and formed by a 10 cm aperture telescope makes the contrast of the image low and the surface details tend to diminish. The main problem with this technique is due to the heating of the eyepiece and of the air column in the telescope tube, which distorts the solar image. As the eyepiece is used to enlarge the solar image and is located near the focal plane of the objective, it tends to become very hot and under no circumstances a cemented eyepiece should be used for this purpose. In case of commercially available telescopes (both refractors and reflectors), often plastic components are located in the path of the rays, which can easily melt if direct unfiltered sunlight falls on them. Although this method has the advantage of being simple and easy for beginners, it is not recommended unless great care is taken during the observations.

Observing with Full Aperture Objective Filter

A full aperture 'Mylar' objective filter made of aluminised Mylar film is very effectively used these days for low and medium resolution simple solar observations. Mylar filters available in the market have transmissions ranging from 0.1% to 0.01% and they effectively reduce light and heat entering the telescope tube. These Mylar filters are rather inexpensive but require great care in handling, as they tend to develop pinholes, which result in multiple images. A certain degree of improvement can be achieved by using full aperture plane parallel glass plate filters with a surface accuracy of about $\lambda/10$, coated with a thin layer of chrome or Inconel (a form of stainless steel), giving about 0.01% transmission. These Inconel filters give a yellowish tinge, while the aluminised Mylar filters give a bluish tinge to the solar image. High quality full aperture glass filters tend to become costlier with increasing diameter, however, small aperture results in loss of spatial resolution. Using a full aperture filter one can safely view the solar image through an eyepiece, or take photographs on film or record digitally.

Observing the Sun with Eyepiece Filters

Eyepiece filters are generally supplied as accessories with inexpensive telescopes. However, these eyepieces are NOT recommended as they are fitted very close to the focal plane in both refractors and reflectors, hence become very hot and some times even crack, causing intense light to leak through the cracks and focus on the eye and permanently damage the retina. *Remember that solar observations made without eliminating intense sunlight is very dangerous to the eye and generally eyepiece filters are not safe.*

Observing with Special Eyepiece Filters

There are special eyepiece filters available in the market, which are also known as 'helioscopes', they reduce the sunlight and heat through reflection, refraction or variable polarisation or a combination of all three. The best helioscopes are capable of variably attenuating the solar image without image distortion. For visual observations practically any filter or helioscope with transmission between 0.01% and 0.001% can be safely used. For photography of the whole or parts of the disk, objective filters with transmission of 0.1% to 0.3% are generally used, in conjunction with a slow speed, fine grain film. But for high resolution images produced by large focal ratio telescopes, a 0.1%–0.3% transmission filter may attenuate light too much and may require longer exposure times, which may distort the image due to seeing fluctuations. Correct exposures have to be decided by trial and error, depending on the film speed, effective f-ratio of the telescope, transmission of the filter, seeing etc. For visual and photographic observations, an optical Herschel wedge has been found to be very useful. This is essentially a thin glass plate with a small wedge angle (about 5°), by which 90–95% of the strong sunlight is reflected and only about 5% is transmitted. Thus a large fraction of the light and heat is rejected, only a small amount, suitable for visual and photographic use is available. Therefore, without loosing the telescope resolution, a Hershel wedge provides a good means to reduce the intense light and heat from the solar image. Any good optical telescope (refractor or reflector) can be used for solar observations, provided proper care is taken to reduce the incoming sunlight to safe levels. Under good to very good seeing conditions, about 2–3 arc second resolution could be achieved with a refracting solar telescopes of 8 to 10 cm optimum aperture. The initial cost of such a refractor will be higher than that of a reflector, but is preferred in spite of the inherent defect due to the chromatic aberration (that different colours have a different focus). A generally asked question, is what would be the ideal aperture of a solar telescope to give the best or optimum spatial resolution? Theoretically, the angular resolution depends on the aperture of the telescope (D) and the wavelength of the light (λ) used, a relation obtained from the Rayleigh criteria is given by $\sim \lambda/D$. The resolution limit of a diffraction limited system is $1.22\lambda/D$ radians. As a rule of thumb, approximately the angular resolution of an optical telescope, working in the visible spectral range, is given in seconds of arc by 120/diameter of the objective in mm.

Thus, theoretically a 100-mm aperture telescope could yield a spatial resolution of 1.2 arc seconds. However, such a resolution is very hard to achieve in practice. The smallest white light features which one may aim to observe are granulation elements, intergranular lanes, penumbral fine filamentary features, bright bridges in sunspots, dark umbral dots and any new feature, they range in size from about less than 1 to 2 arc seconds. But due to the atmospheric turbulence, the solar seeing does not permit one to observe features less than 1 to 2 arc seconds. In addition to the seeing, the film and other parts of the system also degrade the spatial resolution, such that features of this size cannot really be distinguished. An important factor called the Modulation Transfer Function (MTF) has to be considered. This depends on the detector characteristics, the optics and the atmospheric conditions. A beginner may not be able to initially observe solar features of 1 arc second. Nevertheless, to start with a refracting telescope of about 8–10 cm aperture should be quite adequate to observe sunspots, penumbral filaments, pores, faculae, occasionally granulation, and if one is lucky even the rare 'white light' flares.

Modulation Transfer Function

In simple terms the Modulation Transfer Function (MTF) is defined as a quantitative number to indicate the amount of blurriness in an image, formed by an optical system. The properties of such a system can be described by a point-spread function or a line spread function, but the MTF is more convenient. Mathematically it is defined as the ratio of the output modulation of a sinusoidal wave form with spatial frequency ω, to the input modulation of the same frequency. To measure MTF, a regular pattern (such as a sequence of bars) of a certain spatial frequency is imaged through an optical system and the resulting distribution is detected either on a fine grain film or a photoelectric scanner or a CCD camera. The ratio of the amplitudes of the observed to the initial distribution is the MTF for that spatial frequency. If the system is perfect the MTF is unity. The MTF for the over all system is the product of the MTF's of each component, that is objective, film, atmospheric seeing and any other intermediate optics. In astronomical telescopes, especially in reflectors with central obscuration, due to the secondary mirror, the spatial resolution is almost half as compared to refractors. This is essentially due to the degradation of MTF at the secondary. Note that the MTF concept is also applied for temporal frequencies. For more details, the readers are referred to monographs by Smith (1966) and Dainty & Shaw (1974).

2.2 Observation of the Chromosphere

The chromosphere is a region just above the photosphere (or solar surface) extending up to about 2000 km, between the photosphere and the corona. Between the upper layer of chromosphere and the corona (although the demarcation is not sharp) lies the 'transition layer', where the temperature rises very steeply, from about 25 000 to 500 000 K in height difference of just 1000 km. In terms of the density ($\approx 10^{-12}$ g cm^{-3}), the chromosphere is substantially more tenuous than the photosphere ($\approx 10^{-7}$ g cm^{-3}) and the intensity of the emitted radiation is several tens of thousand times less than that of the photosphere. In the higher photosphere and low chromosphere the intense photospheric continuum radiation is absorbed, resulting in strong dark absorption lines of certain atoms and ions. Earlier, this line-forming region was known as the 'reversing layer'. Besides absorbing photospheric radiation, the chromosphere also emits radiation, but due to its low density, the emission is weak and therefore, can not be seen against the strong photospheric background. Hence, the chromospheric emission spectrum can be observed only during the few seconds before or after a total solar eclipse (just before the second and after the third contact).

These spectra taken during the total solar eclipse are called 'flash spectra'. They are formed in the chromosphere which is the region just a few seconds of arc beyond the solar limb (about 1500–2000 km). However, in 1909, Hale and Adams had succeeded in photographing the flash spectrum even outside of solar eclipse, with the 60-foot tower telescope and 30-foot spectrograph of the Mount Wilson Observatory. A description of these observations and spectra, taken in the region 4800–6600 Å, is given by Adams and Burwell (1915). In recent times, a detailed atlas of the flash spectrum of chromospheric emission lines, in the spectral range 3040–9266 Å, has been prepared by Pierce (1968), using the 60-inch solar telescope of the National Solar Observatory, at Kitt Peak. To observe the chromospheric features in front of the solar disk, one has to use the strong chromospheric absorption lines of Hydrogen or ionised Calcium and Magnesium, in which the photospheric continuum is eliminated. This is achieved by using narrow passband filters or spectroscopes, spectrographs and spectroheliographs, which allow one to observe the specific chromospheric spectral lines in which the strong line absorption suppresses the background continuum radiation. These instruments are described in Sect. 3.

2.3 Observation of the Corona

Beyond the chromosphere extends a very tenuous layer of the solar atmosphere, stretching to many solar radii, called solar corona. Because the density of the corona is so low, the emission from it is 100 million times less than the photospheric intensity. Therefore, due to the intense glare of the Sun, the solar corona is not visible outside a total eclipse. The reason for this is that the intense photospheric light is scattered by the dust and air molecules in our Earth's atmosphere, and the intensity of that scattered light is several million times greater than the faint coronal intensity. During a total eclipse, the Moon cuts off the intense photospheric light and only the faint coronal light enters the Earth's atmosphere, which of course is also scattered in our atmosphere. But the intensity of this scattered light is several times less than the coronal emission, thus the faint solar corona extending to several tens of solar radii, becomes visible. In 1930, the French astronomer, Bernard Lyot (1930, 1939) came up with the brilliant invention of an instrument called coronagraph, through which one could see the bright inner corona even without an eclipse.

The coronal emission has three components: the emission line or E-corona, the K-corona, and the F-corona. The coronal emission lines were first observed during the 1868 solar eclipse, and posed one of the major puzzles for more than 70 years for solar physics research. Earlier it was thought that they come from a strange element called 'coronium', until on Grotrian's (1939) suggestion Edlén (1943) correctly identified them as due to the forbidden transitions between low-lying fine structure states of heavy and many times ionised atoms. Edlén (1943) first identified four emission lines in the solar corona, originating from Fe X, Fe XI, Ca XII, Ca XIII. In all Edlén identified 19 of the 24 coronal emission lines known at that time. The best-known lines are the 5303 Å green line of Fe XIV, the 6374 Å red line of Fe X, and the 5694 Å yellow line of Ca XV. Temperatures of more than one million degrees K are needed to produce these high ionisation states. This fact also explains the great height of the corona. If the corona had photospheric temperatures the scale height, which is the height over which the density decreases by a factor $e \approx 2.7$, is only about 150 km. Therefore, at distances of a solar radius (7×10^5 km) the density would be essentially zero. However, with a temperature of the order of two million K the scale height is about 10^5 km which explains why the extension of corona is so large.

The second component of the corona, which is perhaps the main component, extends to several solar radii beyond the solar disk, and is called the K-corona (after the German word 'Kontinuum'). The K-corona arises from the scattering of the photospheric light by the high speed electrons in the corona, which smear out the Fraunhofer lines due to the Doppler shift and make them almost undetectable, except for the strong H and K lines due to ionised Calcium ions. The K-corona is linearly polarised due to the alignment of electrons in a coronal magnetic field and the intensity depends on the electron density. The electron density and the alignment of the coronal magnetic field, are determined from the polarisation and intensity measures, made during the total solar eclipse. The third component, the Fraunhofer, or the F-corona arises due to the scattering by the slow moving dust particles. It extends all the way into the interplanetary medium and there it is observed as the zodiacal light, a faint emission concentrated along the ecliptic. The Fraunhofer lines scattered by the dust particles are clearly seen in the spectrum of the F-corona. Of course, the F-corona is not a part of the solar atmosphere. At many observatories around the globe, mainly in America, Europe and Japan, observations of the E- and K-coronae of the inner corona are regularly made, and these coronal data are available on the web sites.

However, the total solar eclipse observations are still important and it is a 'once in a life time experience' to witness one.

3 Solar Instrumentation

Several types of specialised telescopes and instruments have been developed to observe the Sun. Since the Sun is very bright, the system need not be as fast (small f-number) as stellar telescopes. Solar telescopes and other associated instruments are described in the following subsections.

3.1 Solar Telescopes

Solar astronomers usually use either refractors or reflectors with long focal length, which produce large solar images. In addition to the requirement of light gathering power, magnification and the optical quality, two important constraints are placed on solar telescopes:

1. That the heating of the air column inside the telescope tube should be zero or minimal. When the bright image of the Sun, particularly after focusing by the objective, falls on a lens or mirror in the optical system, small amounts of light and heat will be absorbed at each surface and produce thermal currents in the telescope tube which distort the solar image. Therefore, solar telescopes are designed to minimise or completely eliminate thermal currents in the tube, either by evacuating or by filling the optical path in the telescope with helium gas. However, these measures require major efforts and are necessary for large aperture telescopes, which may be beyond the means of amateur astronomers. A relatively new concept, which lately has come in vogue, is to flush out the hot air from the optical surfaces of a solar telescope by fast flowing air. It is used in the Dutch Open Telescope (DOT) (Hammerschlag & Bettonvil 1998; Rutten et al. 2000; Sutterlin 2001).
2. The second important consideration in solar telescope design is the question of guiding and rigidity of the telescope mount. The surface of the Sun continuously displays dynamic phenomena, where the physical condition in each region differs greatly in spatial and temporal domain. In addition, the scale size could be as small as the resolving power of the telescope or be limited by the seeing. Thus, if we want to study a particular feature on the Sun, we must be able to keep our telescope pointed at that feature, for as long as the observation takes place. This is generally difficult due to the motion of the position of the solar image due to the atmospheric turbulence, seeing, vibrations as well as the shaking and inaccuracy of the telescope drive. Normally, a high quality photoelectric solar guider is attached to a solar telescope, which helps to point the telescope accurately and keep it in position with a fair degree of accuracy of better than $1''$ arc. Through the techniques of active and adaptive optics, solar image motion stabilisation and even correction to the wave front distortions has become possible on a real time basis. Some of these systems will be discussed in the following section.

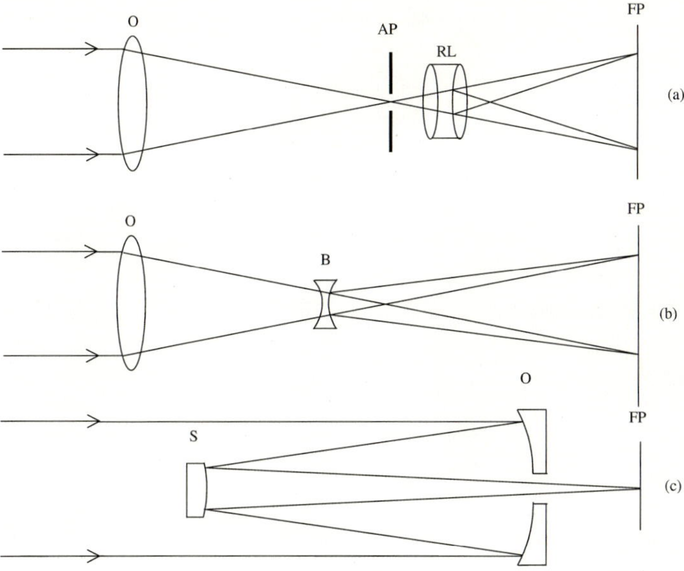

Fig. 1. Optical layout of 3 simple type of solar telescopes, (a) Relay lens RL used to enlarge the image, the image is limited by an aperture stop AP, (b) Solar image enlarged by a negative Barlow lens B, (c) Reflecting Cassegrain telescope with hyperbolic secondary

The simplest design of a solar telescope consists of an objective (convex lens or concave or spherical mirror). Following it an enlarging optics is used, either a negative Barlow lens or a positive projection relay lens. The Barlow lens is placed before the focus, while the relay lens after the prime focus. With an appropriate choice of the focal length of the enlarging lens and the distance from the focus of the objective, the image size and scale can be adjusted. For taking full disk images, it is necessary that the aperture of the enlarging optics should be large enough so that no vignetting (pillow or barrel type distortions) occurs in the solar image. For taking observations of a small region on the Sun, an aperture stop with a suitable hole is placed at the prime focus and only a small region of the Sun is enlarged and focused on the image plane. In this setup, a major portion of Sun's light and heat is reflected out of the telescope, thus maintaining a lower temperature and minimising air turbulence in the telescope tube. Figure 1 shows the optical schematic layouts of 3 simple types of solar telescopes.

Mounting of Solar Telescopes

Normally solar telescopes are either equatorially or fork mounted or make use of the concept of a "Spar". As the Sun's declination remains within $\pm 24°$ of the equator, a solar telescope does not need to cover the whole sky like a stellar telescope, and secondly a solar telescope looks at only one single object, therefore,

Fig. 2. 12-Foot solar spar telescope of the Udaipur Solar Observatory, situated on a small island in the Fatehsagar Lake, Udaipur

Spar mountings are preferred for solar telescopes. A "Spar" is simply a box like framework structure that is pointed towards Sun. Several telescopes and associated optics can be installed on it or inside. Figure 2 shows the typical 12-foot Spar telescope of the Udaipur Solar Observatory, which was originally at CSIRO, Culgoora, Australia.

Heliostats, Coelostats and Siderostats

In some types of solar telescopes, plane mirrors are used to divert the sunlight onto the objective. Depending on the number of mirrors or the type of mounting used, these devices are called heliostats, coelostats or siderostats. A single mirror is placed at some height along the Earth's polar axis in a system called polar heliostat. The sunlight is reflected by this mirror and is diverted onto an objective lens or a mirror, as in the case of National Solar Observatory's 80-inch polar heliostat H coupled with 60-inch spherical objective mirror O (Fig. 3b). However, this system produces a rotating solar image. In the case of a coelostat, there are two mirrors, of these, one tracks the Sun, the light reflected from it is diverted to another 'fixed' mirror, which reflects the light to an objective lens or mirror system to form the solar image. As the face of the tracking mirror lies in and rotates about the polar axis, coelostats give a non-rotating image, in a fixed (usually vertical) direction. An example is the Mount Wilson Observatory's 150-foot tower telescope (Fig. 3a).

In the case of a 'siderostat' arrangement, a single mirror is alt-azimuth mounted and through a clever design, the movement of the mirror is trans-

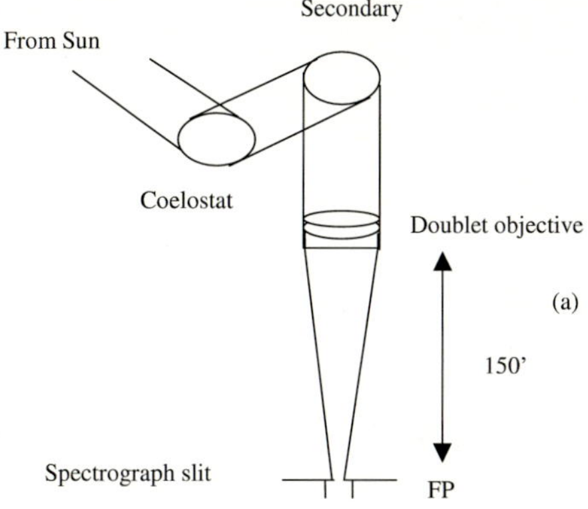

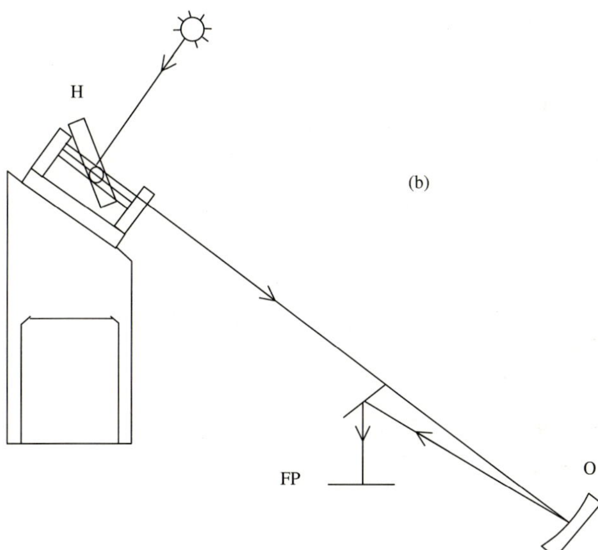

Fig. 3. Optical layout of the (a) 2-mirror coelostat of the Mount Wilson Observatory, and (b) Polar heliostat of the National Solar Observatory at Kitt Peak

formed through a mechanical gear arrangement into an equatorial motion. An 12-inch aperture siderostat is working for the last 100 years at the Kodaikanal Observatory. In all these systems the biggest advantage is that the image plane is fixed, therefore, large and heavy instruments such as big spectrographs, spectroheliographs, filters etc. can be coupled to the telescope. To track the Sun in all these systems, the rotation rate of the tracking mirror is once in 48 hours. In both heliostat and siderostat arrangements, the solar image rotates once in 24 hours around it's centre, while in a coelostat the image does not rotate and is fixed. Therefore, coelostats are generally preferred, but have the disadvantage

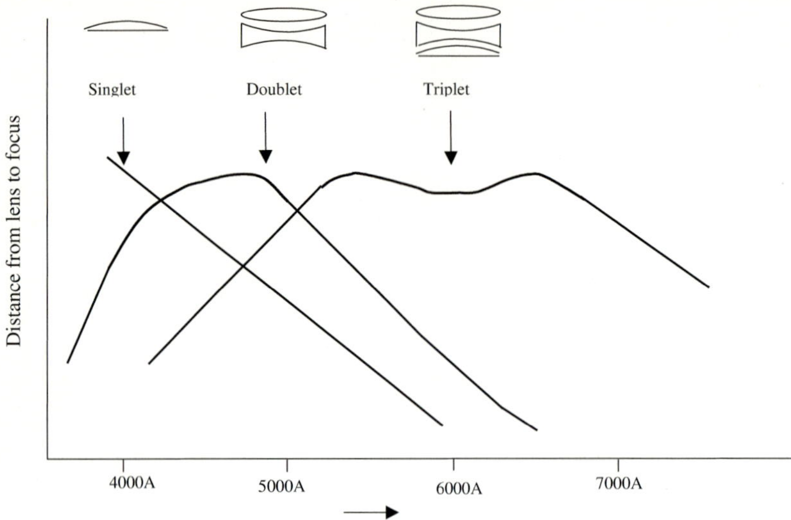

Fig. 4. Typical colour curves of singlet, doublet and a triplet lenses, the X-axis displays the wavelength λ in Å and the Y-axis has an arbitrary scale. For singlets the focal length is shortest for the violet and longest for the red. Doublets and triplets are designed to maintain almost constant or nearly the same focal length for visual and photographic ranges

of using two mirrors and at each mirror surface there is some absorption of light and heating effecting the local seeing. There are various ways to compensate for the rotation of the image, either by rotating the observing table itself, or use an image rotator.

Lens Versus Mirror Objectives

In small and medium size solar telescopes, lens objectives are generally preferred. Lens objectives are normally doublets to correct for the chromatic aberration. Singlet lens objectives are also used in the case of coronagraphs and special purpose solar telescopes dedicated for monochromatic observations, such as the twin 25-cm aperture telescopes of Big Bear Solar Observatory, the Udaipur Solar Observatory's 25-cm refractor and many other dedicated telescopes. To reduce light and heat in the telescope tube, at some observatories, the objective lens is coated to yield a narrow bandpass of few hundred Angstrom wide, but then the observations are limited to a narrow spectral range.

Typical colour curves of singlet, doublet and triplet lenses are shown in Fig. 4. The disadvantage of a singlet lens system is that not all the wavelengths can be focused at the same focal plane. However, secondary colour correctors are now available, which correct colour aberration to some extent. In mirror optics this problem of chromatic aberration is not encountered, but the heating in the telescope tube, due to multiple reflections is a serious problem, for achieving high spatial resolution.

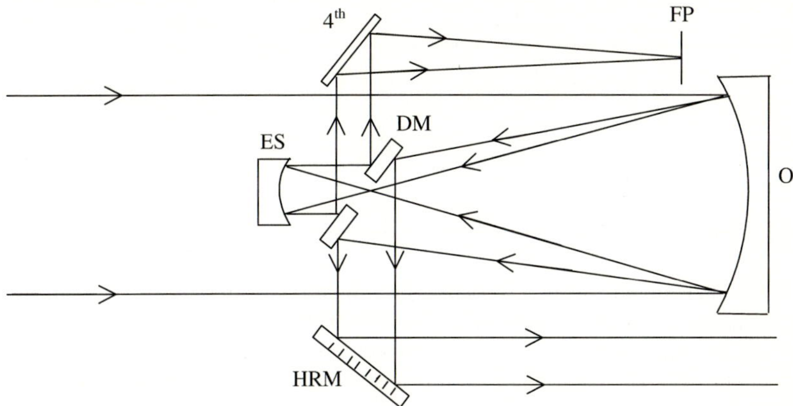

Fig. 5. Optical layout of a Gregorian telescope. FP indicates the focal plane, ES is the elliptical secondary, O is the objective and HRM is the heat rejecting mirror. DM is the aperture stop and third mirror. It is a double mirror where one surface reflects the extra light out of the tube and the other the enlarged image from ES onto the fourth mirror to form a solar image at FP

In the case of advanced solar telescopes, a number of innovative designs has been proposed. One such design of a Gregorian solar telescope is shown in Fig. 5.

3.2 Coronagraphs

As the intensity of the solar corona is several million times less than the photospheric intensity, the faint corona can be seen only during the brief moments of the total solar eclipse, when the Moon cuts off the intense sunlight. Lyot (1933) came up with an ingenious design of an objective lens, which enabled him to observe the solar corona even without a total eclipse. The trick was to have a highly polished, scratch and bubble free singlet objective lens, to take great care to remove stray light and to observe from a high mountaintop where the atmospheric scattered light was minimum (see Fig. 6). For more details of coronagraph, the readers are referred to the review article by Evans (1953).

3.3 Spectrographs, Spectrohelioscopes and Spectroheliographs

A spectrograph is the most important instrument for astrophysical work and especially for solar studies. A typical spectrograph consists of a slit, onto which the solar image is focused. Behind the slit, a collimating lens or a concave mirror is placed to render the beam parallel, which then is followed by a dispersing unit, either a diffraction grating or a prism or a combination of the two. After the grating or prism disperses the light, it is focused by a camera lens or a mirror and recorded on a photographic film, plate or digitally using a CCD chip. Figure 7 shows typical schematic layouts of spectrographs using lens and mirror optics.

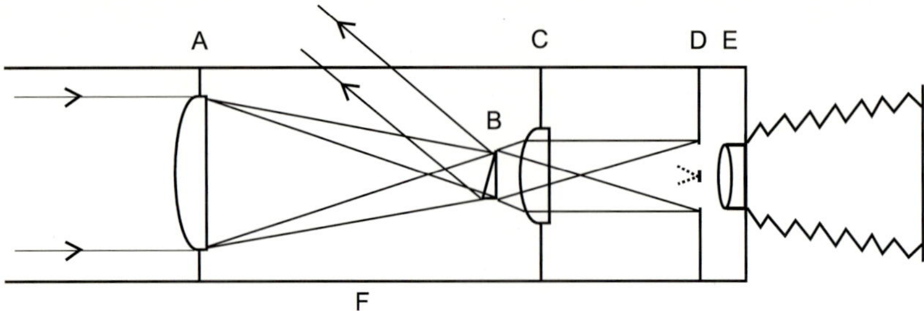

Fig. 6. Coronagraph after B. Lyot (1930). The objective A focuses the solar image on a disk shaped plate B which extends by $10''$ to $20''$ over the solar limb. The photospheric light falling on B is removed from the instrument. The field lens C generates an image of the aperture A at the plane D where a ring shaped aperture stop and a small central disk block stray light. Objective E generates an image of the corona

In a simple Littrow system, a single lens performs the function of both collimator and camera, as shown in Fig. 8. The slit is at the focus of the Littrow lens, producing a parallel beam. This light is diffracted by the grating and refocused by the same Littrow lens at the focal plane, producing spectral lines as images of the slit. This system has several advantages. It is very simple, symmetrical and convenient for the operation, it saves one optical component. But due to the chromatic aberration of the lens system, not all wavelengths can be focused at the same focal plane.

In many spectrographs, mirror systems are used to eliminate the chromatic aberration. In this case one does not need to focus for a selected wavelength range, and the collimator can be of small aperture but large enough to collect the whole solar beam, passing through the full length of the slit. The camera aperture has to be large enough to collect a large spectral range of the dispersed light. Generally, this system is limited in wavelength range by the 'off-axis' spherical aberration (that light rays away from the optical axis have a different focus than those close to the axis), common to all spherical mirrors and the scattered light problem, because the beams go back and forth through the spectrograph. To eliminate the off-axis problem, wide field Schmidt cameras are used. These are cameras with spherical mirrors as objective which employ a correction plate to rectify the spherical aberration.

In case of the solar spectrographs, it is essential that the light beam (image) from the whole length of the slit should fill the collimator and the dispersing unit, grating or prism. Hence, the collimator and the telescope f-ratios have to match for optimum light gathering. Therefore, the collimator has to be much faster (smaller f-number) than normally used for point source images. An illustrative example is shown in Fig. 9.

Echelle Spectrographs

It is sometimes interesting and necessary to observe the whole or a wide range of the solar spectrum as for flares, prominences and other transient phenomena.

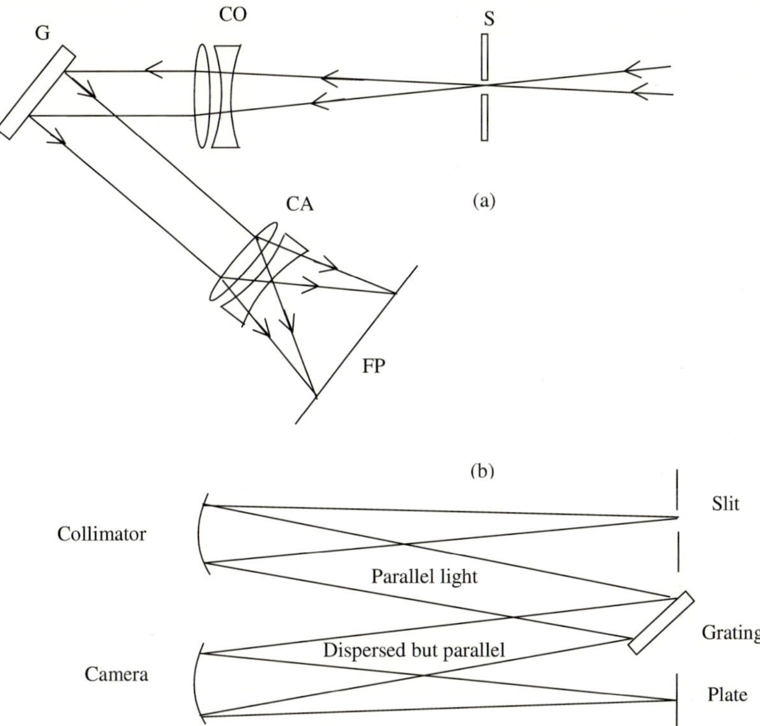

Fig. 7. Typical optical schematic layouts of spectrographs, using lens (a) and mirror optics (b). In (a) S indicates the slit of the spectrograph, CO the collimator, CA the camera, G the grating and FP the focal plane

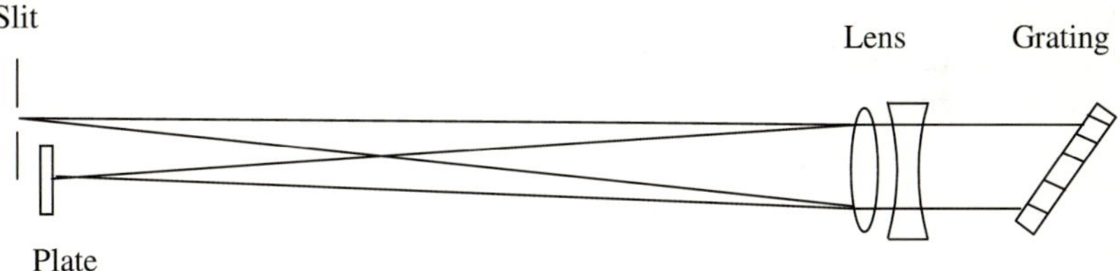

Fig. 8. Optical schematic of a Littrow spectrograph

For example, one may require a wide range of the spectrum to study the spectral changes in the flare development. For this purpose, the most effective systems are 'echelle' type universal spectrographs, as shown in Fig. 10. Echelle gratings consist of a stack of thin glass plates (≈ 40) arranged in a staircase type manner.

In this arrangement almost the entire visible spectrum can be observed by combining a prism and plane grating or using an echelle grating at a high angle (order). At high angles many orders are superimposed on each other, for example the 10th order green falls at the same place as the red of the 11th order and so on. In this case one sorts out the orders with a prism which disperses the light

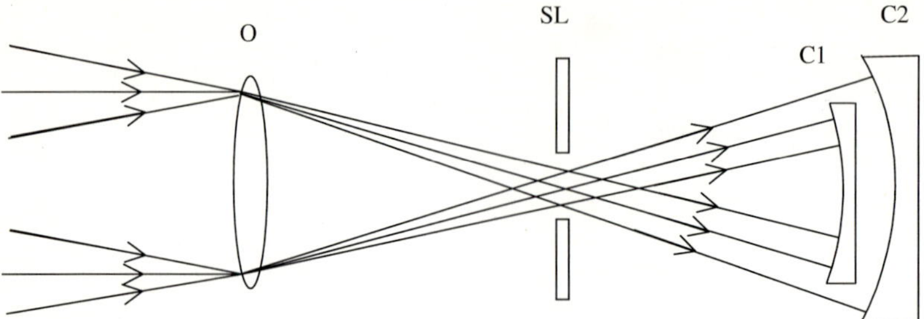

Fig. 9. Illustrating the use of bigger aperture collimators for solar spectrographs as compared to point source images. O indicates the objective, SL is the slit and C1 and C2 are the collimators, C2 has smaller f-ratio and is faster as compared to C1, to accept an extended solar image without vignetting

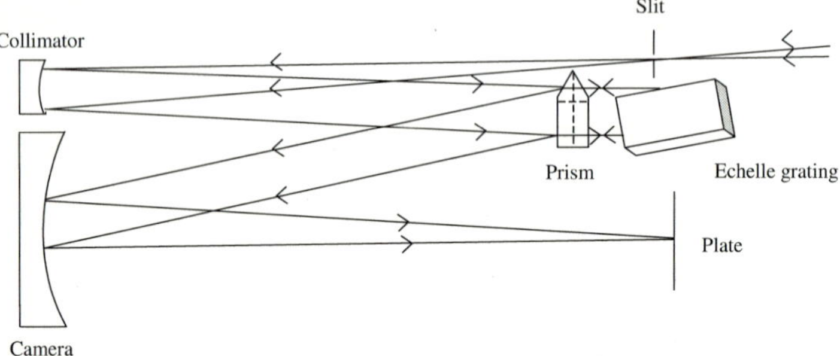

Fig. 10. Optical layout of an echelle spectrograph

perpendicular to the spectrum formed by the grating. This is shown in Fig. 11. Echelle spectrographs have been used for observing solar flare spectra in rapid sequence. The disadvantage of this system is that the photometry of the echelle spectra is extremely complicated for quantitative work, as certain areas of the photographic film may not be correctly exposed due to colour sensitivity of the film emulsion. However, using CCD cameras can solve this problem.

Resolution and Dispersion of a Spectrograph

All modern spectrographs use plane gratings, which diffract light according to the following grating formula

$$n\lambda = d(\sin\theta + \sin\phi) , \quad (1)$$

where n is the order, d is the separation of the grating lines, λ the wavelength, θ the angle of incidence and ϕ the diffraction angle. As n and λ occur as a product in the formula, the orders can not be separated. 6000 Å in the first order falls at the same place as 3000 Å in the second order, and appropriate filters are used to separate these orders. The inverse linear dispersion is the change in dλ per

Fig. 11. Echelle spectrum of a limb flare of May 27, 1959, showing a spectrum which covers the wavelength range from 6000 Å to the Ca II H and K (3933 Å) lines

linear interval $f\mathrm{d}\theta$, where f is the focal length of the spectrograph camera lens, and is obtained by differentiating (1).

In the Littrow case $\phi \approx \theta$ and ϕ is constant,

$$n\,\mathrm{d}\lambda = d\cos\theta\,\mathrm{d}\theta\,, \tag{2}$$

$$\frac{\mathrm{d}\lambda}{\lambda} = \frac{1}{2}\cot\theta\,\mathrm{d}\theta\,. \tag{3}$$

The inverse linear dispersion expressed in Å/mm is given by

$$\frac{1}{f}\frac{\mathrm{d}\lambda}{\mathrm{d}\theta} = \frac{\lambda\cot\theta}{2f}\,. \tag{4}$$

This quantity depends on the focal length of the camera and the cotangent of the incident angle. Since the focal length is fixed, one can get large dispersions by tilting the grating and working in higher and higher orders. However, the efficiency of the grating decreases rapidly. To a certain extent this can be compensated by using 'blazed' gratings available for specific angles of dispersion. Blazed gratings are made by cutting the grooves at a certain angle such that most of the reflected light from the grating is directed in a particular direction, while in other directions or orders the reflected light is decreased. Such blazed gratings have been found to be very useful for working in specified orders.

High dispersion is useful only when the spectral lines to be studied are very narrow and the spectrograph slit is of the right width to resolve the fine solar features, or when line profile studies are made. For example, a telescope of focal length f in mm gives an image scale of $f/206265$ mm/arc second at the focus of the telescope. A 20 cm, $f/30$ telescope yields an image scale of 0.03 mm/arc second. As the solar seeing is rarely better than $1''$ arc, therefore, to keep the slit width less than 0.03 mm is of no advantage.

In the simplest terms, we may define resolution of an optical system as the smallest detectable adjacent points in an image, which do not appear to touch each other. In case of a solar spectrograph, the resolving power depends on the image scale of the telescope as well as the slit width and is given by

$$R = \frac{s}{2000d} \,. \qquad (5)$$

Here s is the slit width in mm, d the diameter of solar image in mm, and 2000 arc seconds is taken as the mean angular diameter of the solar image.

Scattered Light and Ghosts in Spectrographs

Due to multiple reflections from several optical surfaces in a spectrograph, a considerable amount of 'white light' or undispersed scattered light exists in spectrographs, which hinders quantitative measurements of line intensities and line profiles etc. Other inherent defects of grating spectrographs are due to 'ghost' spectra and polarisation at the grating surface. Due to the periodic errors and the imperfect ruling of the diffraction gratings, ghost images are superimposed over the normal spectrum. To remove this defect, double pass spectrographs have been designed and constructed at the Kitt Peak Observatory by A. K. Pierce (1964). A typical optical layout of a double pass spectrograph is shown in Fig. 12.

Diffraction gratings available in the market are replicas of the original ruled gratings on glass substrates. Recently, holographic diffraction gratings are available, they have almost zero scattered light and no ghost images, as these gratings are not made by ruling engines which tend to introduce periodic errors, but through holographic techniques. However, 'blazing' the grating, that is, to divert a major portion of the diffracted light to a particular angle, is not possible in the case of a holographic grating.

Spectroheliographs and Spectrohelioscopes

In 1891, George Ellery Hale (1892) at his private Kenwood Observatory in Chicago invented an instrument to photographically record the monochromatic images of the solar chromosphere by combining the principle of the spectrohelioscope with the photographic plate for a permanent record. Around the same time, in 1891–92, independently H. Deslandres of the Meudon Observatory in France and Evershed constructed spectroheliographs.

The spectrograph breaks up the Sun's light into a spectrum of colours, crossed by thousands of dark Fraunhofer absorption lines which are in fact images of the slit. The image of the Sun falls on the slit of the spectrograph and produces a spectrum of the part of the Sun covering the slit. To observe another portion of

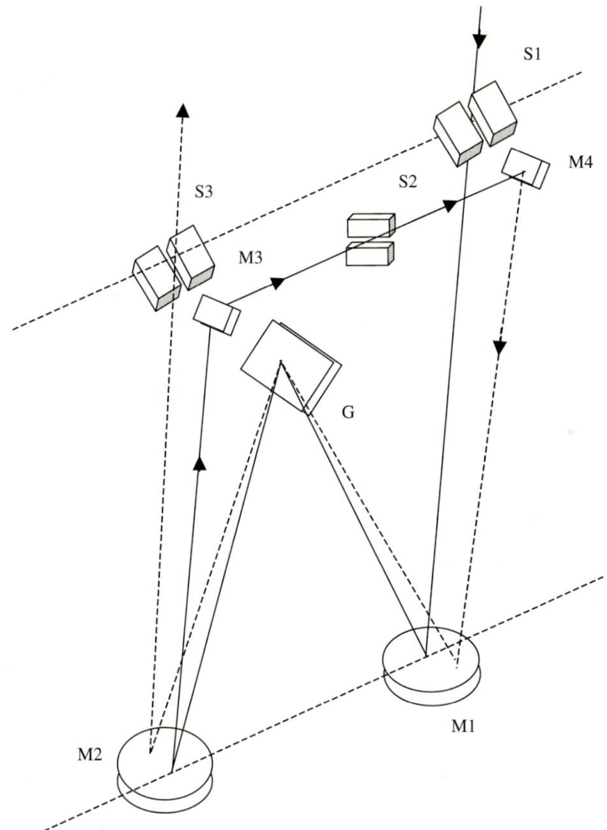

Fig. 12. Optical layout of a double pass spectrograph. S1 is the first slit, M1 and M2 are the collimator and camera spherical mirrors of the spectrograph. The rays shown by the *solid* line form the first spectrum and is picked up by mirror M3 and sent through the limiting slit S2. It is returned back via mirror M4 to the same spectrograph, consisting of M1, the grating G and M2. The rays forming the double pass spectrum is shown by *dotted* lines. As the major portion of the scattered light is eliminated at slit S2, the double pass spectrum formed at S3 is free of scattered light and ghosts. Since the dispersion is doubled the slit width can be double (after A. K. Pierce 1964)

the Sun one will have to either move the slit or the image and by staggering a series of slit images side by side, then a large region of the Sun or the whole solar image can be photographed in one particular line. Figure 13 shows the optical layout of a spectroheliograph. The image of the Sun falls on the slit of the spectroheliograph, which produces a spectrum of the part of the Sun, covering the slit. To observe the Sun in a particular line say the Hα line of Hydrogen at 6563 Å, one moves the image of the Sun across the slit, and observes in the focal plane the spectrum of the solar features in Hα which falling on the slit at that time. To isolate a particular spectral line, a second slit is placed in the focal plane of the spectrograph, and by letting a photographic plate move in this

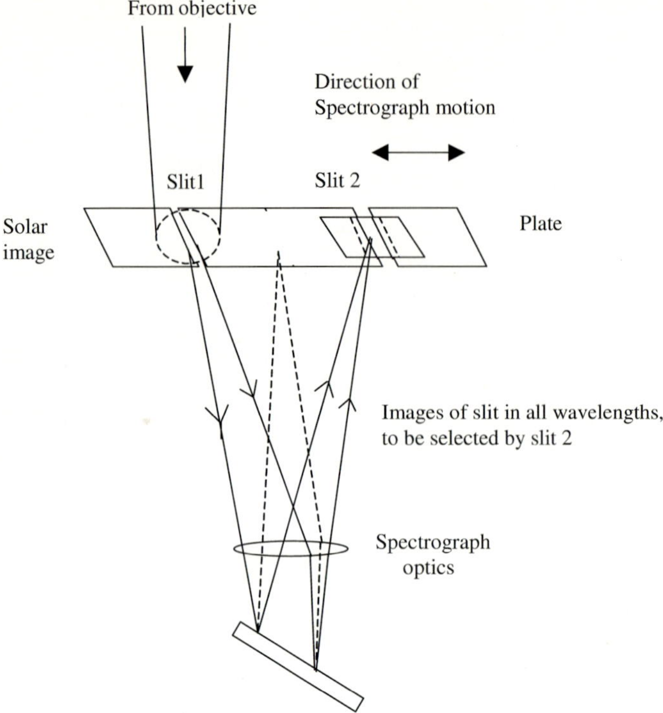

Fig. 13. Optical layout of a spectroheliograph

plane, at the same speed as the solar image on the first slit, a picture of the Sun, as seen in that line is built up.

In this way one obtains monochromatic pictures of the Sun in any desired wavelength. Spectroheliographs have the advantage that the bandpass can be varied and made very small to achieve high spectral purity. However, these have the disadvantage that a spectroheliograph takes considerable time to make a picture of the Sun, and is relatively slow and errors due to seeing and guiding may degrade the image quality.

Spectrohelioscopes

The solar eclipse of 1868 lead to the invention of an instrument called 'spectrohelioscope', for visual observation of the solar prominences in the monochromatic light of Hα. The principle of spectrohelioscope is based on allowing successively small portions of the solar image pass through the first slit of a spectrometer and simultaneously observing through the second slit, which isolates a particular line placed at the focal plane of the spectrometer. The solar image is scanned by an image rotator – "dove" prism, which is just a rectangular glass block placed in front of the first and the second slits. To scan the solar image, this prism is rotated at a speed of about 12 turns per second to visually observe a portion

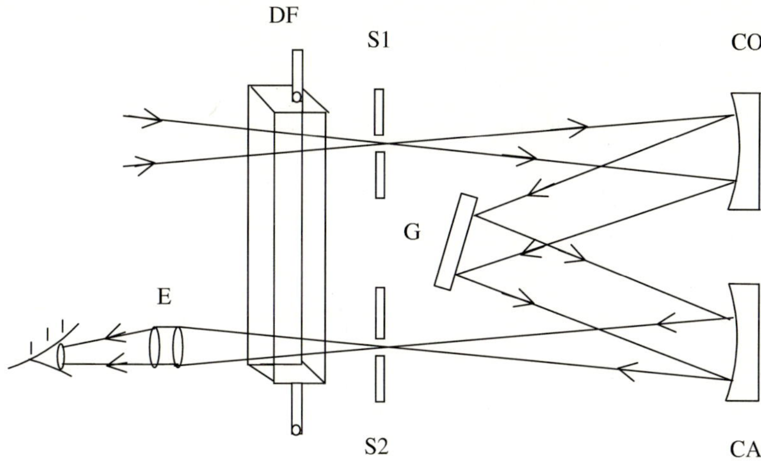

Fig. 14. Optical layout of a spectroheliosope, DF indicates the rotating 'dove' prism, S1 is the first slit, S2 the second slit, G the grating, CO is the collimator and CA the camera mirror

of the solar image falling on the prism, thus by persistence of vision (24 frames per seconds) one sees a two-dimensional monochromatic image of the solar features. Using this technique, in early 1870s, J. Janssen in France and J. Lockyer in England, were first to observe prominences and also chromospheric features outside a total solar eclipse. In the past this simple device had been widely used, all over the world for visually monitoring the solar activity. Spectrohelioscopes are quite easy to build at reasonable cost. Figure 14 shows its optical layout.

3.4 Narrow Band Filters

Until 1938 all the monochromatic observations of the Sun, the chromosphere and prominences were made through spectroheliographs, spectrohelioscopes, and spectrographs or during a total solar eclipse. In 1933, Bernard Lyot (1933) in France and Öhman (1938) in Sweden outlined the construction details of a new type of optical filter making use of the properties of birefringence of calcite and quartz crystals. These filters are generally called Lyot filters.

The principle of a birefringent filter is as follows: if a polarised light is passed through a quartz or calcite crystal, with the crystal face cut parallel to its optic axis, the polarised light is split into 2 rays, the ordinary and the extraordinary rays, travelling in the same plane but with different speeds and phase difference, because the refractive indices for the two rays are different. The ordinary ray travels faster than the extraordinary ray in the case of calcite ($\xi - \omega = -0.17$) and moves slower in quartz ($\xi - \omega = +0.009$) creating a path difference between the two rays which depends on the thickness of the crystal. These two rays interfere and produce interference fringes of bright and dark bands. If ξ and ω

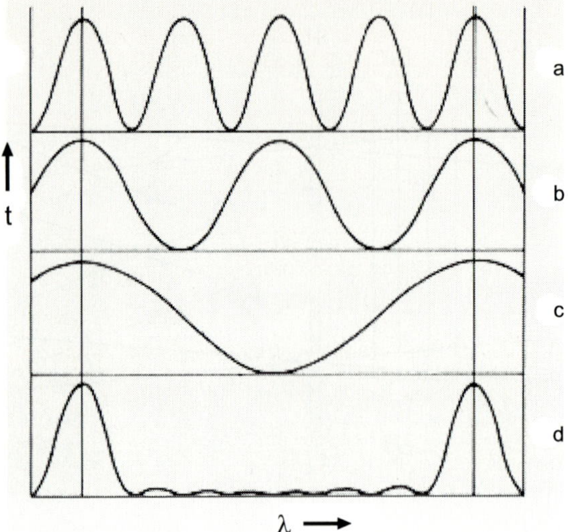

Fig. 15. Transmission curves for a 3-element filter. Shown is the transmission (intensity) as a function of wavelength λ, (a) for the thickest element, (b) and (c) for elements of one half and one quarter thickness, respectively, (d) shows the combined effect of all 3 elements (after Evans 1949). Very low intensity side bands are seen, which can be removed by using additional elements

are the refractive indices of the extraordinary and ordinary rays, respectively, a retardation n will occur for a particular wavelength when the two rays pass through a crystal of thickness d. n is given by the relation

$$n = \frac{d(\xi - \omega)}{\lambda} \,. \tag{6}$$

For some wavelengths, n will be an integer and the plane of polarisation is rotated back to the same plane, for other values of n, circularly or elliptical polarised light is obtained. For half integer values of n, the plane of polarisation is rotated by 90°. A birefringent crystal rotates the plane of polarisation by an amount proportional to the thickness divided by the wavelength. Now if a linear polariser is placed in the emergent beam, to pick out those wavelengths for which n is integer, the transmission of the polaroid - quartz - polaroid sandwich is given by:

$$t = \cos^2 \pi n \lambda \,. \tag{7}$$

The intensity of the emergent light is shown as a function of wavelength in Figs. 15 and 16.

One of the limitation of earlier birefringent filters was due to their narrow acceptance angle or the field of view. This problem was solved by Lyot (1944). He devised a wide field version, in which the narrow pass band elements were split and the two halves were rotated by 90 degrees and separated by a half wave plate. This has an effect of making the optical axes symmetrical and permits the

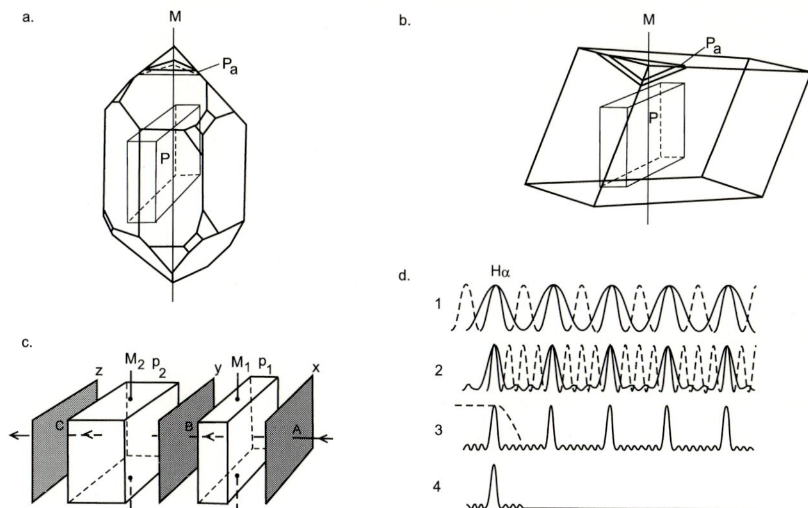

Fig. 16. (a) Quartz (b) calcite crystals, P mark monochromator plates parallel to the optical axis. P_a indicate test plates perpendicular to the axis. (c) assembly using crystal plates P_1 and P_2 (twice the thickness of P_1), and sheets of polaroid X, Y, Z arranged as a monochromator with the optical axes marked by M_1 and M_2. (d) Interference bands in the spectrum are formed by light passing through the monochromator: 1. with two birefringent plates as in (c), 2. with 3 plates, 3. with an interference pattern (3 plates) and transmission curve of red glass (broken line), 4. after combing all 3 plates, showing the resultant pattern as a function of intensity versus wavelength, after light passed through the red glass used to suppress the side bands

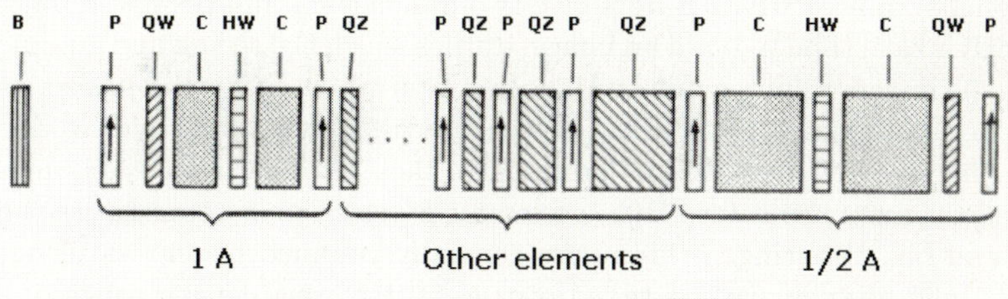

Fig. 17. Arrangement of plates in a Lyot filter. It consists of a series of polaroids, quarter-wave plates and quartz/calcite plates with half-wave plates sandwiched in between

entrance beam up to $f/15$. In addition to this, each element can be tuned in wavelength by placing a quarter-wave plate before the second polaroid. Figure 17 shows an arrangement of series of quartz/calcite plates in a Lyot filter, alternating with polaroid, quarter-wave plate. The thickest elements, used for 0.5 Å and 1.0 Å band pass, are split and half wave plates are sandwiched in between. Each plate is twice as thick as the preceding one, so that the total transmission of the stack is given by:

$$T = \cos^2 \pi n\lambda \, \cos^2 2\pi n\lambda \, \cos^2 4\pi n\lambda \, \cos^2 8\pi n\lambda \, \ldots \, \cos^2 2^{k-1}\pi n\lambda \, . \tag{8}$$

Here k, is the number of plates. The transmitted light exits in the form of a number $(2^k - 2)$ of widely spaced maxima. The width of each maximum is determined only by the thickest element. In narrow bandpass Hα filters, generally calcite is used for 0.5 Å, 1 Å, 2 Å, 4 Å and 8 Å elements, while quartz elements are used for 16 Å and 32 Å bandpass, as it is difficult to work with very thin calcite plates to make elements of more than 8 Å passband. For Hα filters the typical thickness of a 0.5 Å calcite element is about 26 mm and has an aperture of about 30 mm. The thickness of successive elements is reduced by half, for an 8 Å element it is about 1.6 mm. A broad band multi-layer interference blocking filter of about 50 Å bandpass is used to remove the side band peaks which are separated by 32 Å. As the refractive index and thickness of calcite and quartz elements are highly temperature sensitive, the crystal stack with the polaroids, wave plates, broad band blocking filters and the end glass plates, all are placed in a precision temperature controlled chamber, to maintain a constant temperature to within ±0.1 C. There are just a few companies in the world making narrow band birefringent filters. The Halle Nachf. in Germany used to make Hα filters of 0.5 Å pass band. The Nanjing Instrument Factory in China has started making birefringent filters with narrower passband of 0.25 Å and 0.15 Å. The Carl Zeiss Company in Germany used to make narrow band birefringent filters of passband of 0.25 Å, but now have stopped making them.

Birefringent filters have some advantages and disadvantages as compared to spectroheliographs (SHG). Filters give two-dimensional images instantaneously, while SHGs take time to build an image, during which the seeing and guiding may distort the image. Filters made by Halle, are lightweight and small in size, usually about 15 cm in diameter and 25 cm in length, while filters made by Carl Zeiss are slightly bigger. Being small and lightweight, these filters can be easily mounted on small telescopes, but for SHGs a fixed focus telescope is required, as they are heavy and big. Birefringent filters are generally confined to one wavelength. In case of Halle filter, it can be tuned over a limited wave length range of about ±1.0 Å, while the Zeiss filter can be tuned over ±16 Å range. The acceptance angle (field of view) of a birefringent filter is rather limited. Solar beams faster than $f/15$ create non-uniformity over the field of view. Based on Beckers (1973) design and using achromatic wave plates, the Carl Zeiss Company had made a Universal Birefringent Filter (UBF) which covers a wide spectral range from about 4200 Å to 7000 Å, and using blocking filters for specific wavebands one can observe in any line in this spectral range. To make narrow bandpass Lyot filters is a science as well as an art. Henry E. Paul (1980) and Edison Petit (1980) have described practical techniques to make and test birefringent filters.

To observe the chromosphere against the disk, one needs a narrow passband filter of 0.5–0.7 Å. But for prominence observations, seen against the dark sky background, a 3–4 Å passband filter is quite adequate. Because one blocks the strong light of the solar disk by an occulting disk and observes only the outer region of the chromosphere, of course, for even prominence observation through 3–4 Å passband filter, the sky should be very clear, free of haze and scattered light, otherwise the contrast decreases. Another advantage of a broad band filter is that one can observe fast moving prominences with up to 100–150 km s^{-1} line of sight velocities. Lyot filters are widely used for the study of flares, mass ejections and a variety of chromospheric phenomena. For making magnetic and velocity field measurements, narrow birefringent filters with passbands of 0.25 Å or narrower are used for photospheric lines.

Solid Fabry–Perot (FP) etalons are also being used as narrow passband filters. One such filter is available from Daystar Company, USA, and another from Cornoda Filters, Tucson, USA. In the Daystar filter the bi-axial mica sheet is used as substrate, on which dielectric partial reflecting mirror coatings are put, this acts as a solid Fabry–Perot etalon, and produces the standard channel spectrum, with a series of intensity maxima and minima as a function of wavelength. This type of Daystar narrow band filter provides a passband of about 0.5–0.7 Å. The Daystar filters are small, inexpensive and less temperature sensitive. However, they are limited to narrow beams of about $f/20$ and slower, they have low transmission and can not be tuned in wavelength except slightly by temperature changes. They do not have a sharp cut-off and the passband falls exponentially, due to which considerable amount of continuum leaks through in mica etalon filters. But in the case of birefringent filters, the transmission drops steeply to become nearly zero beyond the passband, hence a much cleaner pass band is obtained. The Hα pictures taken through a Lyot filter and through a Daystar filter or other FP-type etalon filters, show chromospheric features quite different in appearance. The cost of Daystar filters is much less than that of Lyot filters.

Another type of narrow band filter has been designed by Cacciani (Agnelli, Cacciani & Fofi 1975) based on the atomic resonance principle and employing the Macaluso–Corbino magneto-optical effect. The principle of this filter is that the light passing through a sodium vapour cell if subjected to a strong longitudinal magnetic field, undergoes resonant scattering in the σ-transitions and is circularly polarised. If this cell is placed between two crossed polaroids, only the light absorbed and re-emitted in the σ-transitions will have its plane of polarisation rotated and pass through the cell. Thus a magneto-optical filter (MOF) isolates the wings of the Na D lines. If a second cell is placed in tandem, it can alternately select the blue or the red wing of the line. The main limitation of the MOF is that it can be used only in resonance lines of Sodium (D_1 & D_2) or Potassium (7699 Å). Due to the simplicity of their construction and being relatively cheap, these cells are used for measuring magnetic fields and line of sight velocity for studying solar oscillations.

3.5 Solar Image Guider

One of the most important requirements for a solar telescope is good pointing stability and guiding of the image. Since the solar surface varies from point to point in position and time, we must keep the telescope pointed at the same place as long as the observations require to study a selected feature. For dedicated and long period observations, it is most important to minimise image jitter due to seeing and inaccuracy in the telescope drive, and this is achieved by the use of a photoelectric servo-guider. Figure 18a shows a schematic of a solar guider. A solar guider consists of four photoelectric cells, which monitor the position of the telescope relative to the Sun and actuate servo motors, in the telescope drive to keep the solar image fixed at one place. If there is no flexure in the telescope, the image will remain fixed, of course, depending on the seeing. For small telescopes, a separate guide telescope, e.g., a 50 mm aperture $f/10$, refractor is quite sufficient, but for bigger telescopes more sophisticated solar guiders are required. Normally, a small image of the Sun is projected on a silicon quadrant cell, with an occulting disk that covers much of the image and allows only the outer edge of the Sun to be detected. The difference signal produced from the two photocells on opposite sides is fed to amplifiers and connected to servomotors attached to the right ascension and declination drives of the telescope. The response time of the telescope is limited by the resonant frequency of the telescope, which may be low if the whole telescope is moved. Therefore, lightweight piezo-electrically-controlled 'tip-tilt' mirror arrangement is being effectively used for fast solar image stabilisation.

Figure 18b shows typical arrangements of a simple solar guider using a quadrant cell and a 'tip-tilt' mirror image stabiliser, using a sunspot image as target area for image stabilisation. The tip-tilt image stabiliser can correct for image shift and image jump. However, this devise cannot correct for wave-front distortion, which results in de-focusing of the image due to the air turbulence and seeing effects. For correcting distortions of the wave-front, the adaptive optics techniques have to be used, details of which are beyond the scope of this review.

4 Solar Observations

Persons starting observational solar astronomy generally ask, how useful are solar observations for serious research, besides just being fun? The answer to these questions depends on the available equipment and resources. A large variety of solar observations can be made with any reasonably good equipment, either in white light or in monochromatic light. Solar observations of any kind are useful to understand the various phenomena occurring on the Sun, on short and long time scales.

4.1 Solar Seeing

All solar observers are familiar with the term "seeing", which is caused by the air turbulence in our atmosphere, and arises from the heating of the ground and the surrounding air by the Sun. The thermal currents are built up as the day progresses, and the solar ray (wave-front) coming through the air column experiences distortions and tilts, resulting in defocusing and shifting of the solar image in the focal plane. There are several criteria to estimate the solar seeing from white light images. The generally used and simple criteria are those given

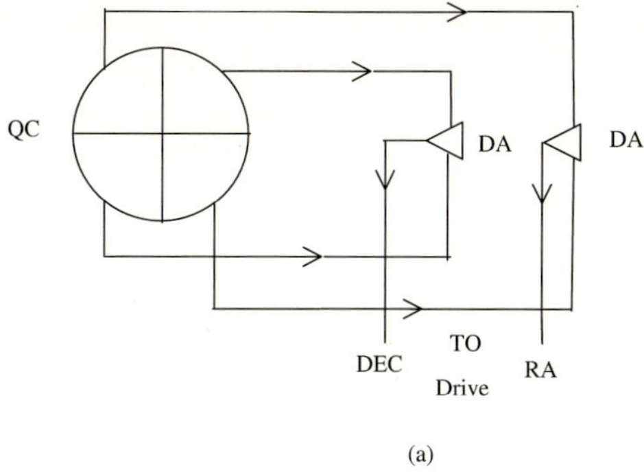

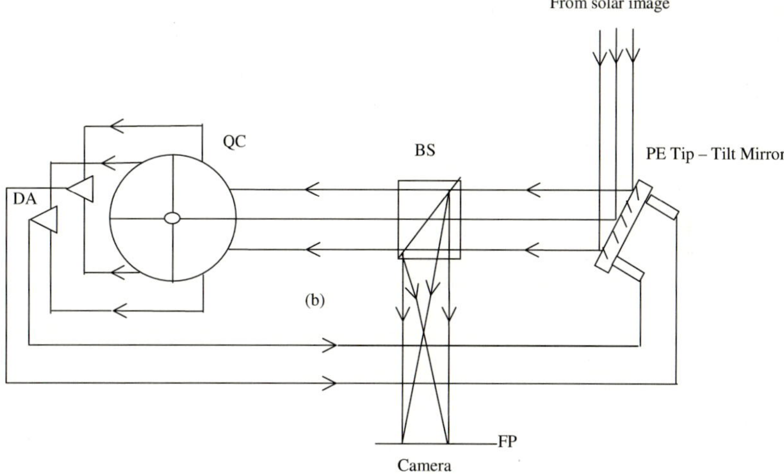

Fig. 18. Shows (a) a simple solar guider and (b) a 'tip-tilt' mirror image stabiliser system, QC indicates guard-photo cell, BS is a beam splitter, FP the focal plane, DA are differential amplifiers

by Kiepenheuer (1964). The seeing scale ranges from 1 to 5, 1 is the best and 5 the worst. These criteria take into account the "Image motion (IM)" at the limb, "Sharpness (S)" and "Quality (Q)".

Image Motion (IM): Qualitatively the magnitude of image motion can be classified as follows:

1. No image motion visible, neither at the limb nor on the disk,
2. Image motion $\leq 2''$ detected only at the limb, no motion on the disk,
3. Image motion $\leq 4''$ visible at the limb and on the disk,
4. Image motion $\leq 8''$ almost prevents distinction between umbrae and penumbrae, solar limb is strongly undulating or pulsating,
5. Image motion $\geq 8''$ reaches diameter of small spots. Solar image is heavily undulating

Under poor to fair seeing conditions, sometimes the solar limb appears boiling, this gives some idea about the degree of air turbulence. Depending on the wind direction, occasionally wavy patterns seems to travel around the limb. These factors are qualitative measures of solar seeing, nevertheless they are very useful to assess the solar seeing during observations.

Sharpness (S): For sharpness one can define the following levels:

1. Granulation is seen very conspicuously, and the structure of penumbral filaments can be recognised,
2. Granulation is well defined, the penumbrae are well visible, sharp boundaries between penumbrae and umbrae are seen,
3. Only traces of the granulation are visible, umbrae and penumbrae are well separated but seen without structure,
4. No granulation structure is detectable, umbrae and penumbrae are distinguishable only in large spots,
5. Granulation is not visible, umbra and penumbrae are indistinguishable even in large spots.

Quality (Q): The image quality can be classified as follows:

E -xcellent – Reserved for days where exceptionally clear details are visible,
G -ood – Average visibility of details on the solar surface,
F -air – Seeing is below average, but the observations are not adversely affected,
P -oor – Considerable image distortion,
W -orthless – Conditions are so bad, that no observation is possible.

From white light observations and making use of the above mentioned 3 main criteria, IM, S and Q and their subdivisions, one can easily estimate the solar seeing in a qualitative manner. Estimation of seeing during observations is very important, as it gives some idea about the quality and reliability of the data. Under exceptionally good seeing conditions, visual observations show much more details and fine structure as compared to a photographic record. The reason for this is that the human eye has the capability to detect fast variations in seeing and rapidly follows the image motion, as the brain freezes the best images, while a photograph integrates over a certain length of time, which results in a blurred integrated image. Visual solar observations by experienced observers are of immense importance even today. The drawings made visually of solar prominences by A. Secchi (1872) at the Vatican Observatory and by Professor Fernley in Oslo in the early 1860 and 70s show extremely fine details, which compare favourably well with the best photographs obtained with modern equipment. Over a hundred of Fernley's drawings have recently been "uncovered" by Jensen, Rustad and Engvold at the University of Oslo.

4.2 Sunspot Observations

One of the simplest observation of the Sun that one can make through a solar telescope, is in white light. On a white light image one sees sunspots, faculae, pores and granulations. The study of sunspots, which are the seat of solar activity, is of immense importance for short and long-term synoptic studies. Here we shall describe white light sunspot observations. The white light observations could either be made visually and by drawing, or recorded photographically or digitally. Generally the most interesting solar photospheric features are 'pores', sunspot umbrae and penumbrae, penumbral filaments, umbral dots, bright points, light bridges, bright rings around sunspots etc., which inspire solar observers to study their growth, decay and inclination. Under exceptional circumstances, perhaps one may be able to observe even white light flares, which are very rare, but very important to understand the energy generation and flare mechanism.

Pores: Pores are very small sunspots without a penumbra, they display rapid change in appearance and number. Generally they mark the position of a new emerging sunspot. Pores have diameters of $1''$ to $5''$ arc, and their lifetime is between a few hours to a day. The intensity in the pores is around 0.2–0.4 of the surrounding photospheric intensity I_{phot}. It is important to study the proper motion of pores because this indicates plasma flows, as well as changes in magnetic flux tubes beneath the solar surface, and provides information about the development of the sunspot group.

Sunspots: Observing sunspots is important, as is the study of their development, their classification, area determination, position and the sunspot number. Initially, sunspots appear on the solar disk as single small dark 'pores', which may grow into larger darker spots with penumbrae. Sunspots appear in pairs and develop over a large area with several small spots in the region. Spots can have different intensities, shape and sizes. Figures 19 and 20 show various types of sunspot groups and their nomenclature.

Void Areas: Under very good seeing conditions, and where granulations are missing, dark areas of about $1''$–$5''$ arc sizes and intensities between 0.7 and $0.75 I_{\mathrm{phot}}$ are seen. These dark areas called 'void' are quite different from pores. While pores are generally round, 'voids' can be irregular, and much smaller in size, and tend to be filled again by granulation, within a few minutes. These areas can change their brightness and then become the first stage in the development of a sunspot.

Sunspot Umbrae: Umbrae are the dark cores of sunspots with an average diameter of about 10 000 km and their colours vary from dark black to reddish brown. It is observed that the darker the umbra the greater is the magnetic

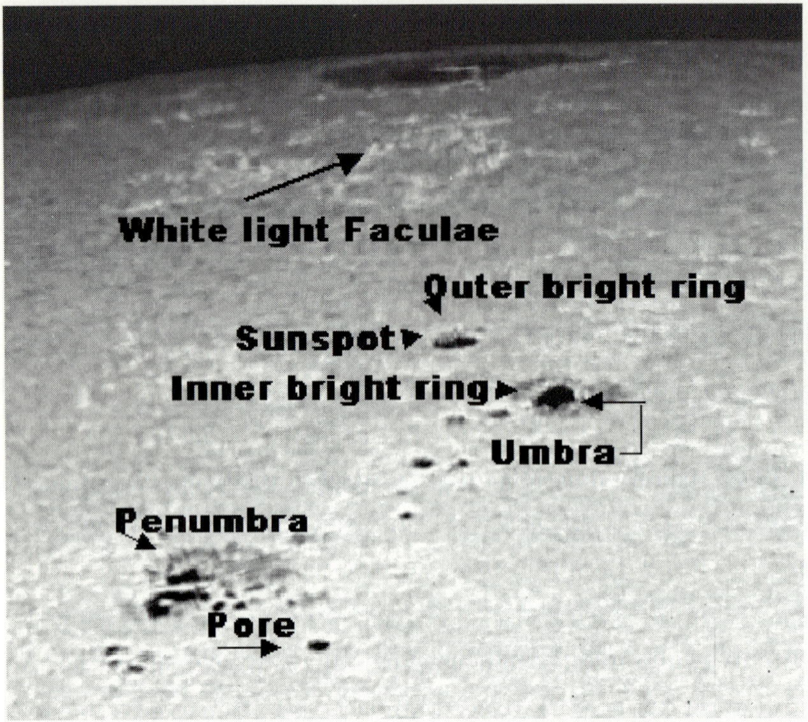

Fig. 19. Terminology of sunspots and sunspot groups

field strength. It is reported, that during the sunspot maximum period, umbrae appear darker compared to their appearance at sunspot minimum. A typical magnetic field strength in umbrae is about 2000 Gauss, however, magnetic field strengths of more than 4000 Gauss have been reported by Livingston (1976). The intensity of umbrae is about $0.1 I_{\rm phot}$, that is, umbrae are one tenth as bright as the photosphere. However, the brightness depends on the wavelength as well as on the seeing, the scattered light and the contrast of the image. The umbral temperature can be calculated from the Stefan-Boltzmann law

$$\frac{I}{I_{\rm phot}} = \left(\frac{T_e}{(T_e)_{\rm phot}} \right)^4 . \tag{9}$$

Assuming a photospheric temperature of 5780 K, the umbral temperature turns out to be around 3300 K.

Umbral Dots: Under good seeing conditions, slightly brighter dots appear in the dark umbra. They have diameters of about $0''.5$ arc and intensities of about $0.13 I_{\rm phot}$. The atmospheric turbulence and the scattered light distort their size and intensity. Determination of their true size is sometimes difficult due to these factors. What are these bright umbral dots, are they the manifestation of

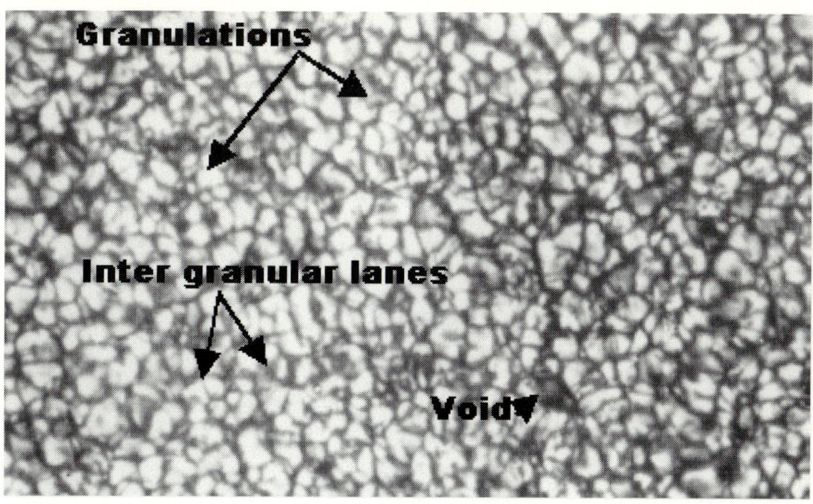

Fig. 20. High resolution white light picture showing the granulation, the intergranular lanes and 'voids'

convection in the umbra, and if so, how does convection occur in the presence of strong magnetic fields? Perhaps convection is not completely suppressed in umbrae.

Light Bridges in Umbrae: In spots, with a fairly large umbral diameter some times bright photospheric emission appears to penetrate the umbra, forming a light bridge, dividing the umbra into two or many parts and surrounded by a common penumbra. Generally, such light bridges appear during the mature stage of the spot's development.

Inner and Outer Bright Rings: The existence of a bright ring around the umbra (inner ring) and the bright outer ring around the penumbra have been debated since long. However, not all spots show bright rings, but there are several visual and photographic observations showing the presence of such rings. Bright rings around the umbrae and penumbrae have an important physical significance related to the energy transport and the inhibition of convection in spots. It is argued that magnetic field in the sunspot tends to block the energy transport by convection and a part of this energy appears in the surrounding region in the form of these bright rings. A detailed study of these rings may throw some light on the understanding of the physical processes of energy transport and the inhibition of convection in spots. The intensity of the bright rings is between 1.03 and $1.07 I_{\text{phot}}$ and corresponding to a temperature of about 50 to 100 K higher, than the photosphere. The bright rings may not completely surround the umbrae and the penumbrae, but can be broken into separate sections.

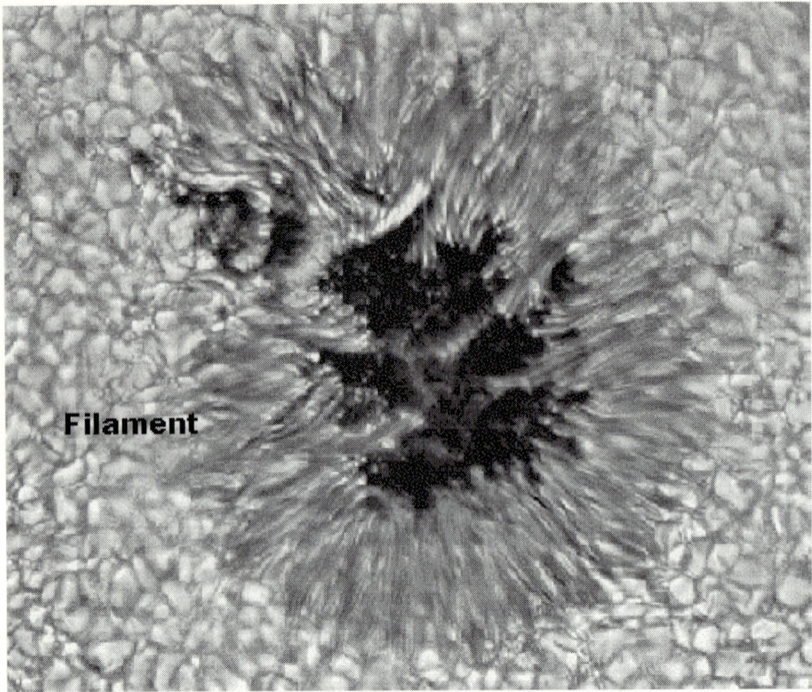

Fig. 21. High resolution white light picture of a sunspot, showing radially oriented penumbral filaments with lengths of 1 to 3″ arc and widths of 0″.3 arc in the penumbra, light bridges in the umbra and umbral dots (courtesy Rimmele)

Penumbrae: Penumbrae are slightly brighter than umbrae and often surround them. Under very good seeing conditions, penumbral bright filaments are seen radially aligned and directed outwards from the umbra (Fig. 21). The radial filaments are generally about 1″–3″ arc in length and about 0″.2–0″.3 arc wide, and have a lifetime of nearly 40–45 minutes. The bright filaments appear to move outwards towards the photosphere with speeds of 7–10 km s^{-1}. Spectroscopic observations of penumbrae reveal outwards mass motion of 1–2 km s^{-1}, which is called the Evershed effect, discovered by Evershed (1909) at the Kodaikanal Observatory. The horizontally aligned penumbral filaments suggest that they are aligned along the magnetic field lines, and emerge vertically out in the umbra and become inclined or horizontal in the penumbra. The first measurements of inclination of magnetic field were made by Hale and Nicholson (1938) in the early 1920s at Mount Wilson Observatory, later Bumba (1960) and other authors determined the inclination of the field lines in sunspots, and found that lines of force emerge vertically from the umbra and become inclined or tilted at an angle of 25–30° to the solar surface near the outer edge of penumbra.

4.3 Development of Sunspots and Sunspot Groups

To observe the development of a sunspot and of sunspot groups is one of the most interesting activity in solar studies. A large variety of phenomena are associated with the various stages of a sunspot's life history. Several authors have given detailed descriptions of sunspot development, for example Newton (1958), Bray and Loughhead (1964), Bumba (1967), Wilson (1968) and McIntosh (1981). Sunspots are seats of strong magnetic field, which emerge from below the surface. Using the white light and Hα pictures, a brief description of the day-to-day development of a large sunspot group is described as follows:

Day 1: A small bright facula in white light near the limb, or a small arch system in Hα, near the disk centre first appears, this indicates that magnetic flux tubes have just reached the 'surface' – the photosphere (Fig. 22).

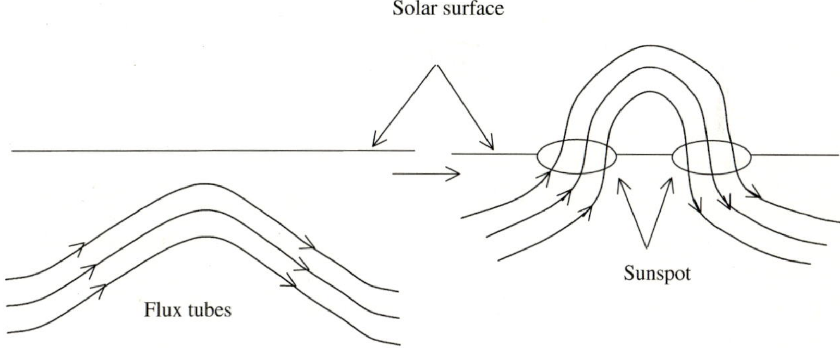

Fig. 22. Flux tubes are emerging from beneath the surface and breaking through the solar surface to form sunspots

Day 2: A small preceding (p-) sunspot appears at the western edge of the facula, it increases in size and brightness contrast and the magnetic field continues to rise.

Day 3: One or more following (f-) spots appear at the eastern edge of the facula, with opposite polarity of the first spot, and the area of the facula increases.

Day 4: Small spots dissolve and coalesce to form large spots. The western 'preceding' p-spot of the group forms a penumbra. The magnetic field distinctly shows a bipolar character.

Day 5–13: The eastern 'following' f-spot forms a penumbra. Then numerous small spots appear between the two main p- and f-spots, until the group attains its greatest extent. The brightness and the extent of Hα faculae or plages increases, and if the magnetic field strength and the 'shear' of field lines are just right, solar flares may be triggered. Generally, the flare activity is highest, during this phase of the sunspot group's development.

Day 14–30: All spots except for the principal p-spot disappear. The Hα plage brightness decreases and it may divide into smaller areas. The magnetic field becomes weaker. Dark Hα filaments appear in the active region and they seem to divide into zones, marking regions of opposite polarity.

Day 30–60: The p-spot too begins to shrink and disappear. The brightness of the Hα plages also decreases. The Hα filaments in the active region increases in length and if it persists without eruption, it may divide the active region into two halves of opposite polarity.

Day 60–100: The Hα plages almost disappear but the photospheric faculae may persist and eventually dissolve. The filament in the region may reach its greatest length and lies almost parallel to the equator.

Day 100–250: No Hα plages or faculae are seen and the filament may break into several small pieces or may even erupt as a whole and form a Coronal Mass Ejection (CME). Some times it is seen to form again in the same location.

It will be clear from the above description that sunspot groups develop rapidly, to their maximum extent and activity in about 10–15 days, but decline slowly over more than 60 days, while the magnetic activity can be seen for more than 100–200 days. Rapid rise and slow decline of active regions, manifests the rapid emergence and relatively slow sinking of flux tubes beneath the surface.

Until now we do not fully understand the details of the formation of sunspots, particularly the formation of the sharp boundaries between the umbra, penumbra and the photosphere, how the bright rings are formed and why sunspots are so dark?

4.4 Classification of Sunspots and Sunspot Groups

There are a large variety of sunspot groups observed on the Sun depending on their polarity, state of umbral and penumbral structures, and area covered. These are classified using different criterion, some of which are described below.

Waldmeier Classification

In 1938 Max Waldmeier introduced a classification system for sunspot groups, which takes into account the polarity, the number of spots and of groups, the formation of penumbrae and the longitudinal extent. According to Waldmeier, the complete classification of a sunspot group consists of a 'letter' for the class and a 'number' indicating the number of spots in the group. For example, a single spot without a penumbra is classified as A1. A spot with 2 umbrae and a common penumbra is classified as J2 or H2, if its diameter exceeds 2.5 heliographic degrees. In case of a bipolar sunspot group with say 8 sunspots and one penumbra, around one of the main spots is classified as C8. The details of this classification are as follows:

A An individual spot or a group of spots without a penumbra or a bipolar structure.

B Group of spots without penumbra in a bipolar arrangement.

C A bipolar sunspot group, the principal spot appears surrounded by a penumbra.
D A bipolar group, the principal spots have penumbrae, length of the group $< 10°$.
E A large bipolar group in which the two principal spots are surrounded by penumbrae and exhibit complex structures and several smaller spots appear between the principal spots.
F A very large bipolar or complex sunspot group, at least 10° long.
G A large bipolar group without small sunspots seen between the principal spots with a length of at least 15°
H Unipolar spot with diameter $> 2.5°$
J Unipolar spot with diameter $< 2.5°$

The Waldmeier sunspot group classification is being widely used by professional and amateur solar astronomers. Taking into account the development of the group, Kleczek (1953) and Künzel (1960) have suggested some modifications to this classification.

McIntosh Classification

The Waldmeier classification is quite popular among amateurs, however, it requires some knowledge about the previous history of the sunspot's development. In addition, it was found inadequate to classify complex groups. To overcome these difficulties McIntosh (1990) introduced a modified version of the Waldmeier system to discriminate between the "active" and "inactive" varieties of sunspot groups and he obtained a good correlation between the complexity of the group and X-ray flares. NOAA (National Oceanic and Atmospheric Administration, Boulder, USA) uses this new system, known as the McIntosh classification in the Solar Geophysical Data (SGD).

In McIntosh classification the initial letter corresponds to the Waldmeier system, but without the G and J class. The unipolar groups are individual spots or individual groups in which the maximum distance between the two spots is less than 3°.

The second letter of the McIntosh system indicates the appearance of the penumbra in the largest spot in the group, for example:

x no penumbra,
r rudimentary (incomplete) penumbra with irregular boundaries,
s symmetrical almost circular penumbra with a filamentary structure directed outward and a diameter less than 2.5°,
a asymmetrical or complex penumbra with filamentary structure and less than 2.5° in diameter,
h symmetrical penumbra like type 's' but with a diameter of more than 2.5°,
k asymmetrical penumbra, like type 'a', but with a diameter of more than 2.5° measured in N–S direction.

The third letter indicates the distribution of the spots within the group, for example,

x indicates individual spot.
o open distribution of spots. The area between the p- and f-spots is free of sunspots, so that the group clearly consists of two parts with different magnetic polarity.
c compact distribution. The area between the main spots appears compactly packed with many large spots, some of them may have only umbrae, but at least one spot may have a penumbra.
i intermediate type between o and c. Some sunspots without a penumbra can be seen between the principal spots.

Figure 23 shows a pictorial representation of the McIntosh classification of sunspot groups. In this system a complex group which exceeds 5° in diameter in which both polarities occur within a penumbra (bipolar group) will be classified as Dkc or Ekc or Fkc etc.

Mount Wilson Magnetic Classification

Since, almost 100 years at the Mount Wilson Observatory, visual measurements of the sunspot's longitudinal magnetic fields are being made daily with the 150-foot tower telescope and the 75-foot spectrograph. Using the magnetic field data, Hale and Nicholson (1938) proposed a magnetic classification, which is based on the magnetic complexity of the spot group. Details of the Mount Wilson magnetic classification are given by Hale and Nicholson (1938) and is described as follows:

The Mount Wilson magnetic classification includes three main classes, designated as Unipolar (α), Bipolar (β), and complex (γ): Unipolar groups are single spots or groups of spots having the same magnetic polarity. They are subdivided as follows:

A = α Groups for which the distribution of calcium flocculi, in the preceding and following part of the group, is fairly symmetrical.
AP = αp Groups situated in the preceding part of an elongated mass of calcium flocculi. All the magnetic measurements in the group are of the same polarity, which corresponds to the preceding spots in that hemisphere for that cycle,
AF = αf Group situated in the following part of an elongated mass of calcium flocculi. All the magnetic measurements in the group are of the same polarity, corresponding to the following spots in that hemisphere for that cycle,

Bipolar groups in their simplest form consist of two spots of opposite polarity. Usually, the bipolar group is a stream of spots, those in the preceding and following parts of the group being of opposite polarity. Bipolar groups are subdivided as follows:

B = β Those in which the preceding and the following members, whether single or multiple are approximately of equal area.

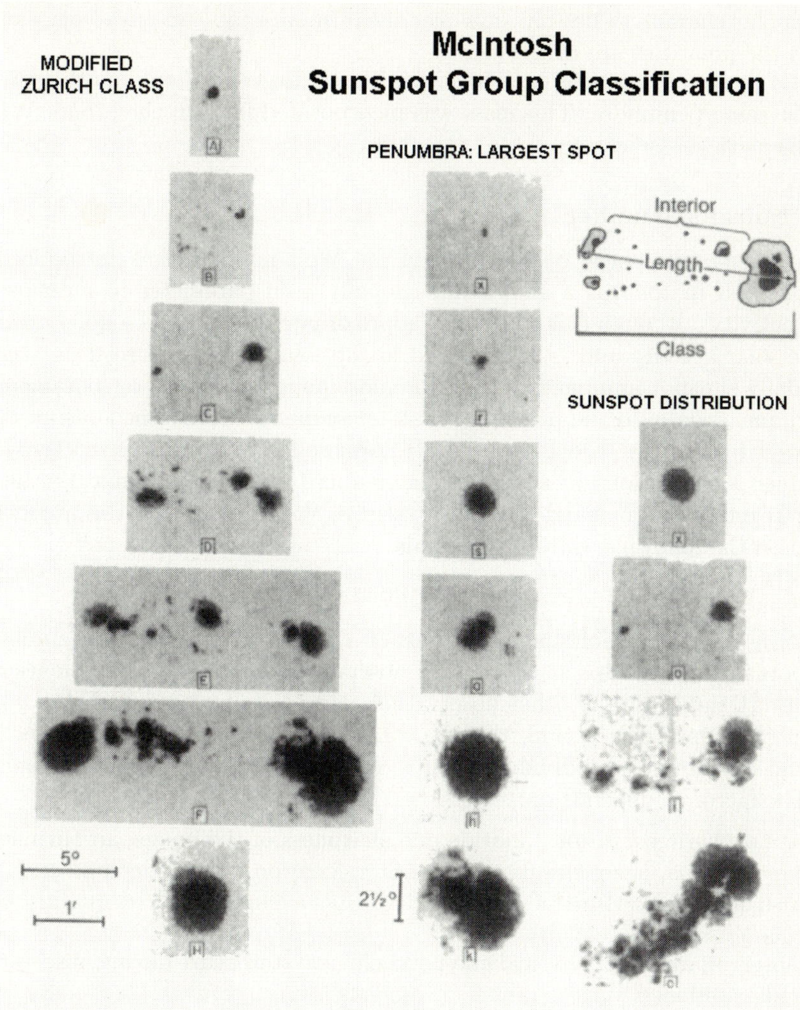

Fig. 23. Development of sunspot groups and illustration of the McIntosh sunspot group classification

BP = βp Those in which the preceding member is the principal component of the group.

BF = βf Those in which the following member is principal component of the group.

BG = $\beta\gamma$ Those in which the bipolar characteristics are shown, but in which there is no marked north-south dividing line between the spots of different polarities.

G = γ Complex groups including spots of both polarities so irregularly distributed as to prevent classification as bipolar groups. A group which has bipolar characteristics but in which one or more spots are out of place as far as the polarities are concerned.

D = δ Spots of opposite polarity are within 2 degrees of one another and in the same penumbra. This class was incorporated later in the Mount Wilson magnetic classification.

4.5 Sunspot Number

The "Sunspot number" also known as the Wolf number, was introduced by Rudolf Wolf in 1848, is a simple and globally used parameter to measure the solar activity, although it is somewhat approximate, but it gives a good estimate of the solar activity and is widely used in the solar–terrestrial relation studies. The daily sunspot number is among the most popular and easiest parameter to determine, even with the help of a small telescope. To verify the sunspot cycle, discovered earlier by H. Schwabe (1844), Wolf collected numerous scattered and dissimilar observations of sunspots and reduced them to a uniform scale. To bring together the old and new observations, Wolf introduced the number R, based on the following empirical formula:

$$R = k(10g + f) \,. \tag{10}$$

Here g is the number of sunspot groups on the Sun and f the total number of all spots in these groups at the time of observation and k is a 'reduction factor' to convert the counts of other observations onto a uniform scale. For $k = 1$, if there is one spot on the Sun, then $R = 11$. For a group of five spots R will be 15, and if there are 5 individual groups, each with one spot, then R would be 55. The factor k, depends on:

1. Atmospheric conditions: motion and sharpness of the image, air turbulence, wind, clouds, haze, elevation of the Sun, location of the telescope etc.
2. Instrument: aperture of the objective, focal length, optical quality, filter, enlargement, projection system,
3. Observer: eyesight, physical and psychological state, care during observations and experience,
4. Level of the solar activity.

In principle, the k-factor can be determined because 3 out of the 4 mentioned parameters are measurable quantities. The greatest influence on the homogeneity of the Wolf number is due to the variation of the atmospheric conditions (visibility). Several authors have tried to refine the determination of the k-factor (Beck 1978; Schindler 1981; Seech and Hinrichs 1977; Wagner 1979). Generally, the sunspot number is considered as a simple measure of solar activity, however, it is not the only index that defines the solar activity. Nowadays, the solar activity is measured by several parameters, the best measurements are from the radio flux emission observed at 10.7 cm, or the UV or X-ray emission through precision recording instruments. These measurements are not subject to obliterating factors as mentioned in the case of sunspot numbers and the k-factor. However, for the sake of continuity, the sunspot number is still used as it gives a good estimate of the magnetic flux on the solar surface.

Sunspot Area Number A

The presence of sunspots alone is not a sufficient condition to characterise the solar activity. The solar magnetic field is the root cause of all solar activity phenomena. If the observations of solar activity is limited to white light and to sunspots only, then the measurements are linked to the magnetic fields and related to the area of the spots. Houtgast and van Sluiters (1948) have empirically shown that the maximum magnetic flux density B_n in Gauss, at the disk centre is related to the area of spot A_I, in millionths of the disk, as follows:

$$B_n = \frac{3700 A_I}{A_I + 60} . \tag{11}$$

This equation applies to stable spots and not to spots which are in developmental phases.

From 1874 to 1976, the Royal Greenwich Observatory has collected and used white light photoheliograms from various observatories around the globe to determine daily the area A_I of each spot visible on the Sun and also corrected for foreshortening. The areas of all spots measured daily are added and divided by the area of the visible hemisphere of the Sun and multiplied by 10^6, to obtain a handy number. The area number A in units of millionth of the visible solar hemisphere (MH) is given by:

$$A = \frac{(\sum A_I \sec\theta) \times 10^6}{2\pi r_0^2} . \tag{12}$$

where θ is the heliocentric angle of the sunspot, that is, the angle between the radius vector and the line of sight at the spot position on the solar disk, and r_0 is the radius of the solar image.

Until 1955 the area number A was published annually in the Greenwich Photo-Heliographic Results. Recently, the Mount Wilson Observatory has started providing these data and they are also available in the weekly Solar Geophysical Data (SGD) published from NOAA, Boulder. As the area determination is mostly done from visual drawings of sunspots, their accuracy is limited and should be used with caution.

Both the area A and the Wolf number R are measures of solar activity, therefore, there should be some relationship between the two parameters particularly when averaged over a certain period. Waldmeier found a linear relation between the annual mean area number A_g from the Greenwich records and the annual mean Wolf number R_z, from Zurich for the period 1874 to 1938, as follows:

$$A_g = 16.5 R_z . \tag{13}$$

However, for a period of increased solar activity, during cycle 18 and 19 the relation between A_g and R_z was found to be non-linear.

Foreshortening

As the Sun is a spherical body, the sunspots on it are not on a flat surface, hence as the Sun rotates, the sunspots appear to rotate with it. Beyond the Sun's centre, say near the limb of the Sun, the spots appear contracted or 'foreshortened'. To correct for this normally a correction factor is applied, by dividing the measured area by the cosine of the angle θ, the angle between the radius vector of the Sun and the line of sight of the position of the sunspot on the disk. Near the extreme edge of the Sun's limb, this correction gives erroneous values, hence this correction factor can be safely applied up to perhaps about 80°. Observations much closer to this are very rare. However, Waldmeier (1978) has observed an H-spot with diameter of 40 000 km up to $\theta = 88°22'$ or $0''.4$ arc from the limb. It will be of interest to investigate the effect of physical foreshortening and interpret its effect on the sunspot's appearance near the limb. Amateur astronomers and beginners could take up the determination of the sunspot area, as these measures are of great importance for long term synoptic data.

The solar group of the British Astronomical Association (BAA) (Dougherty 1981) has developed a very easy method to measure sunspot areas obtained either from a projected image, on a photograph or a drawing by comparing the spot region with circles of known areas. The smallest circle, which just surrounds the spot and the largest that just fits into the spot are found. The spot area is then the average of the two circles. BAA supplies template sheets or those could be made with 15 such circles on a transparent sheet and by overlaying this template on the spots one can easily measure the area with a fair degree of accuracy. For elliptical shaped spots, the template can be modified accordingly. This method is not suitable for large complex and irregular sunspot groups, for such groups counting squares on a grid is the standard method. Now with digital recording of the full disk image, it is possible to write programs, which could measure and calculate areas with good accuracy and speed, and even on a real time basis.

4.6 Position Determination of Solar Features

In addition to the phenomenological observations, it is also important to know and determine the heliographic positions of sunspots, faculae, filaments, plages, flares etc. on the Sun. In this section we present some simple methods to determine heliographic positions, on a full disk photograph or a drawing, obtained from an equatorial or alt-azimuth mounted telescope. Using a stable and properly oriented telescope, an accuracy of measurement of better than 0.3° can be achieved, with full disk digital images available now much better accuracy can be obtained. The heliographic position determination is an essential parameter for most investigations, such as to study of the sunspot development, latitude movement of spots, solar rotation and the proper motion of sunspot groups.

Heliographic Coordinates

Two terms are used in the following paragraph, "location" and "position" to mark a certain solar feature. By "location" we mean the location of a feature

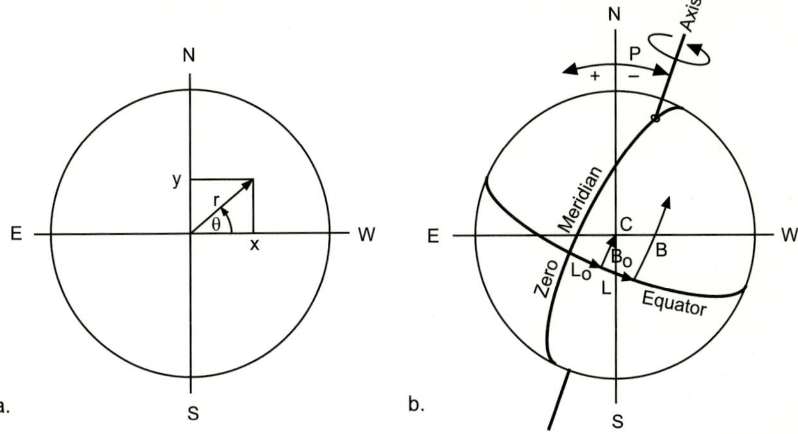

Fig. 24. Explaining the various heliographic coordinates. (a) Cartesian and polar coordinate system, (b) heliographic latitude B, longitude L, and the position angle P of the Solar axis

on the solar image and by "position" we mean location on the Sun itself. The location on the image can be given in either Cartesian (x, y) or polar (r, θ) coordinates. The conversion from the Cartesian to the polar system is given by

$$r = (x^2 + y^2)^{1/2} , \qquad \theta = \tan^{-1}(y/x) , \tag{14}$$

and from the polar to the Cartesian system by

$$x = r\cos\theta , \qquad y = r\sin\theta . \tag{15}$$

Here in (14) it is assumed that the proper branch of $\tan^{-1}$ is taken.

Figure 24 explains various heliographic coordinates, such as L, L_0, B, B_0 and the P angle. The 'position' coordinate of a solar feature, in heliographic latitude B is measured from 0° to +90° from the solar equator to north, and from 0° to −90° to the south, similar to the Earth's latitude. For zero heliographic longitude on the Sun, there is no fixed feature, from which one could count the longitude L. For this purpose a Carrington zero meridian, L_0 is defined as that N–S meridian passing on 1 January 1854 at 12 UT through the ascending node of the solar equator, projected on to the solar disk. The Carrington longitude is measured towards the west from 0° to 360°. The coordinates L_0 and B_0 mark the centre C of the solar disk. Every time when $L_0 = 0$ crosses the solar meridian it marks the beginning of the continuously counted synodic solar rotation, called Carrington rotation. The beginning of Carrington rotation number 1 is assigned as that meridian which crossed the disk centre on 9 November 1853. The Carrington rotation number is given by:

$$\mathrm{CRN} = \mathrm{int}\left(R_0 + \frac{\mathrm{JD} - \mathrm{JD}_0}{27.2753} \right) , \tag{16}$$

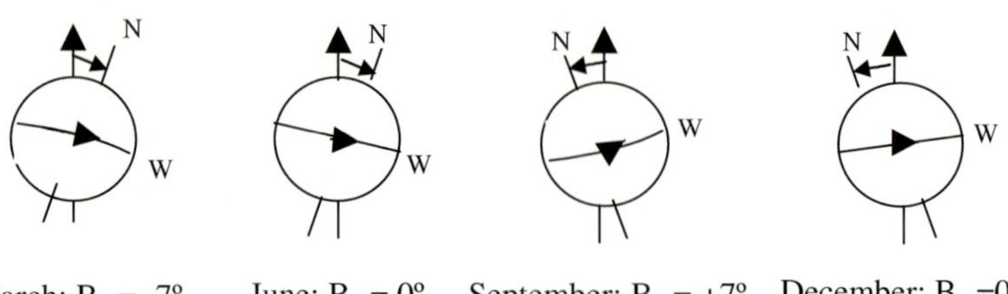

Fig. 25. The effect of the inclination of the Sun's N–S axis

where int(x) given the nearest integer $\leq x$. Here R_0 is the Carrington number for a known day obtained from an almanac, JD_0 the Julian day number for that day obtained from the almanac, while JD is the Julian day number for a particular day for which Carrington rotation is required, and 27.2753 is the Sun's synodic rotation period in days at the equator. Although, the Sun does not rotate like a rigid body, the value of a 'mean' period is assumed for the purpose of defining heliographic longitude. The sidereal rotation rate at the surface of the Sun, is given by the following relation (Newton and Nunn 1951; Howard 1984):

$$\Omega = 14.42 - 2.30 \sin^2 \theta - 1.62 \sin^4 \theta \quad \text{degree/day}\,, \tag{17}$$

where θ is the heliographic latitude.

The sunspots do not appear to traverse over the solar disk in straight lines, but in semi-elliptical paths, this indicated that the Sun's equator or the N–S axis is inclined at an angle to the ecliptic plane. During the course of a year the north and south hemispheres are alternately more inclined towards us. This angle, or the heliographic latitude B_0 of the centre of the solar disk varies between $\pm 7^\circ.25$. During June and December, when $B_0 = 0$, the sunspots appear to traverse in straight line, while at other periods the spots appear to move in elliptical paths, over the solar disk. (see Fig. 25).

The position angle P, between the Sun's N–S rotational axis and the N–S direction of the sky, that is the Earth's N–S axis varies during the year. This angle is determined by superimposing the Earth's equator on to the ecliptic, inclined at an angle of $23^\circ.37$ (Fig. 26). Due to the combined effect of both tilts (B_0 and P), the P angle varies between $\pm 26^\circ.37$. For positive P angles, the solar axis is inclined towards the East, while for negative P, it is inclined towards the West. Daily values of P, B_0 and L_0 are given in the section for Ephemeris for Physical Observations of the Sun in the Astronomical Almanac.

To determine accurately the heliographic coordinates of solar features on the Sun with the help of a small refracting telescope, it is essential to know as precisely as possible the orientation of the solar equator (E_S–W_S) and the solar (N_S–S_S) axis. To achieve this, the following procedure is used.

A solar image is formed by an equatorial mounted telescope of say 100 to 150 mm aperture or even larger and is enlarged to about 150 to 180 mm in

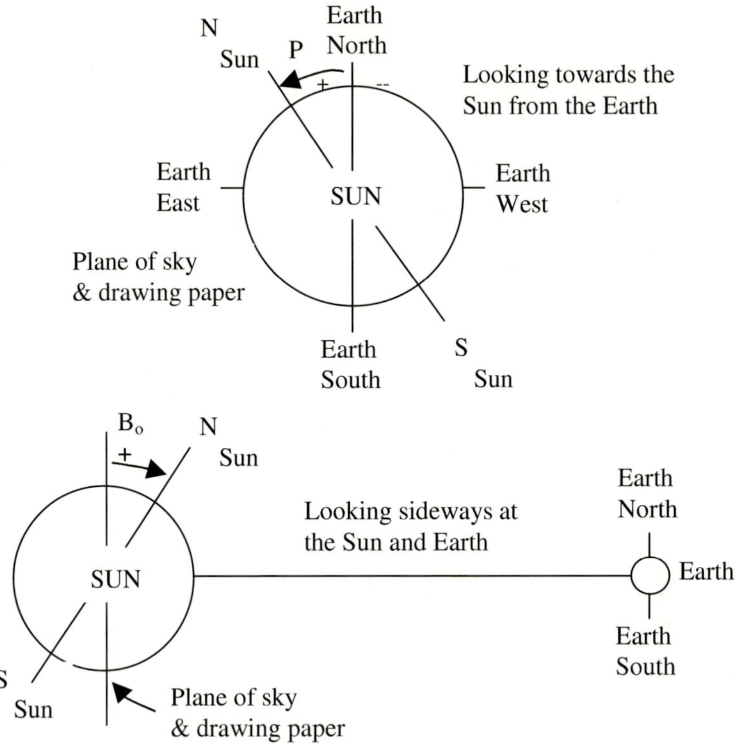

Fig. 26. Showing the tilt of the Sun's north pole (P) in the plane of the sky, 'tipping' of Sun's north pole (B_0) towards the observer on the Earth

diameter by an un-cemented Huygens or Ramsden eyepiece onto a projection screen. A circle of the same diameter, as the solar image is drawn on a paper and placed on the projection screen. The solar image is centred on the circle. Then the telescope drive is switched off and the image is allowed to drift (East to West due to the Earth's rotation). As the limb of the Sun intersects the circle at two points, these two points are quickly marked by a soft pencil and the line joining the two points P_1 and P_2 mark the Earth's North–South axis, as shown in Fig. 27.

To know which is the north and south hemisphere of the Sun, tip the Telescope slightly towards higher declination, the north point of the Sun will disappear last from the field of view. For better accuracy this procedure is repeated several times, before and after taking the observations. The same procedure can be followed for fixed focus telescopes heliostats or coelostats. However, for equatorial telescopes, not every day the North–South axis need to be determined, generally, fiducial marks are permanently made near the focal plane and recorded on the photographic plate or film. Sunspot locations can also be used to determine the Sun's North–South axis on the solar image by the same drift

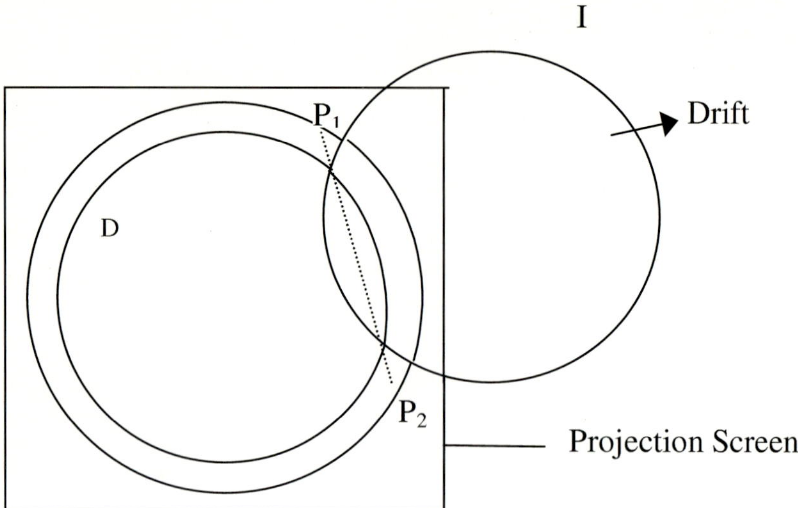

Fig. 27. The method to determine the Earth's North–South direction by the "drift method". D indicates a circle drawn with the size of the solar image on the projection screen, I the drifted solar image. P_1, P_2 are intersection points of the image and circle, marking $N_E - S_E$

method. Photographic techniques give very accurate determinations of the solar axis. This is done as follows: Double exposures of two solar images are made on a single plate, first at one instant and then after letting the image drift for a little while. Thus one obtains two solar images, the line intersecting the two images mark the geocentric Earth's N–S axis. After making the markings, the telescope drive is again switched on and the drawing of the sunspots is made with a soft pencil, care should be taken that the telescope remains sharply focused and that the diameter of the solar image fits the circle accurately. This is important because during the year the Sun's apparent angular diameter changes from $31'\ 29''.6$ arc (in July) to $32'\ 31''.9$ arc (in January). Depending on the accuracy required, one could determine the heliographic coordinates either by using overlay grids, known as Stonyhurst disks or calculated mathematically by measuring the Cartesian or polar coordinates of the solar features and converting them into heliographic coordinates using the formulae given in this section.

Grid Overlay Template Method

To determine the heliographic coordinates of solar features, with fair degree of accuracy, the easiest, quickest and widely used technique is the overlay grid method. The Stonyhurst disks, which have a printed grid with heliographic latitude and longitude, are used as overlay template grids and are available from many sources. Figure 28 shows one such grid. One good source for transparent overlay templates is from NOAA–SESC or could be copied on transparent sheets, from the Solar Geophysical Data No. 489, May 1985, 'Explanation of

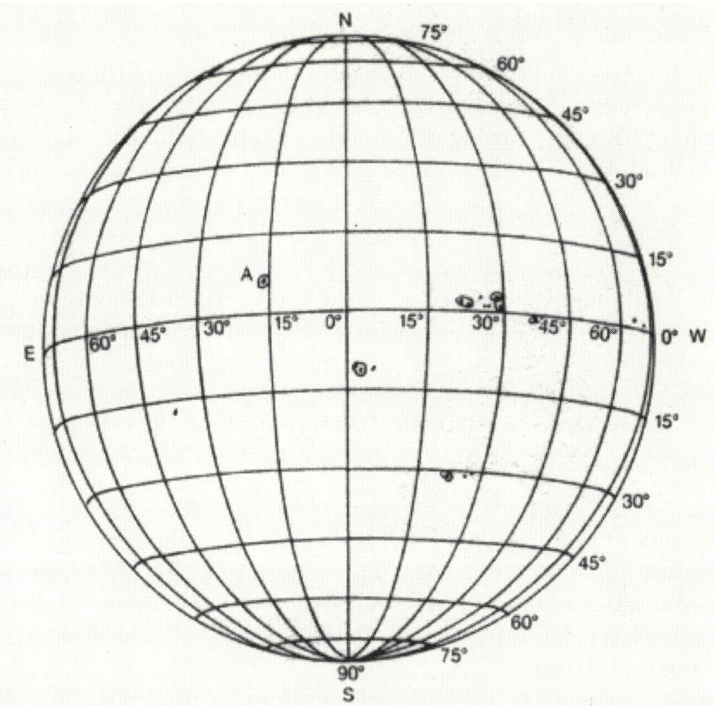

Fig. 28. Overlay of a Stonyhurst disk on a drawing, the heliographic coordinates of the spots are then directly read off from this grid

Data Reports', and enlarged to the appropriate size, to match the solar image formed by the telescope in use. As the value of B_0 varies over a small range, $\pm 7^\circ.25$, generally 8 grids are sufficient ($B_0 = 0^\circ, 1^\circ, 2^\circ, \ldots, 7^\circ$) to yield an accuracy of about $0^\circ.5$. The appropriate grid, for a particular day's B_0 value is overlaid on the solar projection drawing or on the full disk photograph and the central meridian line is turned by the P angle for that particular day, P being the position angle of Sun's N–S axis with respect to the Earth's axis. For positive P the solar N point is towards the East and for negative P angles it is towards the West. For negative B_0 values, that is, when the Sun's South pole is pointing towards the Earth, the same positive disks are turned upside down and the spot positions are measured as usual. Once the B_0 and the P angles are correctly aligned on the drawing with the Stonyhurst disk, the heliographic coordinates B' and L' (angular distance in longitude from the central meridian) are directly read off to better than $0^\circ.5$.

To convert these approximate coordinates into more accurate B and L heliographic coordinates, the following equations may be used

$$\sin B = \cos B_0 \sin B' + \sin B_0 \cos B' \cos L' \,, \tag{18}$$

$$\cot L = \frac{\cos B_0}{\tan L'} - \frac{\sin B_0 \tan B'}{\sin L'} \,. \tag{19}$$

Mathematical Method

This method is based on the measurement of the position coordinates r, θ (polar) which can be determined from the Cartesian coordinates x and y of a solar feature, on a projected drawing or a full disk photograph. The angular distance ρ of a spot from the solar disk centre is measured using the equation

$$\sin\rho = \frac{r}{R} , \tag{20}$$

where R is the radius of the projected image and r is the distance of the spot from the disk centre. To calculate the heliographic latitude B and the heliographic longitude difference l, from the central meridian, the following equations are used

$$\sin B = \cos\rho \sin B_0 + \sin\rho \cos B_0 \sin\theta , \tag{21}$$

$$\sin l = \frac{\cos\theta \sin\rho}{\cos B} . \tag{22}$$

Any of the above two methods provide directly the heliographic latitude B of a solar feature but not the heliographic longitude L. To determine the 'true' heliographic longitude L, from l the following relation between the heliographic longitude L_0 of the central meridian and l is used. L_0 for each day is given in the Astronomical Almanac:

$$L = L_0 + l . \tag{23}$$

The detailed method to calculate the solar heliographic latitude and longitude is described by Duffett-Smith (1988). Ashok Ambastha of the Udaipur Solar Observatory has developed a computer code to compute the heliographic coordinates of sunspots etc., once the date, time of observation and Cartesian x, y position of the spot are given. This programme calculates the P, B_0 and L_0 values also and one does not need to look in the Astronomical Almanac for these data. This code is available from his web site (http://www.prl.ernet.in/~ambastha).

5 Solar Magnetic Fields

Magnetic field plays a dominant role in the solar atmosphere. Sunspots were the first features where magnetic fields were detected on the Sun, but the magnetic fields are not restricted to the sunspots.

5.1 Sunspot Magnetic Fields

From 'iron filing' like structures formed by the magnetic field and the 'vortex' structure seen on a good Hα spectroheliogram of a bipolar sunspot, George Hale in 1907 was motivated to look for magnetic fields in sunspots and he indeed discovered strong magnetic fields. Let me quote his own words about this remarkable discovery, "I applied this test (Zeeman effect) to sunspots on Mount Wilson in June 1908, with the 60-foot tower telescope, and at once found all the

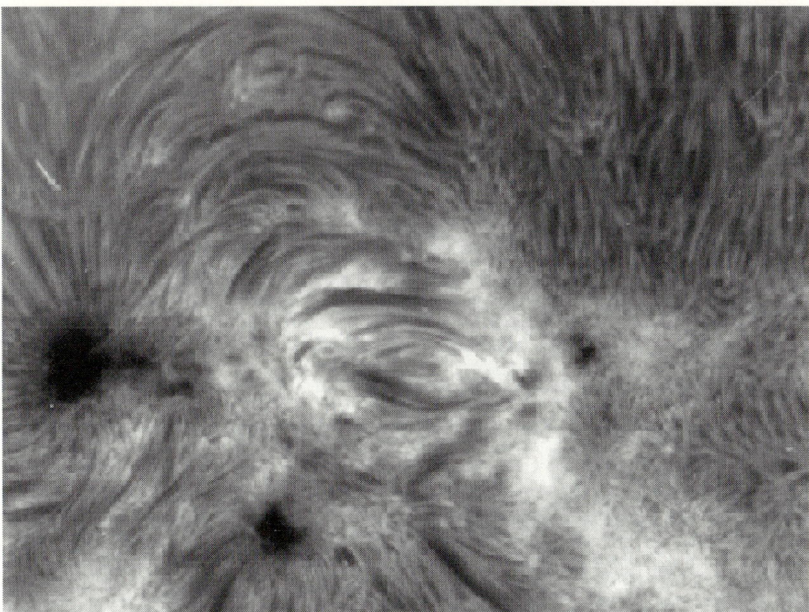

Fig. 29. High resolution Hα filtergram showing fibril structure around a bipolar sunspot group joining the two sunspots (courtesy Big Bear Solar Observatory)

characteristic features of the Zeeman effect. Most of the lines of the sunspot spectrum are merely widened by the magnetic field, but others are split into separate components, which can be cut off at will by the observer" (from *Smithsonian Report for 1913*, pp.145–58). The Zeeman effect was fortunately discovered a few years earlier in 1896, and Hale indeed measured strong magnetic fields of the order of 2000 Gauss. Figure 29 shows a high resolution Hα picture of a bipolar sunspot group, displaying the Hα-fibril structure which interconnects the two sunspots, similar to the distribution of iron filing around a bar magnet. Hale also discovered that sunspots appear in pairs with opposite magnetic polarity.

If the image of a sunspot umbra is placed on the slit of a spectrograph, the sunspot spectral lines broaden or split into two or three components, depending on the strength of the field and whether the magnetic field is longitudinal or transverse, with respect to the observer. For longitudinal magnetic fields, i.e., the field lines are along the line of sight, the spectral lines are split into two oppositely circularly polarised σ-components, but when the field lines are perpendicular to the line of sight the spectral lines are split into three linearly polarised π-components. The wavelength splitting of the spectral lines is given by the relation:

$$\delta\lambda = \frac{\pi e}{M_e}\frac{\lambda^2 gB}{c} = 4.7 \times 10^{-13}\lambda^2 gB \ , \qquad (24)$$

where the factor ($\pi e/M_e c = 4.7 \times 10^{-13}$) is the standard constant, c the speed of light in vacuum, e the electron charge and M_e the electron mass, λ is the

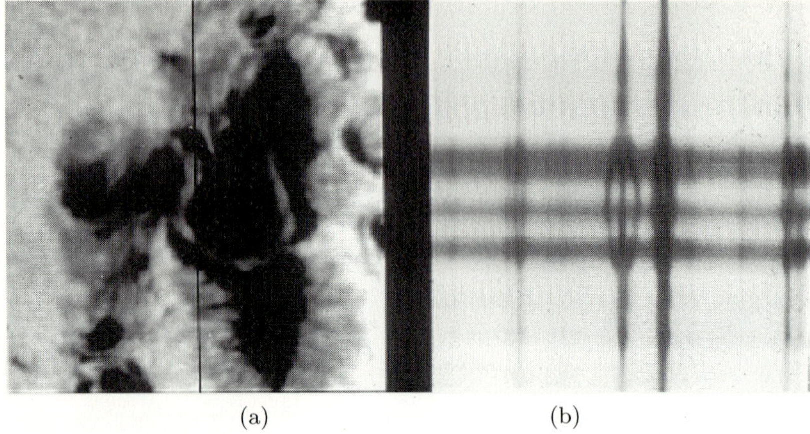

(a) (b)

Fig. 30. (a) Spectrograph slit placed across a large sunspot, (b) splitting of a spectral line at 5250 Å, showing a Zeeman triplet due to the splitting of the line in a sunspot magnetic field of 4130 Gauss (Livingston 1976)

wavelength in Å, g is the Lande g factor, B the magnetic field strength in Gauss. For the Fe I line at 6302.5 Å and $g = 2.5$, a 3000 Gauss field will produce a Zeeman splitting of $\delta\lambda = 0.15$ Å. This quantity can be easily measured with a suitable spectrograph having adequate dispersion and resolving power. Figure 30, shows the spectrum with Zeeman splitting in the 5250 Å line, taken by Livingston at the National Solar Observatory.

At the Mount Wilson Observatory a programme of visually observing the daily sunspot magnetic field and polarity was initiated by Hale in early 1917 and is still being continued. Sunspot magnetic field data can be obtained from their Web Site: http://www.astro.ucla.edu/~obs/intro.html. Field strengths up to 100–200 Gauss can be measured by this method. However, magnetic fields of less than 100 Gauss are difficult to observe, because the splitting of the Zeeman components becomes very small.

5.2 General Magnetic Fields

In 1891 Arthur Schuster, speaking before the Royal Institution on the question of, "is every rotating body a magnet?", remarked about the solar corona, that, "The form of the corona suggests a further hypothesis which, extravagant as it may appear at present, may yet prove to be true. Is the Sun a magnet?". It is very interesting how appearances lead to great discoveries. From the appearance of the corona, the Sun was always suspected to have a general magnetic field like a bar magnet. To detect the general magnetic field, several attempts were made at Mount Wilson Observatory, but all failed, because the fields were much smaller than the sensitivity of the photographic technique being used. Until the father and son team of H. W. Babcock and H. D. Babcock (1952), invented an ingenious photoelectric device to detect small scale general solar magnetic fields (Babcock 1953).

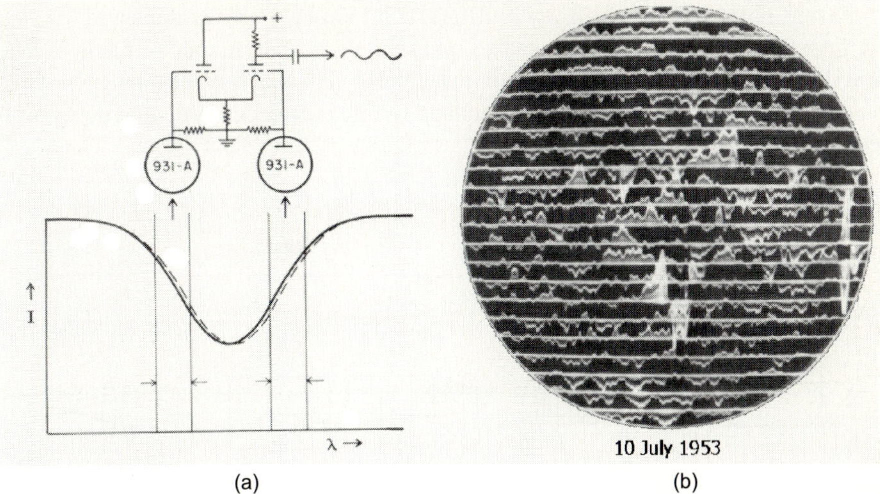

Fig. 31. (a) Principle of the Babcock magnetograph, the line profiles of the two oppositely polarised σ-components are shown by *solid* and *dotted* lines (b) an early magnetogram, positive fields are shown as deviations above the horizontal lines and negative below, the amplitude of the deviation is a measure of the field strength

The basic principle of Babcock's magnetograph was to record photo-electrically the intensity variations in the red and the blue wings of a magnetically sensitive line, as shown in Fig. 31. The solid and the dashed I–λ curves denote the line profiles from a magnetic region, due to the longitudinal magnetic field, when the left and right circularly polarised light is alternately admitted to a high dispersion spectrograph. The two components, which in the normal Zeeman effect are oppositely and circularly polarised, are converted into linearly polarised components by an electro-optical quarter-wave plate, such as KD*P (Potassium Diammonia Phosphate) with its axis inclined at 45° to a plane polariser, placed just before the entrance slit. By alternately changing the voltage from positive to negative on the KD*P crystal at a rate of about 50–60 Hz, the polarity of the quarter-wave plate is changed, thus either the $+\sigma$ or $-\sigma$ component of the magnetic line is allowed to fall through the second slits, on to the two photomultipliers. The intensity difference is measured between the two photomultipliers. The intensity variation of the difference signal is amplified and recorded. The amplitude of the difference signal is roughly proportional to the net magnetic flux in the observed solar region. In an earlier version of Babcock's magnetograph, the difference signals were displayed on a Cathode Ray Tube and was photographed as shown in Fig. 31. However, with improved technology, video and digital recordings, use of fast computers and image grabbers and 2-dimensional narrow passband filters, much higher spatial resolution and sensitivity have been achieved to measure magnetic fields of the order of 4–5 Gauss with a spatial resolution of $2''$–$3''$.

Lately, narrow band birefringent filters, solid Fabry–Perot etalons and Michelson interferometers have been used to measure very weak magnetic fields on the Sun. This filter technique has the advantage that it yields instantaneously a two-dimensional map of the solar magnetic field, while a Babcock type magnetograph needs time to scan the solar image.

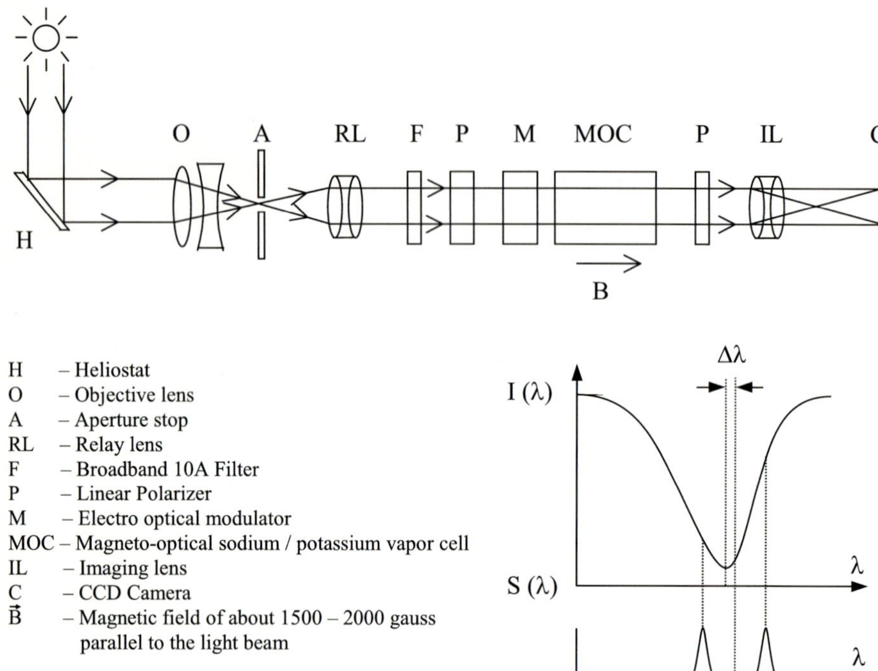

Fig. 32. A Vector Imaging Magnetograph based on a Magneto-Optical Filter (MOF), after Cacciani et al. (1998)

A particularly sensitive method to measure very small line shifts (regardless of whether due to magnetic effects or velocity variations) is to use magneto-optical filters (MOF). These filters consist of a cell filled with cold sodium or potassium vapour to which a 1500 to 2000 Gauss magnetic field is applied (see Fig. 32). The light from the solar absorption line enters the cell parallel to the magnetic field and gets absorbed by the cell's sodium or potassium atoms. But that absorption and the subsequent resonance scattering occurs only at the narrow wavelength positions of their Zeeman-split line components (Fig. 32). By modulating the light input and alternately detecting the scattered light from the two line wings a very sensitive difference signal can be generated.

5.3 Velocity Field Observation

Using the Doppler effect the Babcock magnetograph is also used for measuring solar velocity fields. In this case, another quarter-wave plate is placed after the linear polariser, this second $\lambda/4$ plate converts the linearly polarised components into circularly polarised light. Thus by alternating the sign of the voltage on the KD*P, either the red or the blue component, shifted by the Doppler effect, is allowed to fall on the two photomultipliers. The difference output signal from the two photomultipliers is proportional to the solar velocity amplitude.

5.4 Leighton's Spectroheliographic Technique for 2-D Velocity and Magnetic Field Maps

In the late 1950s Robert Leighton (1959) came up with an idea of using the spectroheliograph of the Mount Wilson Observatory to make 2-D velocity and magnetic field maps of the Sun. The basic idea was to place two slits at the focal plane of the spectroheliograph, each slit centred on the two wings of a line. Simultaneously, two spectroheliograms were made on two photographic plates, one in the blue and the other in the red wing of the line. By photographic subtraction of the two spectroheliograms, a composite picture displaying the Doppler shift or mass motion as dark and bright areas was obtained. Grey areas indicate no motion or velocities less that the sensitivity level of the technique. Using this photographic spectroheliographic technique, Leighton discovered the 5-minute solar oscillations and the supergranulation (Leighton, Noyes & Simon 1962). For making magnetic maps, a polarising optics consisting of a quarter-wave plate and a polaroid sheet, similar to the one used in Babcock's magnetograph, was placed before the entrance of the first slit and two spectroheliograms were made in the two oppositely and circularly polarised components. The two spectroheliograms made in the two polarised components were photographically subtracted to make a magnetic map of the solar region.

5.5 Vector Magnetic Fields

Until now we have described the measurement of only longitudinal magnetic fields, that is magnetic field lines along the line of sight. But to observe fields perpendicular to the line of sight, that is the transverse components, we have to measure the linearly polarised π-components. It is rather difficult to measure accurately the linear polarisation due to the low signal to noise ratio and the linear instrumental polarisation. To measure all the three components of the magnetic field, that is the vector magnetic field, the Stokes I, Q, U and V components (Venkatakrishnan, this volume) have to be determined from the line profiles observed through polarising optics. At several solar observatories around the world, vector magnetic field measurements are being made on a regular basis, using either narrow band filters or Stokes polarimeter.

6 Solar Data from the Internet

Through the international collaboration of solar observatories and data dissemination centres around the world, a number of Websites are operating which give almost on a real time basis solar data extending from radio and optical wavelengths to UV, X-rays and γ-rays. A list of some of the important and major Web sites is given below while new sites are constantly being added.

1. The Solar Data Analysis Centre at Goddard Space Flight Centre, at: http://umbra.nascom.nasa.gov/sdac.html
2. SOHO Observatory Home page at: http://sohowww.nascom.nasa.gov/
3. Solar and Upper Atmospheric Data services at: http://www.ngdc.noaa.gov/stp/SOLAR/solar.html
4. National Oceanographic and Atmospheric Administration (NOAA) at: http://www.SpaceWeather.com
5. Big Bear Solar Observatory at: http://www.bbso.njit.edu/

Acknowledgements

The author is thankful to the Physical Research Laboratory to provide facility to work at Udaipur Solar Observatory and to CSIR for financial support for this work and to Naresh and Anita Jain in helping with the figures.

References

Adams, W. S., & Burwell, C. G. 1915, ApJ, 41, 116
Agnelli, G., Cacciani, A., & Fofi, M. 1975, Sol. Phys., 44, 509
Babcock, H. W. 1953, ApJ, 118, 387
Babcock, H. W., & Babcock, H. D. 1952, PASP, 64, 282
Beck, R. 1978, SONNE, 1, 56
Beck, R., Hilbrecht, H., Reinsch, K., & Volker, P. 1995, Solar Astronomy Handbook (Willmann-Bell, Inc., Richmond)
Beckers, J. M. 1973, Bull. American Astro. Soc., 5, 269
Bray, R. J., & Loughhead, R. E. 1964, Sunspots (Chapman and Hall, London), 9
Bumba, V. 1960, Izv. Crim. Astrophys. Obs., 23, 42
Bumba, V. 1967, in Proc. Enrico Fermi School of Physics, 39, 77
Cacciani, A., Comari, M., Furlani, S., Hanslmeier, A., Messerotti, M., Moretti, P. F., Pettauer, Th., & Veronig, A. 1998, in Three-dimensional structure of solar active regions, eds. C. E. Alissandrakis & B. Schmieder, Second Advances in Solar Physics Euroconference, ASP Conf. Ser. 155, 265
Dainty, J. C., & Shaw, R. 1974, Image Science, Principles, analysis and evaluation of photographic-type imaging process (Academic Press, New York)
Dougherty, L. M. 1981, J. British. Astr. Assoc., 91, 75
Duffett-Smith, P. 1988, Practical Astronomy with your Calculator (Cambridge University Press, Cambridge), 70

Edlén, B. 1943, Z. Astrophys., 22, 30
Evans, J. W. 1949, J. Opt. Soc. Am., 39, 229
Evans, J. W. 1953, in The Sun, ed. G. P. Kuiper (University of Chicago Press, Chicago), 635
Evershed, J. 1909, MNRAS, 69, 454
Grotrian, W. 1939, Naturwissenschaften, 27, 214
Hale, G. E. 1892, Astronomy and Astro-physics, 11, 407 (Goodsell Observatory, Northfield)
Hale, G. E., & Nicholson, S. B. 1938, Publ. Carnegie Inst. No. 498, Washington
Hammerschlag, R. H. & Bettonvil, F. C. M. 1998, New Astron. Rev. 42, 485
Houtgast, J., & van Sluiters, A. 1948, Bull. Astr. Inst. Netherlands, 10, 325
Howard, R. F. 1984, ARA&A, 22, 131
Kiepenheuer, K. O. 1964, in Site Testing, ed. J. Rosch, IAU, Symp. 19, 193
Kleczek, J. 1953, Bull. Astro. Inst. Czech, 4, 9
Künzel, H. 1960, Astro. Nachr., 285, 169
Leighton, R. B. 1959, ApJ, 130, 366
Leighton, R. B., Noyes, R. W., & Simon, G. W. 1962, ApJ, 135, 474
Livingston, W. C. 1976, Sol. Phys., 48, 196
Lyot, B. 1930, Comptes Rendus, Acad. Sci. Paris, 101, 834
Lyot, B. 1933, Comptes Rendus, Acad. Sci. Paris, 197, 1593
Lyot, B. 1939, MNRAS, 99, 580
Lyot, B. 1944, Ann. Astrophys., 7, 31
McIntosh, P. S. 1981, in The Physics of Sunspots, eds. L. E. Cram & J. H. Thomas (Sacramento Peak Observatory, Sunspot NM), 7
McIntosh, P. S. 1990, Sol. Phys., 125, 251
Mitton, S. 1981, Day time Star – the Story of our Sun (Faber and Faber, London/Boston)
Newton, H. W. 1958, The face of the Sun, (Harmondsworth, Middlesex, Baltimore)
Newton H. W., & Nunn, M. L. 1951, MNRAS, 111, 413
Öhman, Y. 1938, Nature 141, 157
Paul, H. E. 1980, in Amateur Telescope Making, Vol. 3, ed. A. G. Ingalls, p. 376
Petit, E. 1980, in Amateur Telescope Making, vol. 3, ed., A. G. Ingalls, p. 413
Phillips, K. J. H. 1992, Guide to the Sun (Cambridge University Press, Cambridge)
Pierce, A. K. 1964, Applied Optics, 3, 1337
Pierce, A. K. 1968, ApJS, 17, 1
Rutten, R. J., Hammerschlag, R. H., Bettonvil, F. M. & Sutterlin, P. 2000, AAS SPD Meeting No. 32, #02.107
Schindler, R. D. 1981, SONNE, 5, 62
Schwabe, M. 1844, Astron. Nachr., 21, 233 (reprinted in 1981, SONNE, 5, 190)
Secchi, A. 1872, Die Sonne, Brannschwing
Seech, A., & Hinrichs, A. 1977, SONNE, 1, 101
Smith, S. F. 1966, Modern Optical Engineering (McGraw Hill, New York)
Sutterlin, P. 2001, A&A, 374, L21
Taylor, P. O. 1991, Observing the Sun (Cambridge University Press, Cambridge)
Taylor, R. J. 1996, The Sun as a star (Cambridge University Press, Cambridge)
Wagner, S. 1979, Solar Observations of the Planetary System Working Group, Report of IAYC, p. 56
Waldmeier, M. 1978, Astr. Mitt. Eidgen. Sternw. Zurich No. 359
Wilson, P. R. 1968, Sol. Phys., 3, 243
Zirin, H. 1988, Astrophysics of the Sun (Cambridge University Press, Cambridge)

Solar Interior and Seismology

H.M. Antia

Tata Institute of Fundamental Research, Homi Bhabha Road, Mumbai 400005, India
email: antia@tifr.res.in

Abstract. Helioseismology is probing the internal structure and dynamics of the Sun with high precision. Frequencies of nearly half a million resonant modes of oscillations have been measured by the ground based Global Oscillation Network Group project and the space based Michelson Doppler Imager. Each of these modes is trapped in a different region of the solar interior and hence its frequency is sensitive to structure and dynamics in the corresponding region. Conversely, by combining the information from these large number of independent modes of solar oscillations it has become possible to infer the structure and dynamics of the solar interior to unprecedented precision. These seismic data have provided a test for solar models and theories of stellar structure, evolution and angular momentum transport. Interesting dynamical phenomena have been inferred from these data which are not understood. Some of these developments are described.

1 Introduction

Scientific development during the middle of the last century provided the necessary framework to describe the internal structure of a star as a self-gravitating sphere of plasma, which generates energy through nuclear fusion occurring in the core where the temperature is high enough to induce the required nuclear reactions. This energy is then transmitted through the star to its surface through radiative processes or through material motions in convective cells. Using the known laws of Physics it is possible to write down the equations governing the stellar structure and evolution. Solution of these equations with appropriate initial and boundary conditions give us the so called standard solar model, described earlier in this volume. However, the calculation of the solar model requires a number of simplifying assumptions and the question still remains as to how do we test these solar models? The study of solar oscillations during the last three decades has given us a tool to study the solar interior in the same way as the study of seismic waves travelling through the Earth have allowed us to study the interior of the Earth. Thus the study of the solar interior using oscillations has been referred to as Helioseismology (e.g., Deubner & Gough 1984; Gough & Toomre 1991). Unlike the Earth, where waves are triggered by a seismic event, like an earthquake, in the Sun these waves are continually present, being excited by the turbulence in the solar convection zone (Goldreich & Keeley 1977), which occupies the outer one-third of the solar body. These oscillations are essentially a superposition of millions of independent modes of oscillations of the Sun. Like a musical instrument, the Sun has a set of discrete frequencies of oscillations, which depend on the structure and dynamics of the solar interior. Since the solar interior is transparent to these waves, they can be observed at the solar surface where they induce oscillatory motions, with characteristic frequencies. These flows are detected through the resulting Doppler shifts in the spectral lines (see Bhatnagar, this volume). The light emitted by a fluid element on the solar surface which is moving outwards will be blue shifted, while that from an element moving inwards will be red shifted. Thus by measuring the shift in spectral lines it is possible to study mass motions on the solar surface and detect possible oscillatory modes.

In the next few sections we begin with a discussion of observations and the basic properties of waves and then describe the inferences that have been obtained using the measured frequencies of solar oscillations.

2 Observations of Solar Oscillations

Solar oscillations were discovered by Leighton, Noyes & Simon (1962) when they measured the velocity at some point on the solar disk using the resulting Doppler shift. They found an oscillatory pattern with a period of around 5 minutes and hence these oscillations are often referred to as five-minute oscillations. The nature of these oscillations was not immediately clear and various theories were put forward to explain them. Ulrich (1970) and Leibacher & Stein (1971) suggested that these oscillations are acoustic modes of solar oscillations, which are trapped in the interior. Subsequent observations by Deubner (1975) confirmed this hypothesis as the power was found to be concentrated in a series of ridges in the k-ω diagram, where k is the spatial wavenumber and ω the temporal frequency of oscillations, exactly as predicted by theoretical models. Once the nature of these oscillations was established it was immediately realised that these can provide information about the solar interior. The main problem with early observations was that the individual modes were not resolved due to limited spatial and temporal resolution. With improvement in instrumentation and longer observations the individual modes could be resolved. Libbrecht, Woodard & Kaufman (1990) gave an extensive table of frequencies for a large range of length scales from observations carried out at the Big Bear Solar Observatory (BBSO). The observations of solar oscillations have been carried out in two different modes, one where the intensity or velocity averaged over the entire solar disk is studied, and another where the spatially resolved image of the Sun is used to study the intensity or velocity at each point on the solar disk. In both cases the observations are repeated at interval of about 1 min to obtain a time series. The observations in integrated light are only sensitive to oscillation modes with large spatial extent, while the spatially resolved observations can study modes with smaller length scales depending on the resolution of the image.

Since the Sun is spherically symmetric to a good approximation the individual modes of oscillations can be expressed in terms of spherical harmonics, $Y_{\ell m}(\theta, \phi)$, where θ is the colatitude and ϕ is the longitude. The spatially resolved observations provide us with the line of sight velocity, v_s, at each point on the solar surface as a function of time. This velocity is then decomposed in terms of the spherical harmonics to get:

$$v_s(\theta,\phi,t) = \sum_{\ell m} A_{\ell m}(t) Y_{\ell m}(\theta,\phi) \,. \tag{1}$$

The amplitude of each component $A_{\ell m}(t)$ should contain all modes with the same ℓ, m values. The Fourier transform of $A_{\ell m}(t)$ then gives the frequencies of oscillations for all these modes. In practice, the decomposition in (1) is not perfect, since we can at best observe only half of the solar surface and over the limited area that is observed the spherical harmonics are not orthogonal. Besides, the Doppler shift only gives the line of sight component of velocity at the solar surface, which will have varying contributions from radial and tangential components of the solar velocity field at different points on the solar disk. Because of these distortions it is not strictly possible to separate out all components and $A_{\ell m}(t)$ will have some contribution from neighbouring values of ℓ, m also, which has to be accounted for when determining the frequencies of individual modes. The maximum value of ℓ that can be studied will depend on the resolution of the observations, while the maximum value of the frequencies that can be studied depends on the time-interval between each observation. For example, an interval of $\Delta t = 1$ minute will give a Nyquist frequency of $1/(2\Delta t) = 8.3\,\text{mHz}$ and only frequencies below this limit should be measured. The frequency resolution in the Fourier transform will be determined by the length of the time series or the duration over which the observations are made. If the observations extend over one full day then we can expect a frequency resolution of $1/86400 \approx 11.6\,\mu\text{Hz}$.

To obtain higher spectral resolution we need observations covering a longer period. From most sites on the Earth it is not possible to observe the Sun continuously for more than 16 hrs. Observations over successive days will necessarily have gaps during the night. These gaps in the data introduce distortions in the power spectrum, which are difficult to disentangle. Thus it is desirable to have continuous observations covering several days or even months. Various techniques have been tried to extend the observing period. The first was observations from the geographic south pole (Grec, Fossat & Pomerantz 1980), which in principle, can give a few months of observations, but in practice because of weather conditions it is difficult to have continuous observations extending over more than a few weeks. Another possibility is to observe from a network of sites spread around the Earth using identical instruments. Many such networks have been operating, like BIrmingham Solar Oscillation Network (BISON) (Chaplin et al. 1996a), International Research on the Interior of the Sun (IRIS) (Fossat 1991), Global Oscillation Network Group (GONG) (Harvey et al. 1996) and Taiwan Oscillation Network (TON) (Chou et al. 1995). Of these the first two observe the Sun in integrated light while others observe the resolved images. The GONG has been operational since 1995 using a network of six stations spread around the Earth. Until recently, GONG was using a 256×256 CCD to observe the solar image, which enables it to study oscillations modes with $\ell \lesssim 250$. The instruments have recently been upgraded to a resolution of 1024×1024, which will enable higher degree modes to be studied. Apart from ground based networks it is also possible to make continuous observations from a suitably located satellite. Observations from satellite have another advantage in that the distorting effects of the atmosphere are eliminated. The most important of these is the Michelson Doppler Imager (MDI) instrument on board the Solar and Heliospheric Observatory (SOHO) satellite (Scherrer et al. 1995), which was launched in December 1995. This satellite which is located at the Lagrangian point between the Earth and the Sun, has been observing the Sun almost continuously, except for a period during 1998–99 when the contact with the satellite was lost.

The GONG and MDI projects have provided accurate helioseismic data over the last seven years. GONG has measured frequencies of about a half million modes with different values of n, ℓ, m (Hill et al. 1996). If the Sun were spherically symmetric, then the frequencies would be independent of m, but due to rotation, magnetic field and other possible aspherical perturbations, the frequencies depend on m. However, since the departure from spherical symmetry is small, it is convenient to express the frequencies in terms of suitable splitting coefficients:

$$\nu_{n\ell m} = \nu_{n\ell} + \sum_{j=1}^{J_{\max}} c_j^{n\ell} \mathcal{P}_j^{\ell}(m) \ . \tag{2}$$

Here, $\nu_{n\ell}$ is the mean frequency for a given n, ℓ multiplet, $c_j^{n\ell}$ are the splitting coefficients and $\mathcal{P}_j^{\ell}(m)$ are orthogonal polynomials of degree j in m. In this expansion $J_{\max}$ is generally much less than 2ℓ, thus reducing the number of data points that are available. Unfortunately, different normalisations for orthogonal polynomials have been used by different workers and there is no unique definition of the splitting coefficients. We use the definition given by Ritzwoller & Lavely (1991). Schou, Christensen-Dalsgaard & Thompson (1994) have used a different normalisation. The MDI project (Rhodes et al. 1997) directly calculates these splitting coefficients rather than the frequencies of individual modes. The mean frequency $\nu_{n\ell}$ which is determined by the spherically symmetric structure of the solar interior can be measured very accurately from seismic observations.

3 Properties of Solar Oscillations

The frequencies of solar oscillations depends on the internal structure and dynamics. To a good approximation the Sun is spherically symmetric. The measured oblateness at the solar surface (Kuhn et al. 1998) is $\lesssim 10^{-5}$, which is comparable to the ratio of centrifugal to gravitational forces. Similarly, the (sidereal) rotation period (25 days) is about 4 orders of magnitude larger than typical period (5 min) of the solar oscillations. Thus to a first approximation we can neglect all departures from spherical symmetry to calculate the mean frequencies of the solar oscillations. The departures from spherical symmetry can be treated as small perturbations to the spherically symmetric model for calculating the splitting coefficients.

Given an equilibrium solar model we can calculate the frequencies by considering small perturbations about the equilibrium structure. Since the equilibrium solar model is spherically symmetric the perturbations can be expressed in terms of the spherical harmonics $Y_{\ell m}(\theta,\phi)$. For example, we can write the pressure as

$$p(r,\theta,\phi,t) = p_0(r) + \sum_{n,\ell,m} A_{n\ell m} R_{n,\ell,m}(r) Y_{\ell m}(\theta,\phi) e^{-i\omega_{n\ell m} t} , \tag{3}$$

Here, n, ℓ, m are the three quantum numbers specifying the eigenmode of oscillations, $\omega_{n\ell m}$ is the frequency of the corresponding mode and $R_{n,\ell,m}(r)$ defines the radial dependence of the eigenfunction, while $p_0(r)$ is the pressure profile in the equilibrium solar model. Similarly, all other scalar quantities can be expanded. The displacement with respect to the equilibrium position due to a single mode of oscillation can be expressed as

$$\boldsymbol{\xi}(r,\theta,\phi,t) = \left(\xi(r) Y_\ell^m(\theta,\phi), \eta(r) \frac{\partial Y_\ell^m}{\partial\theta}, \frac{\eta(r)}{\sin\theta} \frac{\partial Y_\ell^m}{\partial\phi}\right) e^{-i\omega t} , \tag{4}$$

where, ξ and η are respectively, the radial and horizontal components of displacement. Since the typical period of the oscillations turns out to be of order of several minutes, which is much smaller than the thermal time-scale, to a good approximation we can neglect the thermal exchange. Thus the perturbation can be treated as adiabatic. The thermal time-scale varies from over a million years in the core to a few minutes in the photosphere. The adiabatic approximation is good in the interior where the thermal time-scale is large, but breaks down in the layers close to solar surface, where the thermal time-scales may be comparable to the oscillation period. These non-adiabatic effects in the surface layers are traditionally not included in the calculations of oscillation frequencies as we do not have a proper theory to treat all these perturbations, particularly, the convective contribution to the flux (e.g., Balmforth 1992).

Since the observed amplitude of a typical mode of the solar oscillations is very small (≈ 10 cm s^{-1} in velocity or ≈ 5 m in displacement) the resulting equations can be linearised to obtain a system of linear homogeneous differential equations in the perturbed quantities. These equations along with the appropriate boundary conditions (e.g., Unno et al. 1989; Christensen-Dalsgaard & Berthomieu 1991) define an eigenvalue problem with frequency $\omega_{n\ell m}$ as the eigenvalue. A nontrivial solution of these equations is possible only for special values of $\omega_{n\ell m}$ which give the frequencies of the corresponding modes. The modes are identified by the quantum numbers n, ℓ, m, with ℓ, m determining the horizontal variation of perturbation through the spherical harmonic. Here, ℓ is referred to as the degree of the mode, m is the azimuthal order, and n is the radial order which is a measure of the number of nodes in the eigenfunctions in radial direction. The degree l is the number of nodal lines on the surface, while m is the number of nodes along the equator. Figure 1 shows the contour plots of a few spherical harmonics projected on the solar disk. The quantum number n, ℓ, m can take integral values, with $\ell \geq 0$ and $|m| \leq \ell$. The radial order n can take all integral values, positive or negative. The mode with $n = 0$ is the fundamental or f-mode. The $\ell = 0$ modes are referred to as radial modes and in this case the perturbations are spherically symmetric as the entire solar surface oscillates in

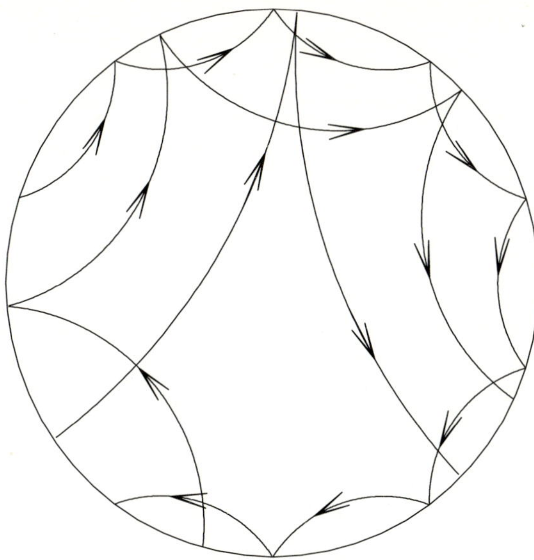

Fig. 3. Schematic diagram showing propagation of acoustic waves corresponding to different values of ℓ. Modes with large ℓ or small horizontal wavelength are trapped in a shallow region just below the surface while modes with low ℓ or large horizontal wavelength penetrate into deeper layers

atmosphere, it is difficult to calculate the frequencies accurately from theoretical solar models as these are affected by uncertainties in the atmospheric models as well as by non-adiabatic effects which are neglected while calculating the frequencies.

The p-modes are conventionally identified with positive values of the radial order, n, and their frequency increases with n. The g-modes or gravity modes are identified with negative values of n and their frequencies decrease with n. In between the p- and g-modes there is the fundamental or f-mode with $n = 0$. For large ℓ, the f-modes are essentially surface gravity modes and their amplitude decreases exponentially with depth. Thus these modes are concentrated in the surface layers. The frequencies of these modes are not very sensitive to internal structure, but depends on the surface gravity. To a good approximation the frequency of f-modes is given by the dispersion relation, $\omega^2 \approx gk$.

Since only f- and p-modes have been observed so far on the Sun, we will restrict our discussion to these modes. Depending on frequency and degree, these modes are trapped in different regions of the solar interior and sample the properties of the corresponding region. Again depending on the value of m these modes sample a different range of latitudes, with $m = \pm\ell$ modes being concentrated near the equator and as $|m/\ell|$ reduces, the mode extends closer to the poles. Thus by studying the characteristics of these p-modes it is possible to study the solar interior as a function of both radius and latitude. The mean frequency of the n, ℓ multiplet is sensitive to the horizontally averaged structure of the Sun

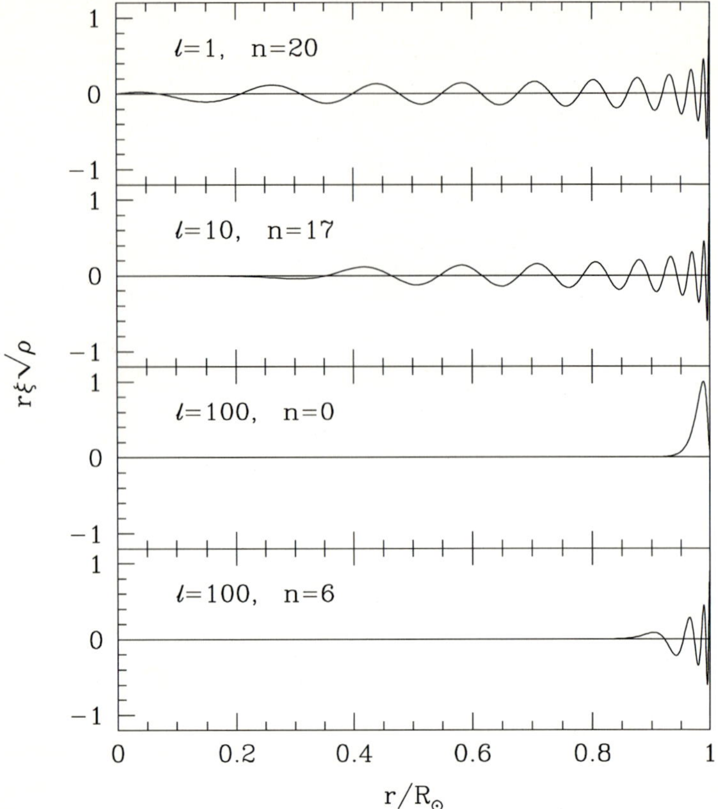

Fig. 4. The eigenfunctions, $r\xi\sqrt{\rho}$, where ξ is the radial displacement, in a solar model for different modes as marked in each panel. The eigenfunctions are scaled to a maximum value of unity

and provides important means of studying the internal structure. Figure 4 shows the eigenfunctions for a few modes. It is seen that while the low degree modes penetrate deep in the interior the high degree modes are concentrated near the surface. Similarly, the f-mode is also concentrated near the surface. Although, ξ for f-modes increase exponentially with radial distance. The scale height of the variation becomes smaller than the density scale height near the surface and hence the kinetic energy density in these modes is trapped in a layer below the surface where the density scale height is half of the scale height of ξ, which depends on the degree ℓ. It can be easily calculated that for $\ell \gtrsim 3500$ the scale height of ξ^2 will never become larger than the density scale height in solar atmosphere and such modes cannot be trapped in the solar interior as the kinetic energy density will keep increasing outwards. Observations of high degree modes (Antia & Basu 1999) have also failed to find f-modes beyond this degree.

Figure 5, shows the mean frequency as a function of degree ℓ obtained using the first 360 days of observation with the MDI instrument. This figure also shows

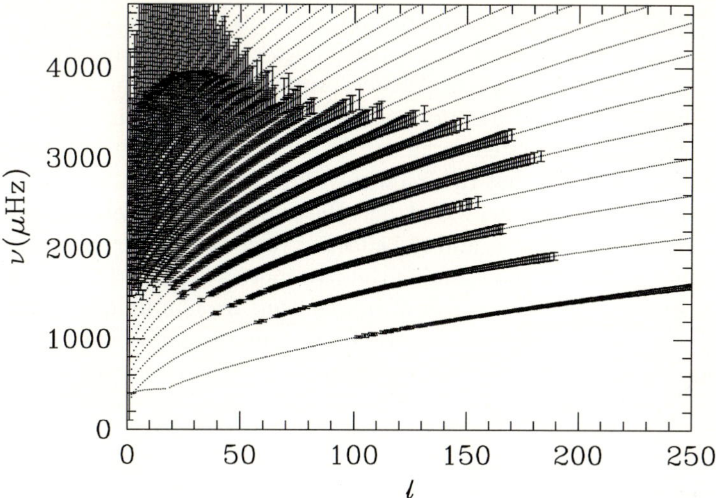

Fig. 5. The frequencies as a function of degree, ℓ for a solar model are shown by dots, which have merged into a line. The crosses with error bars are the observed frequencies from the first 360 days of observations by MDI instrument. The error bars represent 5000σ errors. The lowest line shows f-modes, while the other lines are p-modes

the frequencies computed for a standard solar model. It is clear that there is very good agreement between the observed and calculated frequencies. This excellent agreement confirms the inference that the observed modes are indeed acoustic modes of oscillations of the Sun. Apart from this, it also gives us confidence in the solar model and the underlying theory of stellar structure and evolution. The error bars in this figure are 5000σ errors, which demonstrates the accuracy of the observed frequencies. Note that the theoretical frequencies shown in this figure are from a recent solar model, with improved input physics. With the availability of seismic data there has been significant improvement in the physical input to solar models and that has resulted in the excellent agreement seen in Fig. 5.

4 Seismic Inferences of the Solar Structure

The mean frequency of an n, ℓ multiplet is determined by the horizontally averaged structure of the solar interior. Thus these frequencies provide a diagnostic of these layers. There are two basic ways in which the frequencies can be used to get information about the solar interior, the forward and the inverse techniques. In the forward technique, we construct solar models with some adjustable parameters or with varying physical input and calculate the frequencies for each of these models. These frequencies are then compared with the observed frequencies to find which of these models give the best agreement (e.g., Elsworth et al. 1990b). Most of the early conclusions were obtained using such comparisons. It was realised that the thickness of the outer convection zone should be close to 200 000 km, which was larger than what was previously estimated (Gough 1977).

Similarly, it was noted that an improved treatment of convection due to Canuto and Mazzitelli (1991) led to a significantly better agreement between calculated and observed p-mode frequencies (Basu & Antia 1994a) as compared to the usual mixing-length theory. The main drawback of the forward technique is that none of the solar models computed so far are in perfect agreement with the observed frequencies, in the sense that the frequency difference is much larger than the errors in the observed frequencies. Further, the number of possible parameters that can be modified is very large and the effect of various parameters are correlated, thus making it difficult to determine the parameters separately. As a result, the inferences from direct comparison of frequencies are non-unique as the best fit value of a parameter will depend on the input physics and other parameters that were fixed. Most of the difference between observed and calculated frequencies arises from the treatment of surface layers and hence the direct comparison of frequencies can give misleading information about the internal structure. A small difference in the surface layers can compensate for much larger variation in the interior.

4.1 Inversion Techniques

Because of these limitations of the forward analysis, the inverse technique has been used more often to get information about the solar interior. In this technique, one tries to infer the solar structure directly from the observed frequencies, without involving a solar model. The basic idea in the inverse problem is that since the frequencies are determined by the solar structure, we should in principle be able to infer the solar structure from the observed frequencies. In practice, the inverse problem is ill-conditioned as specification of solar structure, say the sound speed or density as a function of radial distance requires an infinite amount of information, while only a finite set of frequencies can be observed. Thus we do not have sufficient information to calculate the structure in full detail. This problem is normally overcome by restricting the structure functions like the sound speed to a class of smooth functions. The different inversion techniques differ in the manner in which the smoothness is ensured. In general, the smallest scale on which the variation can be reliably determined (which is referred to as the resolution of the inversion technique) depends on the number of modes that are available. The resolution also depends on the depth as the number of modes penetrating to deeper layers reduces with depth.

Most inversion techniques use a reference solar model to calculate the sound speed or density profiles inside the Sun. The advantage of using a reference model is that in that case only the difference in, say, the sound speed between the model and the Sun needs to be determined from the frequency differences. If the reference solar model is a good approximation to the Sun, then these differences are small and the relevant equations can be obtained by neglecting the higher order terms in structural differences. Further, in this case the differences can be expressed in terms of simple functions making it easier to apply the smoothness constraint to define the solution. The inferred profile is not particularly sensitive to the choice of the reference model. In the following part of this subsection, we attempt to give a brief description of the inversion techniques.

For the purpose of inversion the equations of adiabatic oscillations are written in a variational form

$$\omega^2 = \frac{\int_V \boldsymbol{\xi}^* \mathcal{L} \boldsymbol{\xi} \, dV}{\int_V \rho_0 |\boldsymbol{\xi}|^2 \, dV} , \tag{7}$$

where $\boldsymbol{\xi}$ is the displacement eigenfunction and $\mathcal{L}$ is an operator which defines the eigenvalue problem, $\mathcal{L}\boldsymbol{\xi} = \omega^2 \rho_0 \boldsymbol{\xi}$. The operator $\mathcal{L}$ depends on the solar model and the exact form of the operator was given by Chandrasekhar (1964) who also showed that this operator (together with some simple boundary conditions) is Hermitian. The variational principle is applicable for any eigenvalue problem where the operator is Hermitian. The advantage of the variational form is that to first order the perturbations in eigenvalues (due to those in the solar model) can be calculated using the eigenfunctions for the reference model. For this purpose, (7) is linearised by expressing the perturbation of the operator $\mathcal{L}$ in terms of the difference in structure functions and eigenfunctions of the reference model. This yields a complicated equation which can be written as,

$$\frac{\delta \nu_{n\ell}}{\nu_{n\ell}} = \int_0^R \mathcal{K}^{n\ell}_{c^2,\rho}(r) \frac{\delta c^2}{c^2}(r) \, \mathrm{d}r + \int_0^R \mathcal{K}^{n\ell}_{\rho,c^2}(r) \frac{\delta \rho}{\rho}(r) \, \mathrm{d}r + \frac{F(\nu_{n\ell})}{E_{n\ell}} , \tag{8}$$

where the kernels $\mathcal{K}^{n\ell}_{c^2,\rho}(r)$ and $\mathcal{K}^{n\ell}_{\rho,c^2}(r)$ are determined by the eigenfunctions in the reference model. The perturbations $\delta\nu_{n\ell}$, δc^2 and $\delta\rho$ can represent the difference between the Sun and a solar model and $E_{n\ell}$ is the mode inertia defined by

$$E_{n\ell} = \frac{4\pi \int_0^R (|\xi(r)|^2 + \ell(\ell+1)|\eta(r)|^2)\rho_0 r^2 \, \mathrm{d}r}{M(|\xi(R_\odot)|^2 + \ell(\ell+1)|\eta(R_\odot)|^2)} , \tag{9}$$

where ξ and η are the radial and horizontal components of the displacement eigenfunction, M the total solar mass and $\rho_0(r)$ the density profile. In (9) the numerator is the kinetic energy due to perturbation, while the denominator provides a normalisation to make the quantity independent of arbitrary multiplicative factors in the eigenfunction. Equation (8) connects the frequency differences to the differences in sound speed and density between two models or between a solar model and the Sun. This equation can be used in the forward sense to calculate the frequency differences arising from known variations in sound speed and density between two solar models. More often this equation is used to solve the inverse problem when the frequency differences for a set of modes is known, and we need to calculate the differences in sound speed ($\delta c^2/c^2$) and density ($\delta\rho/\rho$). The last term in (8) accounts for the uncertainties arising from the outermost layers which are not resolved by the set of modes that may be available. This term may also account for uncertainties coming from non-adiabatic processes which are not included in calculating the frequencies, since these effects are dominant only in a thin layer close to the solar surface. For each observed mode, (8) defines an integral equation connecting the frequency differences to those in sound speed and density inside the Sun. This set of equations can be solved for differences in the sound speed and density provided the frequencies are known. The inverse problem is in general ill-conditioned and a proper treatment is required to obtain meaningful results. Several inversion techniques have been developed and tested for the purpose of obtaining reliable information about the structure (Gough & Thompson 1991).

There are broadly two classes of inversion techniques, the Regularised Least Squares (RLS) and Optimally Localised Averages (OLA) techniques. In the RLS technique the unknown functions $\delta c^2/c^2$, $\delta\rho/\rho$ and $F(\nu)$ are expanded in terms of a suitable set of basis functions and the coefficients of expansion are determined by fitting to the observed frequencies. Some regularisation is required to obtain meaningful fits. The regularisation is effectively imposed by assuming that the computed solution is smooth in some sense. For example, we can obtain the solution by minimising the function

$$\sum_{n\ell} \frac{d_{n\ell}^2}{\sigma_{n\ell}^2} + \lambda_c \int_0^R \left(\frac{\mathrm{d}^2 \delta c^2/c^2}{\mathrm{d}r^2} \right)^2 \mathrm{d}r + \lambda_\rho \int_0^R \left(\frac{\mathrm{d}^2 \delta\rho/\rho}{\mathrm{d}r^2} \right)^2 \mathrm{d}r \,, \tag{10}$$

where the residual

$$d_{n\ell} = \frac{\delta\nu_{n\ell}}{\nu_{n\ell}} - \int_0^R \mathcal{K}_{c^2,\rho}^{n\ell}(r) \frac{\delta c^2}{c^2}(r)\, \mathrm{d}r - \int_0^R \mathcal{K}_{\rho,c^2}^{n\ell}(r) \frac{\delta\rho}{\rho}(r)\, \mathrm{d}r - \frac{F(\nu_{n\ell})}{E_{n\ell}} \,, \tag{11}$$

and $\sigma_{n\ell}$ is the estimated error in $\delta\nu_{n\ell}/\nu_{n\ell}$. Here, λ_c and λ_ρ are the regularisation parameters. If $\lambda_c = \lambda_\rho = 0$ we get the simple least squares minimisation to calculate the coefficients of expansion. This is generally not satisfactory as the resulting solution shows spurious oscillations due to the magnification of errors. To make the solution acceptable we have to introduce a regularisation. If λ_c and λ_ρ are very large then the resulting solution will be a linear curve for $\delta c^2/c^2$ and $\delta\rho/\rho$ because the smoothing term will dominate. Thus we need to choose suitable values for the regularisation parameters so that the solution is reasonably smooth and also the residuals $d_{n\ell}$ are reasonably small.

On the other hand, in the OLA and related techniques (Backus & Gilbert 1968), a linear combination of equations (8) is constructed such that the resulting kernel is localised in some region and the equations can be used to calculate the unknown quantity in this region. Both these techniques involve the choice of some parameters which control the regularisation or the resolution of inversion. These parameters are usually selected through some experimentation with different values. For more details of inversion techniques the readers can refer to Gough & Thompson (1991) and references therein.

Since the inversion techniques involve heuristics to find the optimal value of parameters for the regularisation, it is necessary to test them before they can be applied to real data. These tests have been carried out through extensive comparisons and through the hare and hound exercise (Antia et al. 1997). In this exercise a member who acts as the hare constructs a solar model with some input physics and calculates the frequencies of p-modes in this solar model. The calculated frequencies are then perturbed by adding random errors with the same distribution as those in the observed frequencies. This set of frequencies along with the error estimates is then supplied to the other members who act as hounds. The hounds only get the frequencies with errors added and are not aware of what model has been used to calculate these. They use the inversion techniques with the same parameters as those used for real data and calculate the sound speed and density profiles from the supplied frequencies. The inverted profiles are then sent to the hare for comparison with the actual model used to calculate the frequencies. Such exercises also test the ability of inverters to chose the parameters in inversion. These tests have validated the inversion techniques. The inversion techniques are able to get a reliable estimate of the sound speed and density over most of the solar radius, except for regions close to the surface and the centre.

Instead of sound speed and density it is possible to use other pairs of independent structure variables to write equations similar to (8). Thus we can also calculate the adiabatic index, $\Gamma_1 = (\partial \ln P / \partial \ln \rho)_S$ or the pressure, using inversion techniques. However, the frequencies of p-modes are most sensitive to variations in the sound speed and hence the sound speed can be determined more precisely. The errors in the inferred density profile are generally larger than those in the sound speed. There is also an additional problem in determining the density, since the total mass needs to be conserved. Thus $\delta\rho/\rho$ must in addition satisfy an integral constraint.

4.2 Inversion Results

One of the important achievements of the inversion technique is the inference of the sound speed and the density in the solar interior (Gough et al. 1996; Kosovichev et al. 1997). The sound speed in the bulk of the solar interior is known to an accuracy of better than 0.1%. From the variation of the sound speed below the convection zone, it was concluded that the opacity of the solar material needs to be enhanced by 15–20%. This was later confirmed by the more up-to-date Livermore opacity calculations (Rogers & Iglesias 1992). Analysis of the sound speed profile just below the bottom of the convection zone also provided evidence of gravitational settling of helium inside the Sun (Christensen-Dalsgaard, Proffitt & Thompson 1993). Helium being heavier than hydrogen is expected to settle towards the centre under the influence of gravity (e.g., Cox, Guzik & Kidman 1989). Although, the estimated time-scale for diffusion is larger than the age of the Sun, a part of the helium settles in the interior causing the surface abundance to be reduced. The incorporation of gravitational settling in the radiative interior, indeed, results in a significant improvement in solar models. Such a diffusion of helium and heavier elements should occur in the interior of other stars as well. During the main sequence phase of stellar evolution, hydrogen burning supplies the required energy to sustain the stellar luminosity. The diffusion of helium in the interior decreases the hydrogen abundance, which in turn reduces the main sequence life-time of stars. The ages of globular clusters are determined by calibrating the observed H–R diagram against theoretical calculations of stellar evolution. The inclusion of the diffusion of helium would naturally reduce

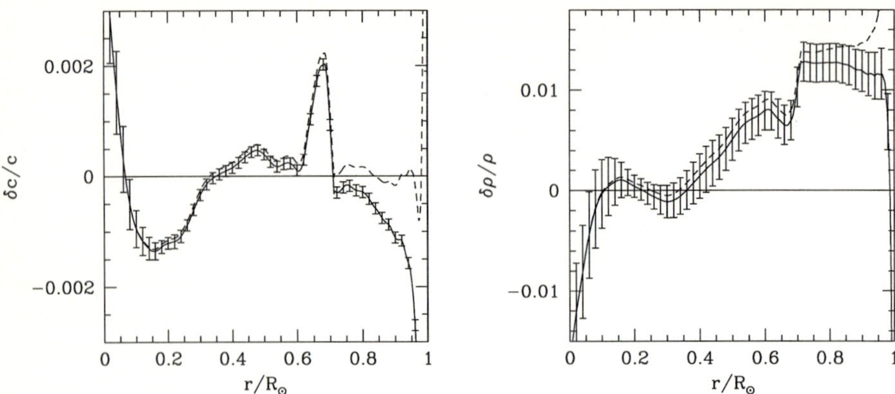

Fig. 6. Relative difference in sound speed and density between the Sun and the standard solar model of Christensen-Dalsgaard et al. (1996) is shown by *solid* lines with error bars. The *dashed* lines show the results obtained when the radial distance in the solar model is scaled by a factor of 1.00018 before taking the difference

the estimated age of globular clusters (Chaboyer et al. 1992). This should help to resolve the age problem in the standard big bang model of cosmology.

Figure 6 shows the plots of the relative difference in sound speed and density between the Sun as inferred by seismic inversions and a standard solar model with gravitational settling of helium and heavy elements (Christensen-Dalsgaard et al. 1996). The agreement between the model and the Sun is quite good with a maximum departure in the sound speed in most of the region being less than 0.2%. The most noticeable discrepancy is the prominent peak just below the convection zone, which has been identified to be due to a sharp gradient in the helium abundance profile in the solar model. This discrepancy can be alleviated if a moderate amount of turbulent mixing, possibly due to a rotationally induced instability, is included (Richard et al. 1996; Brun, Turck-Chièze & Zahn 1999). As will be seen in the next section, this region exhibits strong shear due to variation in rotation rate and is referred to as the tachocline (Spiegel & Zahn 1992). This shearing motion in the tachocline is probably responsible for mixing. The dip in the sound speed difference around $0.2R_\odot$ is not yet understood. It may be due to uncertainties in the nuclear reaction rates. The nuclear energy generation mainly takes place in the region $r < 0.2R_\odot$ while the ^{3}He abundance has a prominent peak around $r = 0.27R_\odot$ due to competition between nuclear reactions producing and destroying ^{3}He. Close to the centre, the uncertainties in the inversion results are quite large as only a few of the modes penetrate to this region. Moreover, the sound speed is rather large in this region and as a result the frequencies of p-modes, which in some sense are determined by the sound travel time, are not so sensitive to conditions in these layers. On the other hand, the frequencies are very sensitive to the structure of the outermost layers where the sound speed is low.

The figure shows a significant discrepancy in the outer region, particularly, in the convection zone, where there is a sharp dip. Inside the convection zone the temperature gradient is expected to be equal to the adiabatic gradient (except in the outermost layers), which in turn is determined by the equation of state of the solar material. We do not expect much uncertainty in the equation of state in this region and hence it is surprising to find this discrepancy in the sound speed. In fact, it turns out that if the radial distance r in the solar model is scaled by a factor of 1.00018 before taking the difference, most of this discrepancy in the sound speed is wiped off as can be seen by the dashed line in Fig. 6. Thus it appears that this discrepancy is due to an error in the solar radius. Since the scale height reduces as we approach the surface, the relative variation in sound speed due to an error in radius increases. The error in radius is most probably due to uncertainties in the treatment of the surface layers which are affected by convection as well as radiative transfer. In the lower part of the convection zone, the temperature gradient is almost equal to the adiabatic gradient and convection does not introduce much uncertainty, but close to the surface where the density is low, convection is not efficient in transporting energy and the temperature gradient becomes significantly higher than the adiabatic value. Since there is no accepted treatment of convection the structure of these layers is uncertain.

Similarly, in the atmosphere where the fluid is optically thin, we need a more sophisticated treatment of radiative transfer to model these layers. It is quite possible that a combination of these effects results in an error of 0.02% or 140 km in the solar radius. It may be noted that the surface in a solar model is generally defined to be the layer where the temperature equals the effective temperature. There are alternative definitions of the solar surface in terms of optical depth, but all these definitions agree with each other within about 30 km. Thus if the position of this layer is incorrectly determined in the solar model the radius will be incorrect. In fact, a better treatment of convection (Canuto & Mazzitelli 1991) and use of semi-empirical atmospheric model (Vernazza, Avrett & Loeser 1981) removes most of this discrepancy in a static solar model constructed using the composition profile obtained from evolutionary calculations (Brun et al. 2002). It should be noted that this discrepancy has nothing to do with uncertainties in the solar radius. Irrespective of what radius is used in the solar model, a similar scaling of r is found to remove the discrepancy in the sound speed and density.

The sound speed in ionisation zones is affected by the variation in the adiabatic index, Γ_1 which in turn is determined by the chemical composition. In particular, the dip in Γ_1 inside the second helium ionisation zone may be effectively used to determine the helium abundance in the solar convection zone. The inferred sound speed profile can be employed to compute the quantity,

$$W(r) = \frac{r^2}{GM_\odot}\frac{\mathrm{d}c^2}{\mathrm{d}r} \approx \frac{m(r)}{M_\odot}\left(\frac{P_0}{\rho_0 g_0}\frac{\mathrm{d}\Gamma_1}{\mathrm{d}r} - \Gamma_1(1-\frac{1}{\gamma\chi_\rho})\right) , \qquad (12)$$

where the second expression is valid only in most of the convection zone, where the temperature gradient is close to the adiabatic gradient. Here, G is the gravitational constant, $m(r)$ is the mass within a sphere of radius r, $\gamma = C_p/C_v$ is the ratio of specific heats and $\chi_\rho = (\partial \ln P/\partial \ln \rho)_T$. In the lower part of the convection zone, $\Gamma_1 \approx \gamma \approx 5/3$ and $W(r) \approx -2/3$. Figure 7 displays $W(r)$ for

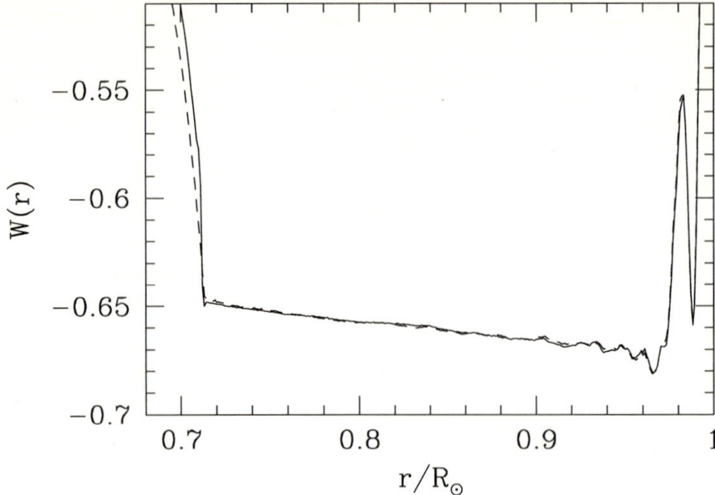

Fig. 7. The function $W(r)$ for a solar model is shown by the *solid* line, while the *dashed* line represents the same for the Sun using the seismically inferred sound speed profile

a solar model as well as for the Sun using the inferred sound speed profile. The slow rise in the curve as one approaches the base of the convection zone is due to the variation in $m(r)/M_\odot$. The total mass in the convection zone is about 2.5% of the total solar mass. The peak around $r = 0.98R_\odot$ in this curve is due to the dip in Γ_1 inside the He II ionisation zone. This peak can be calibrated to measure the helium abundance (Gough 1984; Däppen et al. 1991) which is found to be 0.249 ± 0.003 (Basu & Antia 1995). This value is somewhat less than what was adopted in earlier standard solar models and the discrepancy has been attributed to the diffusion of helium from the convection zone to the radiative interior. The dip in Γ_1 inside the ionisation zone is also determined by the equation of state and the inferred sound speed in this region provides a test for the equation of state (Basu & Christensen-Dalsgaard 1997). It is found that early equations of state which were widely used in stellar evolution calculations are not good enough to model the solar interior. In particular, the Coulomb corrections to the equation of state are quite important in this region (Christensen-Dalsgaard & Däppen 1992). More sophisticated equations of state, like the MHD (for Mihalas, Hummer, Däppen) (Däppen et al. 1988) or OPAL (Rogers, Swenson & Iglesias 1996) equation of state, are found to yield better agreement with helioseismic data. Furthermore, the OPAL equation of state is found to be in better agreement with solar data as compared to the MHD equation of state (Basu & Antia 1995). Even these equations of state are found to be slightly discrepant in the core and this discrepancy has been attributed to the neglect of relativistic corrections for electrons (Elliott & Kosovichev 1998).

Apart from the peak in the He II ionisation zone, there is a prominent discontinuity in the gradient of $W(r)$ at the base of the convection zone (Fig. 7). This discontinuity arises because the temperature gradient changes from the adiabatic value inside the convection zone to the radiative gradient below it. As a result, the second derivative of the temperature and hence that of the sound speed is discontinuous at the base of the convection zone. This discontinuity can be used to locate the position of the base of the convection zone (Christensen-Dalsgaard, Gough & Thompson 1991). The sound speed as well as the frequencies of p-modes are very sensitive to the depth of the convection zone and seismic inversions, therefore, enable a very accurate determination of its thickness. Using recent data the depth of the convection zone is estimated to be $(0.2865 \pm 0.0005)R_\odot$ (Basu 1998). Further, the position of the base of the convection zone in solar models depends on the opacity of the solar material. We can thus estimate the opacity at the base of the convection zone (Basu & Antia 1997) and it has been found that the current opacity tables from OPAL (Iglesias & Rogers 1996) with the inferred chemical composition (Grevesse, Noels & Sauval 1996) are consistent with helioseismic data to within an estimated error of 3%.

Although in a solar model we assume a sharp boundary between the convection zone and radiative interior, in actual practice the convective eddies inside the convection zone are expected to penetrate beyond the theoretical boundary of the convection zone, where the adiabatic gradient equals the radiative gradient. This is referred to as convective overshoot. Such penetration has been observed in laboratory convection as well as in the atmosphere of the Earth and the Sun. However, the conditions at the base of the solar convection zone are quite different from these situations and it is not clear if the same amount of penetration is possible at the base of the convection zone. In the absence of any theory of convection, the extent of the overshoot beyond the convection zone is treated as a free parameter in calculations of stellar structure and evolution. A significant overshoot, particularly from convective cores in massive stars can alter the course of evolution as it will give rise to mixing beyond the boundaries of the convection zone. Although, the Sun doesn't have a convective core, the conditions at the base of the convection zone are not very different from those at the outer edge of the convective cores and we may expect a similar extent of the overshoot in the two cases. Thus it is important to measure the extent of the overshoot below the solar convection zone using seismic data. The discontinuity in the derivatives of the sound speed at the base of the convection zone introduces an oscillatory component (Gough 1990) in the frequencies as a function of radial order n. The amplitude of this signal is controlled by the magnitude of the discontinuity, which in turn depends on the extent of the overshoot below the solar convection zone. Thus by measuring the amplitude of this oscillatory signal we can determine the extent of the overshoot below the convection zone (Monteiro, Christensen-Dalsgaard & Thompson 1994; Basu, Antia & Narasimha 1994). The measured oscillatory signal is found to be consistent with no overshoot and on basis of this result an upper limit of 1/20 of the local pressure scale height has been set (Basu 1997) for the overshoot distance. This is, of course, too small ($\approx 2800\,\mathrm{km}$) to affect the stellar evolution calculations significantly. It is also found that the amplitude of the oscillatory signal depends on the treatment of diffusion of helium and heavy elements below the convection zone (Basu & Antia 1994b). If there is a sharp gradient in the composition profile below the base of the convection zone, then the amplitude of the oscillatory signal in the frequencies is increased, but the measured amplitude from the observed frequencies is consistent with no gradient in the composition profile at the base of the convection zone. It would thus appear that the region immediately below the base of the convection zone is mixed. This inference is also confirmed, as noted earlier, by the bump in Fig. 6, just below the base of the convection zone.

4.3 Inversion for Temperature and Chemical Composition

The primary inversions which have provided information about the physical quantities like the sound speed, density and adiabatic index in the solar interior are based on the equations of mechanical equilibrium. The equations of thermal equilibrium are not used because on time scales of several minutes, no significant energy exchange is expected to take place in moving elements, except in the outermost layers. The frequencies of solar oscillations are, therefore, largely unaffected by the thermal processes in the interior and the equation of adiabatic oscillations does not involve the temperature and chemical composition directly. However, once the sound speed and density profiles in solar interior are deduced through primary inversions, it is possible to employ the equations of thermal equilibrium to determine the temperature and chemical composition profiles inside the Sun (Gough & Kosovichev 1990; Takata & Shibahashi 1998; Antia & Chitre 1998) provided the input physics like the opacity, the equation of state and the nuclear energy generation rates are known. These equations may be written as:

$$L(r) = -\frac{64\pi r^2 \sigma T^3}{3\kappa\rho}\frac{\mathrm{d}T}{\mathrm{d}r} , \tag{13}$$

$$\frac{\mathrm{d}L(r)}{\mathrm{d}r} = 4\pi r^2 \rho\epsilon , \tag{14}$$

where $L(r)$ is the total energy generated within a sphere of radius r, ϵ is the rate of nuclear energy generation per unit mass, κ is the Rosseland mean opacity, T the temperature and σ is the Stefan-Boltzmann constant. In addition, the equation of state provides a relation connecting the inferred sound speed and density with temperature and chemical composition profiles. These equations can be solved for $L(r), T(r)$ and $X(r)$ provided $Z(r)$ is known.

In general, the computed luminosity resulting from these inferred profiles do not match the observed solar luminosity. The discrepancy between the computed and measured solar luminosity can, in fact, provide a test of the input physics, and using these constraints it has been demonstrated that the nuclear reaction cross-section for the proton-proton reaction, S_{11} needs to be increased slightly to $(4.15 \pm 0.25) \times 10^{-25}$ MeV barns (Antia & Chitre 1998). Similar conclusions have been obtained by constructing solar models with different nuclear reaction rates (Degl'Innocenti, Fiorentini & Ricci 1998; Schlattl, Bonanno & Paterno 1999). This pp reaction cross-section has a controlling influence on the rate of the nuclear energy generation and the neutrino fluxes, but it has never been measured in the laboratory and all estimates are based on theoretical computations. The current estimate by Adelberger et al. (1998) is $(4.00 \pm 0.08) \times 10^{-25}$ MeV barns, which is somewhat lower than the helioseismic estimate. Much of the uncertainty in the helioseismic estimate arises from the uncertainty in the heavy element abundance, Z in the solar core. Antia & Chitre (1999) have estimated the proton-proton cross-section as a function of Z to find that the current best estimates for Z or the pp cross-section need to be increased slightly to match helioseismic constraints.

Part of the uncertainty in the nuclear reaction cross-section could also come from the treatment of electron screening (Weiss, Flaskamp & Tsytovich 2001). The screening arises because the nuclear cross-sections generally refer to bare nuclei, while in stellar material as well as in the laboratory the target nuclei have electrons surrounding them. Further, there is a significant difference between electrons surrounding stellar material which is almost completely ionised and the laboratory targets where electrons are bound in atoms. Thus the measured cross-section in the laboratory is to be corrected for screening by these electrons to get the cross-section for bare nuclei. This quoted value in turn has to be corrected for a different kind of screening in stellar interiors (Gruzinov & Bahcall 1998). For the pp reaction the theoretical value again refers to bare protons which has to be corrected for electron screening in the stellar interior. Most of the helioseismic estimates of the pp reaction cross-section are based on the formulation of Graboske et al. (1973) for electron screening. However, this treatment is probably not applicable to the solar interior because they assume complete electron degeneracy (Dzitko et al. 1995). If instead we try the weak screening formulation of Salpeter (1954) or an intermediate screening due to Mitler (1977) the estimated cross-section for the pp reaction is closer to the theoretical estimate. Figure 8 shows the helioseismically estimated value of S_{11} as a function of heavy element abundance, Z in the solar core using the treatment of electron screening due to Mitler (1977). This figure can be compared with similar figure in Antia & Chitre (1999) using the electron screening treatment due to Graboske et al. (1973). It is clear that with improved treatment of electron screening there is very little departure from theoretical estimate for S_{11}. The helioseismically estimated value of $S_{11} = (4.07 \pm 0.07) \times 10^{-25}$ MeV Barns, is within 1σ of the theoretically estimated value.

The solid line in Fig. 8 shows the estimated S_{11} when the effect of Z is included only in the opacities as was also done by Antia & Chitre (1999). However, Z will also affect the nuclear energy generation rate through the CNO cycle reactions. In the standard solar model the CNO cycle contributes less than 2% of the total luminosity. But when Z is increased, the central temperature increases and CNO reactions become more effective in producing energy. Further, the increased abundance of CNO will also enhance the rate of energy generation. Thus if for simplicity we assume that the abundances of CNO increase in the

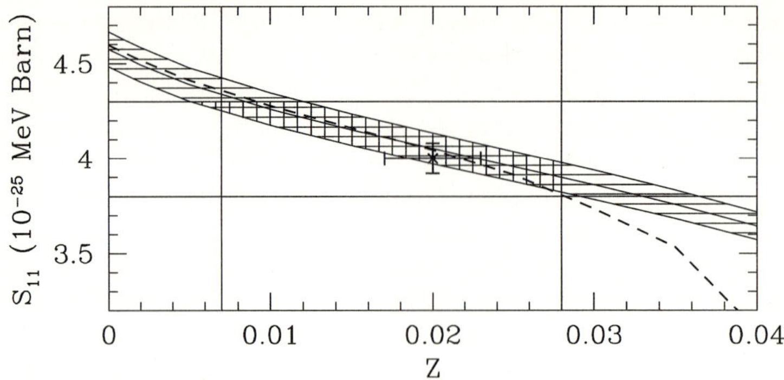

Fig. 8. The cross-section of the pp nuclear reaction as estimated from seismic constraints is shown as a function of heavy element abundance in the solar core. The shaded area shows the 1σ error bars and the *solid* line shows the best estimate. The point with error bars shows the current best estimates for Z and S_{11}. The vertical lines denote the limits on the central Z values obtained by Fukugita & Hata (1998) and the horizontal lines mark the limits on S_{11} as obtained by various calculations so far. The region with vertical shading indicates the area that is consistent with all data. The *dashed* line shows the same curve when the effect of Z is also included in the nuclear energy generation through the CNO cycle

same ratio as Z we can estimate the nuclear energy generation due to changes in heavy element abundances. These results are shown by the dashed line in Fig. 8. Interestingly, we find that as Z is increased to 0.04 the estimated S_{11} drops very sharply, because for this value of Z a large fraction of the solar luminosity can be accounted for by the CNO reactions and hence to maintain the solar luminosity the cross-section for the pp reaction has to be reduced significantly. This is clearly unacceptable and hence these calculations effectively put an upper limit of 0.04 on Z in the core (Antia & Chitre 2002). Similar limits have been obtained by Fukugita & Hata (1998) using solar neutrino fluxes, which are also shown in the figure.

The inferred helium abundance profile is in good agreement with that in the standard solar model with diffusion of helium and heavier elements, except in layers just below the solar convection zone. This is the region where the solar rotation rate has a sharp gradient in radial direction. The inferred helium abundance profile, for example, shown in Fig. 9 is essentially flat in this region. This indicates the presence of some sort of mixing process, possibly by rotationally induced instability which has not been properly accounted for. Solar models including mixing in the tachocline region (Brun, Turck-Chièze & Zahn 1999) are in good agreement with the helioseismically inferred composition profile. The mixing in this region can also explain the anomalously low lithium abundance in the solar envelope. It is known that the lithium abundance in the solar envelope is about a factor of 100 lower than the cosmic abundance inferred from meteorites (Vauclair 2000). Lithium can be destroyed by nuclear reactions at temperatures exceeding 2.5×10^6 K. At the base of the solar convection zone, the temperature

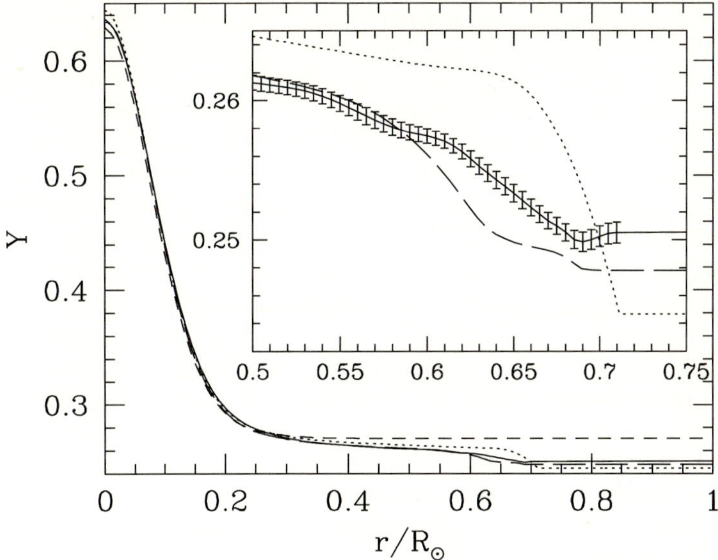

Fig. 9. Fractional helium abundance by mass in the Sun as obtained from inversions is shown by the *solid* line, while the *short-dashed* line represents the same for a solar model without diffusion and the *dotted* line shows that for a model incorporating diffusion of helium and heavy elements. The *long-dashed* line shows the helium abundance in the solar model of Brun, Turck-Chièze & Zahn (1999) which includes mixing in the tachocline region. The inset shows a blow-up of the region close to convection zone base

is still not high enough to burn lithium, but if the mixing extends a little beyond the solar convection zone to a radial distance of $0.68R_{\odot}$ the temperature becomes high enough to explain the low abundance of lithium. This is exactly the region where the inferred composition profile is flat, indicating the possible operation of a mixing process. The exact mechanism for this mixing is not yet understood.

Once the temperature and chemical composition inside the Sun is inferred seismically, it is possible to construct a seismic model using these profiles. We can also calculate the neutrino fluxes in these seismic models, which turn out to be close to those in the standard solar models. These seismic models can be used to study neutrino properties through the observed flux of solar neutrinos at the Earth (e.g., Chitre, this volume).

5 Rotation Rate in the Solar Interior

The solar surface rotation has been extensively studied through observations of sunspots and other tracers (Howard 1984). These observations have established that the Sun does not rotate like a rigid body. Instead the rotation rate varies with latitude resulting in the well known pattern of differential rotation with equatorial regions rotating faster than the polar regions. Doppler measurements at the surface have also confirmed this differential rotation (Ulrich et al. 1988). Thus one expects that the rotation rate varies with radial distance inside the Sun. The current profile of the rotation rate in the solar interior is the result of how the Sun is spun down during the course of its evolution from a fast rotating initial state to a relatively slow rotation at the present epoch (Skumanich 1972; MacGregor & Charbonneau 1994; Schrijver 1994). The gradual loss of angular momentum over the main-sequence life of the Sun is due to magnetic coupling to the solar wind (Rosner & Weiss 1985; Mestel & Weiss 1987; Charbonneau & MacGregor 1993). The exact mechanism for this angular momentum loss and transport in the solar interior is not fully understood and the inferred profile of the rotation rate in the solar interior at the current epoch provides a strong constraint on these theories (Talon & Zahn 1998). Before the advent of helioseismic data it was generally believed that the solar core rotates faster than the surface (e.g., Pinsonneault et al. 1989). This belief arises because the loss of angular momentum is supposed to happen at the surface.

Knowledge of the rotation rate in the solar interior is also crucial for the test of the General Theory of Relativity using the measured precession of the perihelion of the planet Mercury. If the solar interior is rotating rapidly then it will cause some distortion in the solar structure due to the centrifugal force which will give rise to a quadrupole moment. If the quadrupole moment of the Sun is sufficiently large it can explain part of the measured precession of the planet mercury from purely Newtonian effects, thus introducing a discrepancy between the measured value and the prediction by general relativity (Dicke & Goldenberg 1974; Dicke, Kuhn & Libbrecht 1985).

A rapidly rotating solar core can also reduce the central temperature of the Sun thus lowering the solar neutrino fluxes (Ulrich 1969; Roxburgh 1974), which may have some bearing on the solar neutrino problem (e.g., Haxton 1995). This arises because the resulting centrifugal force effectively decreases gravity thus reducing the pressure gradient and consequently the temperature gradient. Any significant lowering of the solar neutrino flux would require a rotation rate in the solar core that is at least a few hundred times the surface value. Such a high rotation rate would cause an observable distortion at the solar surface.

5.1 Inversion for Rotation Rate

Seismic observations in recent times have provided us a tool to determine the rotation rate in the solar interior. As explained earlier, rotation lifts the degeneracy of frequencies with the same n, ℓ and introduces frequency splittings. The magnitude of the splittings is determined by the rotation rate in the region where a given mode is trapped. Since each mode of solar oscillation is trapped in a different region, it is possible to infer the rotation rate in the solar interior as a function of radial distance and latitude by studying the splitting coefficients for all these modes.

In order to understand how rotation might affect the observed frequencies, let us consider a simple situation where the rotation rate Ω is uniform, that is, the Sun is rotating like a rigid body. Let us choose a spherical polar coordinate system (r, θ, ϕ) with the axis coinciding with rotation axis. If we consider another frame that is rotating with the Sun, then in this frame the coordinate $\phi' = \phi - \Omega t$. A perturbation with frequency ω_0 in the rotating frame has the form $\cos(m\phi' - \omega_0 t)$. In the inertial frame it will translate to $\cos(m\phi - \omega_m t)$, where $\omega_m = \omega_0 + m\Omega$ can be considered as the frequency as seen in the inertial frame. Thus a frequency ω_0 in an inertial frame gets split into $2\ell + 1$ components separated by the rotation rate Ω. This is purely geometrical effect. In practice, there will be additional effects coming due to the Coriolis force in the rotating frame. For the Sun an additional complication is caused by the fact that the rotation rate is not uniform and hence we cannot define a frame that is rotating with the Sun. Thus a more sophisticated treatment is needed to estimate the rotational splittings.

As mentioned earlier, the effect of rotation on solar oscillations frequencies is small and hence can be treated as a small perturbation to the non-rotating model. The first order contribution from rotation arises from the Coriolis term and the resulting splitting varies linearly with the rotation rate in the solar interior. The first order splitting from rotation turns out to be an odd function of m and affects only the odd splitting coefficients, $c_1, c_3, \ldots$ in (2). Further, these splitting coefficients are sensitive only to the north-south symmetric component of the rotation rate and hence only this component can be inferred from the measured splittings of the global modes. Local helioseismic techniques (Hill 1988; Duvall et al. 1993) can be employed to determine the antisymmetric component as well as other large scale flows in the outer part of the convection zone. We will be concerned only with the symmetric component of the rotation rate. Surface observations indicate that the antisymmetric component of the rotation rate is rather small and in fact, it has not been reliably measured.

For a slowly rotating Sun the effect of rotation can be treated as a small perturbation over the eigenvalue problem for a spherically symmetric model (Lynden-Bell & Ostriker 1967)

$$\mathcal{L}\boldsymbol{\xi} + \rho_0\omega^2\boldsymbol{\xi} = -2\mathrm{i}\omega\rho_0(\mathbf{v}\cdot\boldsymbol{\nabla}\boldsymbol{\xi}) , \tag{15}$$

where $\mathcal{L}$ is the operator defining the eigenvalue problem for a spherically symmetric star (e.g., (7)) and $\mathbf{v} = \boldsymbol{\Omega} \times \boldsymbol{r}$ is the rotational velocity, while Ω is the rotation rate. Equation (15) is written in an inertial frame and if the right hand side is replaced by zero then it reduces to the normal linear adiabatic equation for stellar oscillations in a nonrotating star. The centrifugal term being of second order in Ω would be much smaller and is neglected in this approximation. It turns out that the centrifugal term and the accompanying distortion in the equilibrium state gives a small contribution to the frequencies which is an even function of m (Gough & Thompson 1990), while the first order contribution from rotation has only odd powers of m. Thus these two contributions can be separated and for the purpose of calculating the rotation rate in the solar interior we need not worry about the even order terms in m arising from the centrifugal force.

The term on the right hand side of (15) represents the perturbations arising due to rotation, and to first order its effect on the frequencies can be estimated using the variational principle (Chandrasekhar 1964) to get the change in frequency of an eigenmode due to rotation

$$\delta\omega = \frac{-\int \mathrm{i}\rho_0 \boldsymbol{\xi}^* \cdot (\mathbf{v}\cdot\boldsymbol{\nabla}\boldsymbol{\xi})\,\mathrm{d}^3r}{\int \rho_0 \boldsymbol{\xi}^* \cdot \boldsymbol{\xi}\,\mathrm{d}^3r} , \tag{16}$$

where $\boldsymbol{\xi}$ is the eigenfunction for the nonrotating solar model. Using (4), this expression can be simplified to obtain the change in frequency of the mode specified by the quantum numbers n, ℓ, m. Further, the even order contribution can be suppressed by taking the difference in frequency of modes with $\pm m$ to get the rotational splittings

$$D_{n\ell m} = \frac{\nu_{n\ell m} - \nu_{n\ell -m}}{2m} = \int_0^1 \int_0^1 \mathrm{d}r\,\mathrm{d}\cos\theta K_{n\ell m}(r,\theta)\Omega(r,\theta) , \tag{17}$$

where r is the fractional radius and θ the colatitude. Here, $K_{n\ell m}(r,\theta)$ are the mode kernels which depend on the eigenfunctions in the spherically symmetric model (Pijpers 1997). Instead of considering the frequency difference between modes with $\pm m$ we can use the splitting coefficients as defined by (2) to obtain an equation similar to (17). Ritzwoller & Lavely (1991) have shown that if $\mathcal{P}_j^\ell(m)$ are the orthogonal polynomials over a discrete set, then the latitudinal dependence of the rotation rate can be expanded in terms of independent functions to separate out the radial and latitudinal dependence. This separation helps in improving the efficiency of the inversion process.

Equation (17) can be used to compute frequencies of individual modes which can then be compared with the observed values (the forward technique) or alternately the measured frequencies or splitting coefficients can be inverted to infer the rotation velocity as a function of r and θ. The diagnostic power of the solar oscillations arises from the fact that the kernels corresponding to different modes are peaked in different regions of the solar interior. The latitude dependence is determined by ℓ, m, while the radial dependence is determined by n, ℓ. The modes with low ℓ penetrate deeper in the solar interior and the corresponding kernels are significant in the deep interior, while modes with high ℓ are trapped in the shallow layer just below the solar surface and these kernels have significant values only in these layers.

Early helioseismic data on rotational splittings included only the sectoral modes with $m = \pm\ell$ (Duvall & Harvey 1984), which are trapped mainly in the region around the equator. Duvall et al. (1984) showed that there was little variation of rotation rate with depth. More complete data (Brown 1985; Libbrecht 1989) showed that the differential rotation observed at the surface persists throughout the convection zone, while below the convection zone, there was little evidence for differential rotation (Brown & Morrow 1987; Brown et al. 1989; Christensen-Dalsgaard & Schou 1988). With better quality data from GONG and MDI now becoming available, inversions have been very successful in inferring the rotation rate in solar interior (Thompson et al. 1996; Schou et

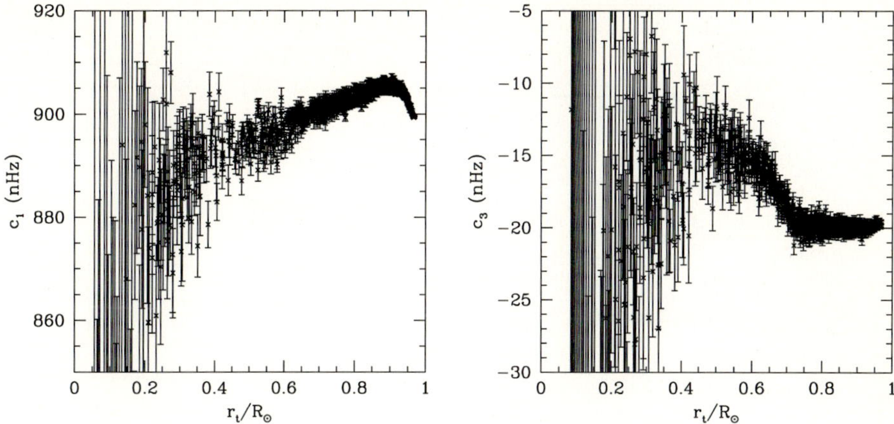

Fig. 10. Splitting coefficients c_1 and c_3 from GONG data as function of the position of the lower turning point

al. 1998). For illustration Fig. 10 shows the first two odd splitting coefficients as a function of the position of the lower turning point of the mode, which is the layer where the acoustic modes suffer total internal reflection. The splitting coefficient c_1 which determines the latitudinally independent component of the rotation rate increases slightly for modes penetrating below the surface before decreasing again. This implies that the rotation rate should increase with depth just below the surface. On the other hand, the splitting coefficient c_3 is almost constant for modes trapped in the convection zone, while below that there is a sharp decline in its value. This feature is attributed to the tachocline.

Equation (17) can be used for inversion to determine the rotation rate in the solar interior. Once again we can use either the RLS or some variant of the OLA technique to perform inversions (Gough & Thompson 1991). In this case, we directly obtain the rotation rate as a function of both latitude and radius and the inversion is referred to as 2D inversion. The number of data points as well as the number of basis functions required to approximate the rotation rate in the RLS technique will be large and significant computing resources are required. On the other hand, if the expansion proposed by Ritzwoller & Lavely (1991) is used then the 2D problem is decomposed into a series of 1D problem, which require much less effort and the inversion is very efficient. This process is often referred to as 1.5D inversion. However, there is significant uncertainty in the inversion of higher order coefficients when they are handled separately and the 1.5D inversion may not give very reliable results at high latitudes or in the deep interior, unless the smoothing or other parameters in inversion are chosen carefully. Nevertheless, all inversion techniques have been tested through extensive hare and hound exercises (Schou et al. 1998) and have performed well on these tests. With improvement in computing resources 2D inversion techniques are being preferred.

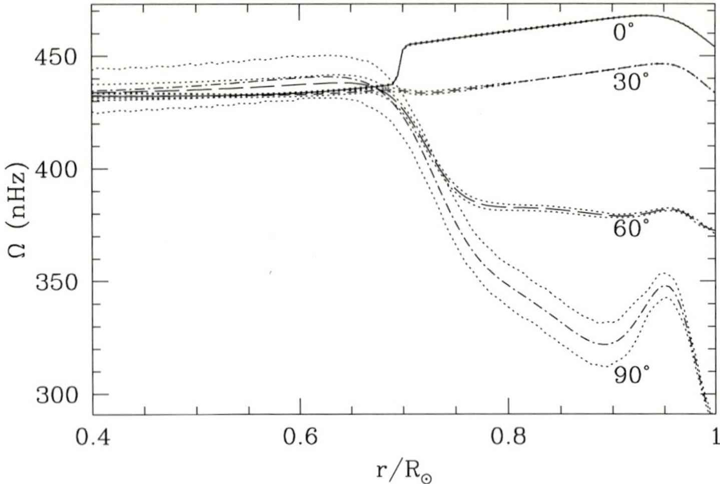

Fig. 11. Rotation rate at various latitudes as a function of radial distance, as inferred from MDI data using a 2D RLS inversion technique. The *solid*, *short-dashed*, *long-dashed* and *dot-dashed* lines show the rotation rate at latitudes of $0°, 30°, 60°, 90°$ respectively. The *dotted* lines show the respective 1σ error limits

5.2 Inversion Results

Figure 11 shows the rotation rate as a function of radial distance at various latitudes inferred using a 2D RLS inversion of MDI data. Figure 12 shows the results from a 2D inversion of GONG data in the form of contours of constant rotation rate. From these results, it is clear that the differential rotation observed at the solar surface continues through the convection zone. While in most of the radiative interior the rotation rate is essentially independent of latitude. Near the base of the convection zone which is located at a radial distance of $0.713R_\odot$ (Christensen-Dalsgaard, Gough & Thompson 1991; Basu & Antia 1997) there is a sharp transition from differential rotation in the convection zone to solid body like rotation in radiative interior. This shear layer has been named tachocline (Spiegel & Zahn 1992) and as pointed out earlier, it is the region where some mixing process is operating. Apart from this there is another shear layer just below the surface where the rotation rate appears to increase with depth, reaching a maximum value around $r = 0.95R_\odot$ or at a depth of 35 Mm from the solar surface. It is not very clear if this shear layer continues at higher latitude. Some inversion results using MDI data (Schou et al. 1998) appear to suggest that the gradient in the rotation rate reverses its sign around a latitude of 60°. Inversion of GONG data does not show any change in the sign of the shear with latitude. The existence of this shear layer is also confirmed by the local helioseismic technique which employs high degree modes that are more sensitive to the surface region (Basu, Antia & Tripathy 1999).

Another interesting feature that has emerged in some inversion results using the MDI data is the possible existence of a jet like feature of rapid rotation

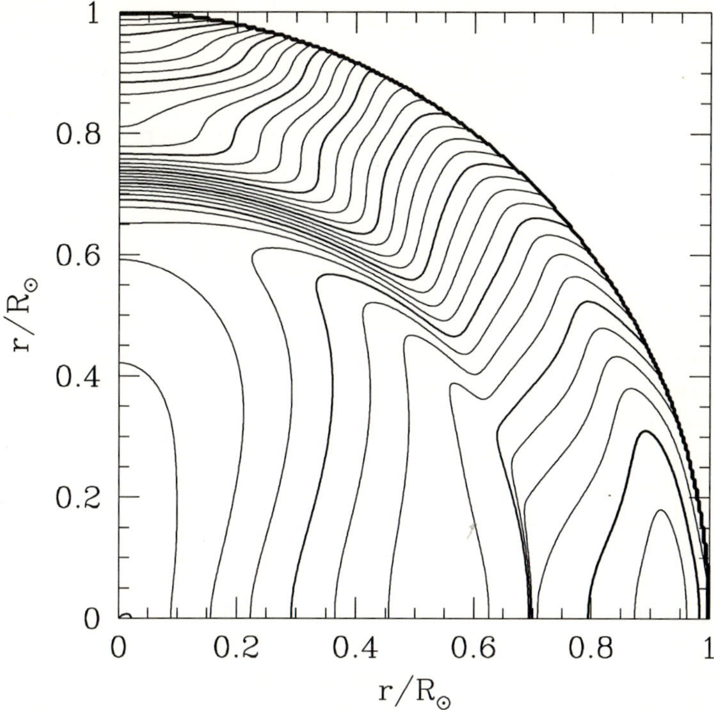

Fig. 12. Contours of constant rotation rate as obtained by 2D inversion technique using GONG data. Due to the symmetry of the inversion results, the rotation rate has been shown for just one quadrant only. The contours have been drawn at intervals of 5 nHz., and the thick *solid* contour corresponds to a rotation rate of 440 nHz. The x-axis represents the solar equator while the y-axis represents the rotation axis

around latitude of 75° and a radial distance of $\approx 0.95R_{\odot}$. This feature causes the bump in the polar rotation rate shown in Fig. 11. However, this feature does not show up in inversions using the GONG data (Fig. 12). More detailed analysis of the results (Howe et al. 1998) has shown that this difference is due to the analysis procedure used in reducing the data to calculate the frequency splitting. The reality of this feature has not yet been demonstrated. The series of contours close to the centre are not likely to be real as the errors in inferred profiles in that region is much larger than the contour spacings. Leaving aside these questionable features the rotation rate in the radiative interior is roughly constant.

The inferred rotation rate at the solar surface is found to be in good agreement with Doppler observations, but is somewhat lower than the value inferred from magnetic features. The magnetic features appear to rotate at a rate close to that found at $r = 0.97R_{\odot}$. This is generally interpreted to mean that magnetic features are anchored at a layer below the solar surface where the rotation rate is higher. Traditionally observations of the rotation rate at the solar surface are fitted to a three term relation:

$$\Omega = A + B\cos^2\theta + C\cos^4\theta \ . \tag{18}$$

The helioseismically inferred rotation profile has many more terms included and an attempt has been made to identify these terms with the zonal flows reported at solar surface (Kosovichev & Schou 1997). Ideally, zonal flows should be identified with the time varying component of the rotation rate and we will discuss this aspect in the next section. There is a pronounced dip in the rotation rate in the polar region as compared to what is obtained using this three term relation (Schou et al. 1998). The cause of this polar dip is not understood.

The rotation rate in the solar core is somewhat uncertain as only very low degree modes penetrate into the core and the rotational splittings of these modes is not determined very reliably. The best measurement of low degree splittings is obtained from the observations of integrated sunlight by the BISON and IRIS networks or the Global Oscillations at Low Frequencies (GOLF) instrument on board SOHO. Most inversion results show that the solar core is rotating slower than the equatorial rotation rate at the surface (Elsworth et al. 1995; Tomczyk, Schou & Thompson 1995; Chaplin et al. 1996b; Thompson et al. 1996; Corbard et al. 1997; Charbonneau et al. 1998). However, IRIS data has yielded a much higher value for the rotation rate in the core (Lazrek et al. 1996; Gizon et al. 1997). The cause for this discrepancy is not understood.

The inferred rotation rate in the solar interior is found to be quite different from what was predicted by theoretical calculations using the existing ideas of angular momentum transport in the stellar interior (Gilman & Miller 1986; Glatzmaier 1987; Brummell, Hurlburt & Toomre 1998). These numerical simulations of convection in rotating spherical shells indicated that the rotation rate is nearly constant on cylinders aligned with the rotation axis. Thus it is clear that these theories need to be suitably revised. Recent numerical simulations (Elliott, Miesch & Toomre 2000) have yielded a rotation rate that is closer to the helioseismically inferred profile, but there are a number of parameters in their model which have been adjusted to get such a rotation rate.

As mentioned earlier, the rotation rate in the solar interior has a bearing on the test of the general theory of relativity using the measured precession of perihelion of planet Mercury. As is well known general relativity predicts a precession by $43''$ per century which has been observed. However, if the solar core were rotating sufficiently rapidly it would distort the Sun and introduce a substantial quadrupole moment, J_2, which would yield some precession of the orbit purely from the Newtonian effects. Thus it is necessary to ensure that the rotation rate in the solar interior is sufficiently small, not to disturb the agreement between observation and theoretical prediction from general relativity.

Using the inferred rotation rate in the solar interior it is possible to infer global quantities like angular momentum, kinetic energy and quadrupole moment (Pijpers 1998). These values are

$$\text{Angular Momentum:} H = (190.0 \pm 1.5) \times 10^{46} \text{ gm cm}^2 \text{ s}^{-1} \ , \tag{19}$$

$$\text{Kinetic Energy:}T = (253.4 \pm 7.2) \times 10^{40}\ \text{gm cm}^2\ \text{s}^{-2}\ , \tag{20}$$

$$\text{Quadrupole Moment:}J_2 = (2.18 \pm 0.06) \times 10^{-7}\ . \tag{21}$$

It may be noted that a major contribution to J_2 comes from outside the core and hence the uncertainty in the rotation rate in solar core does not affect its value significantly. This value of J_2 will yield a precession of the perihelion of the planet Mercury by about 0.03 arc sec/century, which is smaller than the errors in the measurement, thus maintaining consistency of the general theory of relativity.

5.3 The Tachocline

The tachocline is the shear layer near the base of the convection zone, where the rotation rate changes from the differential rotation inside the convection zone to a solid body like rotation in the radiative interior. This layer is thought to be the seat of the solar dynamo and hence the study of this region is of great interest. Although the existence of this layer has been known from early inversion results, the exact location of the tachocline and its thickness is difficult to determine from inverted profiles, because the regularisation used in the inversion technique always tends to smooth out the variation in rotation rate. Thus forward modelling techniques have been employed to study this region (Kosovichev 1996). For example, we can parameterise the rotation rate in the tachocline and calculate the resulting rotational splittings. The parameters can then be determined by comparing these calculated splittings with the observed values.

We can assume the rotation rate at a fixed latitude to be given by (Antia, Basu & Chitre 1998)

$$\Omega_{\rm tac}(r) = \begin{cases} \Omega_c + B(r-0.7) + \frac{\delta\Omega}{1+\exp[(r_d-r)/w]} & \text{if } r \le 0.95\ , \\ \Omega_c + 0.25B - C(r-0.95) + \frac{\delta\Omega}{1+\exp[(r_d-r)/w]} & \text{if } r > 0.95\ , \end{cases} \tag{22}$$

where Ω_c, B, C are the three parameters defining the smooth part of rotation rate while $\delta\Omega$, r_d and w define the tachocline. Here B is the average gradient in the lower part of convection zone, while C is the gradient in the near surface shear layer. The latitude dependence in these six parameters can be accounted for by expanding each of them in terms of latitude. These parameters can be determined by a non-linear least squares fit to the observed splittings of modes with the lower turning point near the location of the tachocline. However, this involves a nonlinear minimisation problem in several variables and it is difficult to find the global minimum which gives the best estimate for the parameters. Various techniques have been employed for this purpose including genetic algorithms (Charbonneau et al. 1997) and simulated annealing (Antia, Basu & Chitre 1998). These techniques generally require considerable computer time. Another alternative is to fix the parameters Ω_c, B, C defining the smooth part from the known inversion results and determine the three parameters defining the tachocline separately using their characteristic signature in observed splittings for the modes with lower turning point near the tachocline. For this purpose, one needs a series of calibration models with different values of the parameters to calibrate the signal (Basu 1997). This technique is very efficient but has the disadvantage that each parameter has to be determined separately by calibration and its value may be affected by the other parameters. Another possibility is to modify the inversion techniques to allow steep variations in some regions (Corbard et al. 1998, 1999). All these techniques have yielded results that are roughly consistent with each other.

The mean properties of the tachocline can be approximately determined by considering only the splitting coefficient c_3, which has the dominating influence on this shear layer. Using c_3 the tachocline is found to be centred at a radial distance of $(0.705 \pm 0.003)R_\odot$ (Basu 1997) which is just below the base of the convection zone. The thickness of the tachocline is more difficult to determine and its value will also depend on the form of the function used in defining the rotation rate. Using the form given by (22), the mean half-width is found to be $(0.010 \pm 0.003)R_\odot$ (Basu 1997). This would imply that most of the tachocline is located below the base of the convection zone. Alternative models for rotation rate in the tachocline have also been used (Kosovichev 1996) and these differences should be accounted for while comparing the widths obtained by different workers.

Since the rotation rate has a strong dependence on latitude inside the convection zone, it would be interesting to determine whether the thickness or location of the tachocline has also a latitude dependence. Inversion results generally tend to show larger thickness at higher latitudes and it also appears to suggest that the tachocline is shallower at higher latitude. However, this may be an artifact of the smoothing, since the jump in rotation rate across the tachocline is much larger at high latitudes. Thus it is necessary to confirm this trend from the forward modelling approach. The initial results were inconclusive as there is no clear trend with latitude seen in either the depth or the thickness of the tachocline (Antia, Basu & Chitre 1998). However, with accumulation of more data it is possible to combine data at different times to improve the accuracy and it turns out that there is a distinct latitudinal variation in depth of the tachocline (Charbonneau et al. 1999; Basu & Antia 2001). It is found that the tachocline is prolate (i.e., the equatorial diameter is less than the polar diameter) and the difference between the tachocline position at 0° and 60° latitude is about $(0.020 \pm 0.003)R_\odot$ which is statistically significant. There is also some increase in thickness of the tachocline with latitude, by about $(0.006 \pm 0.002)R_\odot$, which is less significant. At the same time it is found that there is no significant latitudinal variation in the position of the base of the convection zone. Any possible variation would be less than $0.0002R_\odot$ (Basu & Antia 2001), which is two orders of magnitude smaller than the variation in the tachocline position. Although the latitudinal resolution of the tachocline studies is not very high, there is a possibility that the apparent latitudinal variation is due to the fact that the tachocline actually consists of two different parts, one at low latitudes ($< 30°$) where the rotation rate increases with radial distance and a second one at higher latitudes where the rotation rate decreases with radial distance. It is possible that these two parts are located at slightly different depths and have different thickness, while there is no latitudinal variation within each of these parts (Basu & Antia 2003). This essentially means that apart from a sharp variation with radial distance, there is a discontinuity in latitude also in the tachocline region.

While the properties of the tachocline can be established from helioseismic data, it is not clear how such a shear layer arises inside a star. Stars are expected to be born with much larger angular velocities and during the course of evolution the angular momentum is gradually lost from the surface, thus slowing down the rotation. How exactly, this slowdown gives rise to the tachocline is not understood. The strong shear in the tachocline can induce instabilities which can mix the region (Charbonnel et al. 1994; Chaboyer, Demarque & Pinsonneault 1995; Richard et al. 1996; Brun, Turck-Chièze & Zahn 1999). As mentioned earlier, structure inversion results support the presence of mixing in this region.

5.4 Meridional Flow

Surface observations using Doppler shifts have demonstrated that in addition to rotation, there is a large scale flow in the north-south direction, referred to as the meridional flow (Hathaway et al. 1996). The meridional flow is found to be from the equator to the poles in both the hemispheres. The magnitude of this flow is known to be of the order of 20 m s^{-1}, as compared to the rotation velocity of 2 km s^{-1}. Because of this small magnitude, it is difficult to measure the flow velocity reliably. The global modes of oscillations described above are not sensitive to meridional flows and hence cannot be used to study them. However, local helioseismic techniques like Ring diagram (Hill 1988) and time-distance analysis (Duvall et al. 1993) can be employed to study the meridional component of a large scale flow. These techniques use high degree modes and hence are not sensitive to the deep interior, but can be used to study the outer convection zone. From these studies (Giles et al. 1997; Braun & Fan 1998; Schou & Bogart 1998; Basu, Antia & Tripathy 1999) the meridional component is known to penetrate to a depth of at least 10% of the solar radius. The meridional velocity also depends on the latitude and the dominant component of this flow has a variation of the form $v_0 \sin(2\theta)$, where θ is the latitude. The amplitude v_0 is about 25 m s^{-1} near the surface and has only a weak variation with depth. Higher order components with amplitudes of a few m s^{-1} have also been detected in the meridional velocities (Hathaway et al. 1996; Basu, Antia & Tripathy 1999). From conservation of mass arguments one would expect that in deeper layers the direction of flow must reverse, but this reversal has not been seen up to a depth of about 10% of solar radius (Braun & Fan 1998). Meridional flow plays a crucial role in the operation of the solar dynamo (Nandy & Choudhuri 2002). With improved data and techniques of local helioseismology we hope to learn more about this flow.

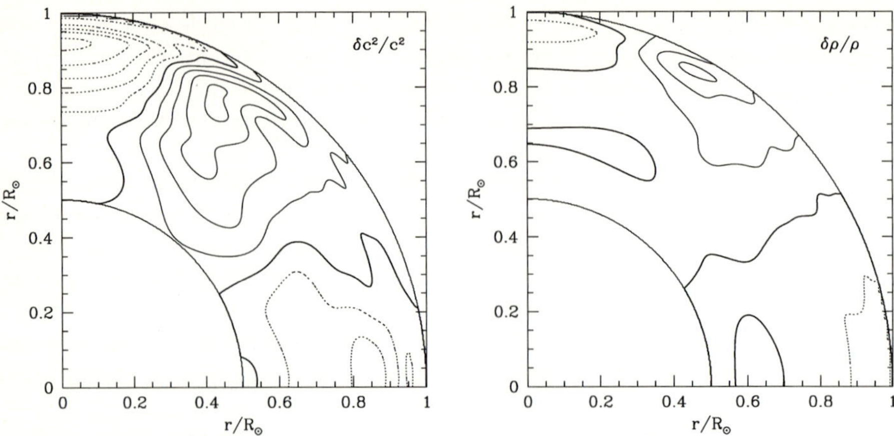

Fig. 13. The aspherical component of sound speed and density in the Sun as inferred by inversion of GONG data averaged over the duration for which data is available. The left panel shows contours of constant $\delta c^2/c^2$ and the right panel shows those for $\delta\rho/\rho$. The contour spacing is 2×10^{-5} with *solid* contours showing positive values, and *dashed* contours showing negative values. The thick line shows the zero contour. The x-axis represents the solar equator, while the y-axis represents the rotation axis

6 Asphericity in Solar Structure

As mentioned earlier, to first order, rotation affects only the odd splitting coefficients. While the even splitting coefficients are determined by second order effects of rotation, magnetic field and any other latitudinal dependence in solar structure. Since the rotation rate can be inferred using the odd splitting coefficients, the inferred profile can be used to estimate its contribution to the second order effects. These can then be subtracted from the observed even coefficients to estimate the magnetic field strength (Gough & Thompson 1990) or other latitudinal variations in solar structure. Unfortunately, it is not possible to separate out the effect of magnetic fields and other aspherical perturbations to solar structure. Ignoring the magnetic field, it is possible to generalise the inversion technique to estimate possible latitudinal variations in the solar structure (Antia et al. 2001a) from the observed even splitting coefficients. The inversions of both GONG and MDI data show that most of the contribution to even splitting coefficients arise from near surface effects. There is also a strong temporal variation in these coefficients as discussed in the following section. A temporal average over the duration for which data are available shows a distinct peak inside the convection zone (Fig. 13) at latitude of about 60°. This peak with a maximum magnitude of about 10^{-4} in sound speed, extends through the convection zone. Although the figure shows some extension below the convection zone also, this could be due to the finite resolution of the inversions and may not be real. It is clear that departures from spherical symmetry are fairly small but much larger than the expected errors in inversion ($\approx 10^{-5}$), when we take the average over all data sets. The aspherical component of density is much smaller

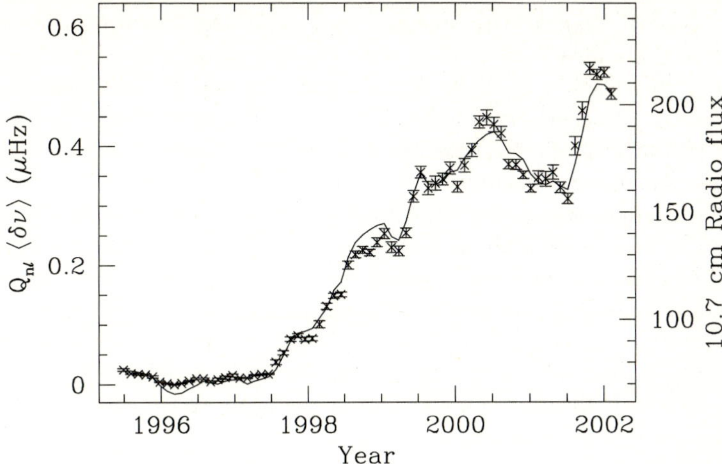

Fig. 14. The mean frequency shift of p-modes as a function of time using GONG data is shown by the *solid* line. The crosses with error-bars show the mean 10.7 cm radio flux on a scale marked on right hand side

and its significance is not very clear. The origin of this feature is not clear as it could be due to magnetic field or due to the asphericity in solar structure (Antia, Chitre & Thompson 2003). If this signal is due to magnetic fields, then the field strength could be about 70 kG at $r = 0.9R_\odot$ near the peak in $\delta c^2/c^2$, while in the tachocline region the field strength may be 200 kG or smaller.

7 Temporal Variations in the Solar Interior

With the accumulation of helioseismic data over the past 7 years covering the rising phase of cycle 23, it has become possible to study temporal variations in the solar interior. Early studies had already established the temporal variation in p-mode frequencies with solar cycle (Libbrecht & Woodard 1990; Elsworth, et al. 1990a). It is found that p-mode frequencies shift by up to 0.4 μHz during the solar cycle and that the frequencies are larger during the phase of maximum activity. The frequency variation is found to be well correlated to the solar activity indices (Dziembowski et al. 1998; Bhatnagar, Jain & Tripathy 1999; Howe, Komm & Hill 1999). Figure 14 shows the mean frequency shift as a function of time in the GONG data. Also shown on a suitable scale is the variation in the 10.7 cm radio flux, which is believed to be an index of solar activity. It can be seen that the two variations are reasonably well correlated. Further, if the frequency shift is scaled for differences in mode inertia, then the scaled shift is found to be a function of frequency alone, i.e., modes with different degree ℓ but same frequency have similar scaled frequency differences. Thus the modes which penetrate deeper and hence have larger inertia show smaller variation in frequency. This appears to suggest that the frequency variations are caused by some perturbations residing in the outer layers of the Sun. In principle, it is possible to apply the inversion technique to the observed frequency shifts to study the temporal variations in the solar structure. But such studies have not shown any significant variation in the interior (Basu & Antia 2000b), thus confirming that the cause for most of the variations is confined to the outermost layers. In particular, no temporal variation has been detected in the depth of the convection zone which can be determined very accurately from the seismic data. Similarly, there is no signature of temporal variation in sound speed or density near the base of the convection zone (Eff-Darwich et al. 2002), where the solar dynamo is believed to operate. Even the non spherically symmetric component of the sound speed does not show any significant temporal variation in the interior (Antia et al. 2001a) and the observed variations in the even splitting coefficients are well correlated with the corresponding component of magnetic field at the surface. It is likely that the expected variations in solar structure from the dynamo are too small to be detected helioseismically.

7.1 Temporal Variations of the Solar Radius

The frequencies of f-modes, which are surface gravity modes, are largely independent of stratification in the solar interior and are essentially determined by the surface gravity. These frequencies which have now been measured reliably by GONG and MDI data provide an important diagnostic of the near-surface regions, including the turbulence and the magnetic field. These frequencies also provide an accurate measurement of the solar radius (Schou et al. 1997; Antia 1998). Using these frequencies the solar radius can be determined to an accuracy of 1 km and thus possible variations in the solar radius of this magnitude can be determined by studying temporal variations in frequencies of f-modes. Unfortunately, systematic errors in calibrating the solar radius from f-mode frequencies are much larger, being of the order of 100 km (Tripathy & Antia 1999). If these systematic errors are independent of time we may still be able to study the temporal variation in solar radius using f-mode frequencies. The value of solar radius inferred from f-mode frequencies is found to be about 200 km lower than the usually accepted value of 695.99 Mm. This discrepancy most probably arises because of difference in the definition of the solar surface. Direct observations of the solar disk use the point of inflection in the limb profile as the definition of surface. This point probably corresponds to a layer about 500 km above the layer where optical depth is unity (Brown & Christensen-Dalsgaard 1998). The helioseismic estimate of the solar radius is based on the calibration of f-mode frequencies in a solar model to the observed frequencies and hence corresponds to the layer where the temperature equals the effective temperature, which is the definition used in solar models. After accounting for all these variations there is still a discrepancy of about 200 km between the two measurements of the solar radius. The cause of this discrepancy is not understood, but it is likely to be due to uncertainties in the treatment of surface layers in theoretical models.

Direct observations of the Sun have given conflicting results on the variation of the solar radius with time (Delache, Laclare & Sadsaoud 1985; Wittmann, Alge & Bianda 1993; Fiala, Dunham & Sofia 1994; Laclare et al. 1996; Noël

1997; Emilio et al. 2000). The reported change in the measured angular radius varies from 0 to $1''$, which implies a change of up to 700 km in the radius. Such large changes will affect the frequencies by 0.1%, which are much larger than the estimated errors in these frequencies. Helioseismic data collected over the last six years has shown that the variation in solar radius over this time is less than 5 km (Dziembowski, Goode & Schou 2001; Antia et al. 2001b). This is an order of magnitude less than the reported variations from direct measurements.

However, there is considerable discrepancy between different helioseismic studies of solar radius variations. It has been shown that the variation in f-mode frequencies with time is more complicated than what was assumed in earlier studies (Antia et al. 2001b). The observed temporal variations in the f-mode frequencies can be resolved into two components: an oscillatory component with a period of 1 year and another slowly varying component which appears to be correlated to the solar activity. The oscillatory component is probably an artifact of the data analysis, since its period agrees exactly with the orbital period of the Earth. Both these components have a strong dependence on frequency and are therefore unlikely to arise from radius variation. Any possible variation in radius should yield relative frequency variations which are independent of frequency. A detailed analysis of these frequency differences suggests that the perturbing influence is localised in the outer 1% of the solar radius. Any possible variation in the solar radius should be less than a few km over the solar cycle (Antia et al. 2001b). On the other hand, using data from MDI, Dziembowski, Goode & Schou (2001) have claimed that the solar radius is reducing at the rate of 1.5 km per year during the rising phase of cycle 23. However, they have not removed the oscillatory component in their study. A close look at their results shows that apart from the oscillatory trend with period of one year in their estimated radius variation, there is a reduction of the solar radius by 3–4 km during 1998.4–1999.4. This happens to be the period when SOHO had lost contact with the control station and was subsequently recovered. It is very likely that this shift is not of solar origin, but due to changes in instrumental characteristics during recovery of SOHO. Apart from this there is no evidence for any systematic variation in the solar radius.

It can be easily shown that even a variation in solar radius by 1 km during the solar cycle will release (or absorb) a large amount of energy through variation in the gravitational potential energy, which would be more than the observed solar luminosity. Thus any possible variation in the solar radius must be confined to the rarefied outermost layers. The observed f-modes are localised in the region between a depth of 1 000–12 000 km below the solar surface and thus the limits on radius variation obtained using f-mode frequencies presumably apply to these layers. So far there is no evidence to suggest that even the radius in these layers has changed. An upper limit on radius variation during solar cycle would be a few km.

Estimating the radius variation with the solar cycle can provide useful constraint for models to explain the luminosity variation with the solar cycle (Gough 2001). In particular, the ratio of the radius variation to the luminosity variation, $W = (\Delta R/R)/(\Delta L/L)$ depends on the theoretical model of luminosity variations. With an upper limit of a few km on possible radius variation, or $\Delta R/R < 3 \times 10^{-6}$. This would yield $W < 0.003$ as the ratio of radius to luminosity variation. Such a small value should favour models involving changes in the outer layers to explain the observed luminosity variations.

7.2 Temporal Variations of the Rotation Rate

While no significant temporal variation has been seen in the solar internal structure, the rotation rate has been found to vary with time. Rotation is believed to play an important role in the operation of the solar dynamo and it is natural to look for possible variations in the rotation rate with time. Surface observations indicate that there is indeed some variation in the rotation rate over the solar cycle (Howard & LaBonte 1980). These observations have shown the existence of zonal bands of slow and fast rotation, which migrate slowly from high to low latitude during the solar cycle. These have also been referred to as torsional oscillations. These migrating flow bands appear to be correlated with the migrating magnetic activity bands well-known from the butterfly diagram (Snodgrass 1991). However, the connection between the zonal shear flow and activity bands is not understood (Schüssler 1981; Wilson 1987; Küker, Rüdiger & Pipin 1996). Helioseismic studies which can infer the variation of the zonal flow pattern with depth and time may help us in understanding this connection.

In order to study temporal variations in the rotation rate we first take the temporal average of the rotation rate at each latitude and depth and then subtract this average from the rotation rate at any given epoch (as obtained by inversion of observed frequencies during that period). The residual in rotation rate would contain the temporally varying component of the rotation. Since the helioseismic data sets cover only about half of the solar cycle, the temporal average does not represent a long term average, but the residuals will still give the time-varying component. Figure 15 shows this residual rotation rate as a function of latitude at a radial distance of $0.98R_\odot$, at different times (Antia & Basu 2000). It is clear that the pattern is changing with time and there appears to be some equatorward movement in peaks representing maximum velocity in the period covered. Similar results have been found using MDI data (Schou 1999; Howe et al. 2000a) and using local helioseismic techniques (Basu & Antia 2000a). It can be seen that the amplitude of this zonal component is around a few nHz, which is much smaller than the total rotation rate of about 460 nHz at the equator. Nevertheless, the variation is significant as it is much larger than the error estimates, which are about 0.2 nHz at low latitudes. Because of the smoothing used in the inversion technique and because the errors increase with depth, it is difficult to estimate the depth to which these flows penetrate, but a detailed analysis appears to suggest that this zonal flow pattern persists up to a depth of, at least, $0.1R_\odot$ (Howe et al. 2000a; Antia & Basu 2000). This depth is somewhat larger than the depth of the outer shear layer though much smaller than the depth of the convection zone. It implies that the zonal flow pattern is not confined to the outer surface layers but penetrates into the interior unlike

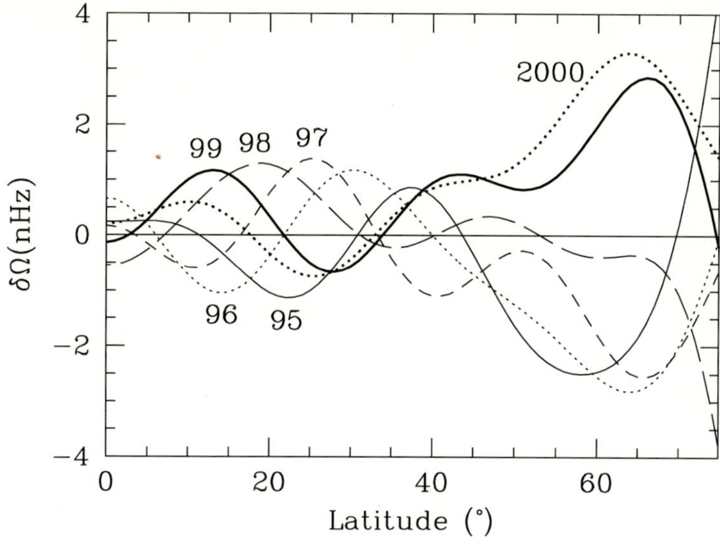

Fig. 15. The zonal flow rotation rate at a depth of $0.98R_{\odot}$ at different times: July 1995 (*solid line*), 1996 (*dotted line*), 1997 (*short-dashed line*), 1998 (*long-dashed line*), 1999 (*thick solid line*) and 2000 (*thick dotted line*)

the structural variations which are confined to the outermost layers. Theoretical models (Covas et al. 2000) of mean field dynamos suggest that the zonal flow pattern should penetrate up to the base of the convection zone. It is possible that because of the increase in errors with depth the pattern is not visible in the inversion results. Recently, there has been some evidence that the zonal flow pattern may actually be penetrating to the base of the convection zone (Vorontsov et al. 2002; Basu & Antia 2003).

In order to get a better idea of the time variation in this zonal component of the rotation rate, we show in Fig. 16 contours of constant rotation velocity residuals at a depth of $0.02R_{\odot}$ as a function of latitude and time. This figure is based on GONG data for each month. The bands of faster and slower than average rotation can be clearly seen in this figure. These bands appear to be moving towards the equator at low latitude. At latitudes above 50° there is some tendency of the contours of constant $\delta\Omega$ to migrate poleward (Antia & Basu 2001; Vorontsov et al. 2002). Similar migrations have been seen in magnetic patterns (Leroy & Noens 1983; Makarov & Sivaraman 1989). This poleward movement may be crucial for the magnetic field reversal during the solar cycle. Similarly, studies of some axisymmetric mean field dynamo models (Covas et al. 2000) also show zonal flow patterns with bands migrating poleward at high latitude. The rotation rate in the polar region appears to be decreasing with time during 1995–99 after which it has started increasing again (Antia & Basu 2001). Thus the minimum in the polar rotation rate occurs distinctly before the maximum in the solar activity.

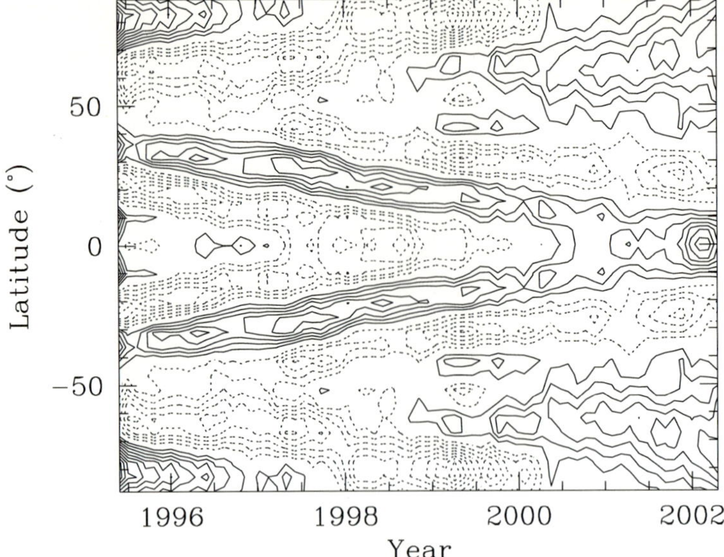

Fig. 16. The contours of time varying component of rotation velocity $\delta\Omega r\cos\theta$ (where θ is the latitude) at a depth of $0.02R_\odot$ below the solar surface are shown as a function of latitude and time. The *solid* contours are for positive velocity, while *dotted* contours denote negative values. The contours are drawn at intervals of 1 m s^{-1}

No significant temporal variation in the rotation rate has been seen below the convection zone, including in the tachocline region (Basu & Antia 2001). Howe et al. (2000b) have reported an oscillatory pattern with a period of 1.3 years, in the equatorial rotation rate at $r = 0.72R_\odot$. However, no such pattern has been seen in other inversion results using the same data sets (Antia & Basu 2000; Corbard et al. 2001) and its significance is not clear.

8 Summary

Precise data from recent helioseismic projects, like GONG and MDI have enabled us to infer the internal structure of the Sun to remarkable detail. The standard solar model with improved physical input is in good agreement with the seismically inferred structure. The rotation rate in the solar interior has been determined as a function of both depth and latitude from seismic inversions. The inferred profile of the rotation rate shows that the differential rotation persists through the convection zone and there is a strong shear layer, referred to as the tachocline, just below the base of the convection zone below which the rotation rate is essentially independent of latitude and depth. The tachocline should play a crucial role in the operation of solar dynamo. The seismic data also allows us to study the temporal variation in the solar interior. The frequencies of solar oscillations are known to vary with solar activity, but this variation is due to changes in the outer surface layers. There is no evidence so far of any significant variation in the solar structure in deeper layers. On the other hand, the rotation rate shows a distinct pattern of temporal variation identified with zonal flows, which penetrate up to a depth of at least $0.1R_\odot$ and possibly up to the base of the convection zone. Although, seismic data have shown unexpected results on the solar dynamics, the origin of many of these observed features, like the tachocline and the zonal flows is still not properly understood. Continued seismic observations over the entire solar cycle will hopefully improve our understanding.

In addition to the global modes of oscillations considered in this article, in recent years local helioseismology is providing more information about the dynamics of the near-surface layers. These techniques can also provide information about the variation of flows with longitude as well as about the north-south antisymmetric component of the rotation rate and the meridional (north-south) component of flow velocities. These near-surface inferences coupled with global techniques will continue to probe the dynamics of the solar interior.

Acknowledgements

This work utilises data obtained by the Global Oscillation Network Group and from the Solar Oscillations Investigation / Michelson Doppler Imager on the Solar and Heliospheric Observatory (SOHO). GONG is managed by the National Solar Observatory, which is operated by AURA, Inc. under a cooperative agreement with the National Science Foundation. The data were acquired by instruments operated by the Big Bear Solar Observatory, High Altitude Observatory, Learmonth Solar Observatory, Udaipur Solar Observatory, Instituto de Astrofísico de Canarias, and Cerro Tololo Interamerican Observatory. SOHO is a project of international cooperation between ESA and NASA.

References

Adelberger, E. G., Austin, S. M., Bahcall, J. N., et al. 1998, Rev. Mod. Phys., 70, 1265

Antia, H. M. 1998, A&A, 330, 336

Antia, H. M., & Basu, S. 1999, ApJ, 519, 400

Antia, H. M., & Basu, S. 2000, ApJ, 541, 442

Antia, H. M., & Basu, S. 2001, ApJ, 559, L67

Antia, H. M., & Chitre, S. M. 1998, A&A, 339, 239

Antia, H. M., & Chitre, S. M. 1999, A&A, 347, 1000

Antia, H. M., & Chitre, S. M. 2002, A&A, 393, L95

Antia, H. M., Basu, S., Christensen-Dalsgaard, J., Elliott, J. R., Gough, D. O., Guzik, J. A., & Kosovichev, A. G. 1997, in proceedings of IAU Symp. 181: Sounding Solar & Stellar Interiors, posters volume, eds. J. Provost, F.-X. Schmider, Nice Observatory, p. 61

Antia, H. M., Basu, S., & Chitre, S. M. 1998, MNRAS, 298, 543

Antia, H. M., Basu, S., Hill, F., Howe, R., Komm, R. W., & Schou, J. 2001a, MNRAS, 327, 1029

Antia, H. M., Basu, S., Pintar, J., & Schou, J. 2001b, in Proc. SOHO 10/GONG 2000 Workshop on Helio- & Asteroseismology at the Dawn of the Millennium, Ed. A. Wilson, ESA SP-464, 27

Antia, H. M., Chitre, S. M., & Thompson, M. J. 2003, A&A, 399, 329
Backus, G. B., & Gilbert, J. F. 1968, Geophys. J. Roy. Astr. Soc., 16, 169
Balmforth, N. J. 1992, MNRAS, 255, 632
Basu, S. 1997, MNRAS, 288, 572
Basu, S. 1998, MNRAS, 298, 719
Basu, S., & Antia, H. M. 1994a, J. Astrophys. Astron., 15, 143
Basu, S., & Antia, H. M. 1994b, MNRAS, 269, 1137
Basu, S., & Antia, H. M. 1995, MNRAS, 276, 1402
Basu, S., & Antia, H. M. 1997, MNRAS, 287, 189
Basu, S., & Antia, H. M. 2000a, Sol. Phys., 192, 469
Basu, S., & Antia, H. M. 2000b, Sol. Phys., 192, 449
Basu, S., & Antia, H. M. 2001, MNRAS, 324, 498
Basu, S., & Antia, H. M. 2003, ApJ, 585, 553
Basu, S., & Christensen-Dalsgaard, J. 1997, A&A, 322, L5
Basu, S., Antia, H. M., & Narasimha, D. 1994, MNRAS, 267, 209
Basu, S., Antia, H. M., & Tripathy, S. C. 1999, ApJ, 512, 458
Bhatnagar, A., Jain, K., & Tripathy, S. C. 1999, ApJ, 521, 885
Braun, D. C., & Fan, Y. 1998, ApJ, 508, L105
Brown, T. M. 1985, Nature, 317, 591
Brown, T. M., & Christensen-Dalsgaard, J. 1998, ApJ, 500, L195
Brown, T. M., & Morrow, C. A. 1987, ApJ, 314, L21
Brown, T. M., Christensen-Dalsgaard, J., Dziembowski, W. A., Goode, P., Gough, D. O., & Morrow, C. A. 1989, ApJ, 343, 526
Brummell, N. H., Hurlburt, N. E., & Toomre, J. 1998, ApJ, 493, 955
Brun, A. S., Turck-Chièze, S., & Zahn, J.-P. 1999, ApJ, 525, 1032
Brun, A. S., Antia, H. M., Chitre, S. M., & Zahn, J.-P. 2002, A&A, 391, 725
Canuto, V. M., & Mazzitelli, I. 1991, ApJ, 370, 295
Chaboyer, B., Deliyannis, C. P., Demarque, P., Pinsonneault, M. H., & Sarajedini, A. 1992, ApJ, 388, 372
Chaboyer, B., Demarque, P., & Pinsonneault, M. H. 1995, ApJ, 441, 865
Chandrasekhar, S. 1964, ApJ, 139, 664
Chaplin, W. J., Elsworth, Y., Howe, R., Isaak, G. R., McLeod, C. P., Miller, B. A., van der Raay, H. B., Wheeler, S. J., & New, R. 1996a, Sol. Phys., 168, 1
Chaplin, W. J., Elsworth, Y., Isaak, G. R., McLeod, C. P., Miller, B. A., & New, R. 1996b, MNRAS, 283, L31
Charbonneau, P., & MacGregor, K. B. 1993, ApJ, 417, 762
Charbonneau, P., Christensen-Dalsgaard, J., Henning, R., Schou, J., Thompson, M. J., & Tomczyk, S., 1997, in Sounding Solar & Stellar Interiors, Posters Volume, eds. J. Provost J. & F.-X. Schmider, IAU Symp. 181 (Kluwer, Dordrecht), 161
Charbonneau, P., Tomczyk, S., Schou, J., & Thompson, M. J. 1998, ApJ, 496, 1015
Charbonneau, P., Christensen-Dalsgaard, J., Henning, R., Larsen, R. M., Schou, J., Thompson, M. J., & Tomczyk, S. 1999, ApJ, 527, 445
Charbonnel, C., Vauclair, S., Maeder, A., Meynet, G., & Schaller, G., 1994, A&A, 283, 155
Chou, D.-Y., Sun, M.-T., Huang, T.-Y., et al. 1995, Sol. Phys., 160, 237
Christensen-Dalsgaard, J., & Berthomieu, G. 1991, in Solar interior & atmosphere, eds. A. N. Cox, W. C. Livingston & M. Matthews, Space Science Series (University of Arizona Press, Tucson) 401
Christensen-Dalsgaard, J., & Däppen, W. 1992, Astron. Astroph. Rev., 4, 267

Christensen-Dalsgaard, J., & Schou, J. 1988, in Seismology of the Sun & Sun-like Stars, eds. V. Domingo & E. J. Rolfe, ESA SP-286, 149
Christensen-Dalsgaard, J., Gough, D. O., & Thompson, M. J. 1991, ApJ, 378, 413
Christensen-Dalsgaard, J., Proffitt, C. R., & Thompson, M. J. 1993, ApJ, 403, L75
Christensen-Dalsgaard, J., Däppen, W., Ajukov, S. V., et al. 1996, Science, 272, 1286
Corbard, T., Berthomieu, G., Morel, P., Provost, J., Schou, J., & Tomczyk, S. 1997, A&A, 324, 298
Corbard, T., Berthomieu, G., Provost, J., & Morel, P. 1998, A&A, 330, 1149
Corbard, T., Blanc-Féraud, L., Berthomieu, G., & Provost, J. 1999, A&A, 344, 696
Corbard, T., Jiménez-Reyes, S. J., Tomczyk, S., Dikpati, M., & Gilman, P. 2001, in Helio- and Astero-seismology at the Dawn of the Millennium, ed. A. Wilson, ESA SP-464, 265
Covas, E., Tavakol, R., Moss, D., & Tworkowski, A. 2000, A&A, 360, L21
Cox, A. N., Guzik, J. A., & Kidman, R. B. 1989, ApJ, 342, 1187
Cutler, C. & Lindblom, L. 1996, Phys. Rev., D54, 1287
Däppen, W., Mihalas, D., Hummer, D. G., & Mihalas, B. W. 1988, ApJ, 332, 261
Däppen, W., Gough, D. O., Kosovichev, A. G., & Thompson, M. J. 1991, in Challenges to theories of the structure of moderate-mass stars, eds. D. O. Gough & J. Toomre, Lecture Notes in Physics, vol. 388, (Springer, Heidelberg) 111
Degl'Innocenti, S., Fiorentini, G., & Ricci, B. 1998, Phys. Lett., B416, 365
Delache, P., Laclare, F., & Sadsaoud, H. 1985, Nature, 317, 416
Deubner, F.-L. 1975, A&A, 44, 371
Deubner, F.-L., & Gough, D. O. 1984, ARA&A, 22, 593
Dicke, R. H., & Goldenberg, H. M. 1974, ApJS, 27, 131
Dicke, R. H., Kuhn, J. R., & Libbrecht, K. G. 1985, Nature, 316, 687
Duvall, T. L., Jr., & Harvey, J. W. 1984, Nature, 310, 19
Duvall, T. L., Jr., Dziembowski, W. A., Goode, P. R., Gough, D. O., Harvey, J. W., & Leibacher, J. W. 1984, Nature, 310, 22
Duvall, T. L., Jr., Jefferies, S. M., Harvey, J. W., & Pomerantz, M. A. 1993, Nature, 362, 430
Dziembowski, W. A., Goode, P. R., DiMauro, M. P., Kosovichev, A. G., & Schou, J. 1998, ApJ, 509, 456
Dziembowski, W. A., Goode, P. R., & Schou, J. 2001, ApJ, 553, 897
Dzitko, H., Turck-Chièze, S., Delbourgo-Salvador, P., & Lagrange, C. 1995, ApJ, 447, 428
Eff-Darwich, A., Korzennik, S. G., Jiménez-Reyes, S. J., & Pérez Hernández, F. 2002, ApJ, 580, 574
Elliott, J. R., & Kosovichev, A. G. 1998, ApJ, 500, L199
Elliott, J. R., Miesch, M., & Toomre, J. 2000, ApJ, 533, 546
Elsworth, Y., Howe, R., Isaak, G. R., McLeod, C. P., & New, R. 1990a, Nature, 345, 322
Elsworth, Y., Howe, R., Isaak, G. R., McLeod, C. P., & New, R. 1990b, Nature, 347, 536
Elsworth, Y., Howe, R., Isaak, G. R., McLeod, C. P., Miller, B. A., New, R., Wheeler, S. J., & Gough, D. O. 1995, Nature, 376, 669
Emilio, M., Kuhn, J. R., Bush, R. I., & Scherrer, P. 2000, ApJ, 543, 1007
Endler, F., & Deubner, F.-L. 1983, A&A, 121, 291
Fiala, A. D., Dunham, D. W., & Sofia, S. 1994, Sol. Phys., 152, 97
Fossat, E. 1991, Sol. Phys., 133, 1
Fukugita, M., & Hata, N. 1998, ApJ, 499, 513

Giles, P. M., Duvall, T. L., Jr., Scherrer, P. H., & Bogart, R. S. 1997, Nature, 390, 52
Gilman, P. A., & Miller, J. 1986, ApJS, 61, 585
Gizon, L., Fossat, E., Lazrek, M., et al. 1997, A&A, 317, L71
Glatzmaier, G. A. 1987, in The Internal Solar Angular Velocity, eds. B. R. Durney & S. Sofia (Reidel, Dordrecht) 263
Goldreich, P., & Keeley, D. A. 1977, ApJ, 212, 243
Gough, D. O. 1977, in Energy balance and hydrodynamics of the solar chromosphere and corona, eds. R. M. Bonnet & P. Delache, IAU Colloq. 36 (G. de Bussac, Clermont-Ferrand) 3
Gough, D. O. 1984, Mem. Soc. Astron. Ital., 55, 13
Gough, D. O. 1990, in Progress of seismology of the Sun & stars, eds. Y. Osaki, & H. Shibahashi, Lecture Notes in Physics, vol. 367, (Springer, Berlin) 283
Gough, D. O. 2001, Nature, 410, 313
Gough, D. O., & Kosovichev, A. G. 1990, in Inside the Sun, eds. G. Berthomieu & M. Cribier, Proc. IAU Coll. 121, (Kluwer, Dordrecht) 327
Gough, D. O., & Thompson, M. J. 1990, MNRAS, 242, 25
Gough, D. O., & Thompson, M. J. 1991, in Solar interior & atmosphere, eds. A. N. Cox, W. C. Livingston & M. Matthews, Space Science Series (University of Arizona Press, Tucson) 519
Gough, D. O, & Toomre, J. 1991, ARA&A, 29, 627
Gough, D. O., Kosovichev, A. G., Toomre, J., et al. 1996, Science, 272, 1296
Graboske, H. C., DeWitt, H. E., Grossman, A. S., & Cooper, M. S. 1973, ApJ, 181, 457
Grec, G., Fossat, E., & Pomerantz, M. 1980, Nature, 288, 541
Grevesse, N., Noels, A., & Sauval, A. J. 1996, in Cosmic abundances, eds., S. S. Holt & G. Sonneborn, ASP Conf. Ser., 99, 117
Gruzinov, A. V., & Bahcall, J. N. 1998, ApJ, 504, 996
Harvey, J. W., Hill, F., Hubbard, R., et al. 1996, Science, 272, 1284
Hathaway, D. H., Gilman, P., Harvey, J. W., et al. 1996, Science, 272, 1306
Haxton, W. C. 1995, ARA&A, 33, 459
Hill, F. 1988, ApJ, 333, 996
Hill, F., Stark, P. B., Stebbins, R. T., et al. 1996, Science, 272, 1292
Howard, R. 1984, ARA&A, 22, 131
Howard, R., & LaBonte, B. J. 1980, ApJ, 239, L33
Howe, R., Antia, H. M., Basu, S., Christensen-Dalsgaard, J., Korzennik, S. G., Schou, J., & Thompson, M. J. 1998, in Structure & Dynamics of the Interior of the Sun & Sun-like Stars, eds. S. Korzennik & A. Wilson, ESA SP-418, 803
Howe, R., Komm, R., & Hill, F. 1999, ApJ, 524, 1084
Howe, R., Christensen-Dalsgaard, J., Hill, F., Komm, R. W., Larsen, R. M., Schou, J., Thompson, M. J., & Toomre, J. 2000a, ApJ, 533, L163
Howe, R., Christensen-Dalsgaard, J., Hill, F., Komm, R. W., Larsen, R. M., Schou, J., Thompson, M. J., & Toomre, J. 2000b, Science, 287, 2456
Iglesias, C. A., & Rogers, F. J. 1996, ApJ, 464, 943
Kosovichev, A. G. 1996, ApJ, 469, L61
Kosovichev, A. G., & Schou, J. 1997, ApJ, 482, L207
Kosovichev, A. G., Schou, J., Scherrer, P. H., et al. 1997, Sol. Phys., 170, 43
Kuhn, J. R., Bush, R. I., Scheick, X., & Scherrer, P. 1998, Nature, 392, 155
Küker, M., Rüdiger, G., & Pipin, V. V. 1996, A&A, 312, 615
Laclare, F., Delmas, C., Coin, J. P., & Irbah, A. 1996, Sol. Phys., 166, 211
Lazrek, M., Pantel, A., Fossat, E., et al. 1996, Sol. Phys., 166, 1

Leibacher, J. W., & Stein, R. F. 1971, Astrophys. Lett., 7, 191
Leighton, R. B., Noyes, R. W., & Simon, G. W. 1962, ApJ, 135, 474
Leroy, J.-L., & Noens, J.-C. 1983, A&A, 120, L1
Libbrecht, K. G. 1989, ApJ, 336, 1092
Libbrecht, K. G., & Woodard, M. F. 1990, Nature, 345, 779
Libbrecht, K. G., Woodard, M. F., & Kaufman, J. M. 1990, ApJS, 74, 1129
Lynden-Bell, D., & Ostriker, J. P. 1967, MNRAS, 136, 293
MacGregor, K. B., & Charbonneau, P. 1994, in Cool Stars; Stellar Systems; and the Sun, ed. J.-P. Caillault, Astron. Soc. Pac. Conf. Ser., 64, 174
Makarov, V. I., & Sivaraman, K. R. 1989, Sol. Phys., 123, 367
Mestel, L., & Weiss, N. O. 1987, MNRAS, 226, 123
Mitler, H. E. 1977, ApJ, 212, 513
Monteiro, M. J. P. F. G., Christensen-Dalsgaard, J., & Thompson, M. J. 1994, A&A, 283, 247
Nandy, D., & Choudhuri, A. R. 2002, Science, 296, 1671
Noël, F. 1997, A&A, 325, 825
Pijpers, F. P. 1997, A&A, 326, 1235
Pijpers, F. P. 1998, MNRAS, 297, L76
Pinsonneault, M. H., Kawaler, S. D., Sofia, S., & Demarque, P. 1989, ApJ, 338, 424
Rhodes, E. J., Jr., Kosovichev, A. G., Schou, J., Scherrer, P. H., & Reiter, J. 1997, Sol. Phys., 175, 287
Richard, O., Vauclair, S., Charbonnel, C., & Dziembowski, W. A. 1996, A&A, 312, 1000
Ritzwoller, M. H., & Lavely, E. M. 1991, ApJ, 369, 557
Rogers, F. J., & Iglesias, C. A. 1992, ApJS, 79, 507
Rogers, F. J., Swenson, F. J., & Iglesias, C. A. 1996, ApJ, 456, 902
Rosner, R., & Weiss, N. O. 1985, Nature, 317, 790
Roxburgh, I. W. 1974, Nature, 247, 220
Salpeter, E. E. 1954, Australian J. Phys., 7, 373
Scherrer, P. H., Bogart, R. S., Bush, R. I., et al. 1995, Sol. Phys., 162, 129
Schlattl, H., Bonanno, A., & Paternó, L. 1999, Phys. Rev., D60, 113002
Schrijver, C. J. 1994, in Cool Stars; Stellar Systems; and the Sun; ed. J.-P. Caillault, Astron. Soc. Pac. Conf. Ser., 64, 328
Schou, J. 1999, ApJ, 523, L181
Schou, J. & Bogart, R. S. 1998, ApJ, 504, L131
Schou, J., Christensen-Dalsgaard, J., & Thompson, M. J. 1994, ApJ, 433, 389
Schou, J., Kosovichev, A. G., Goode, P. R., & Dziembowski, W. A. 1997, ApJ, 489, L197
Schou, J., Antia, H. M., Basu, S., et al. 1998, ApJ, 505, 390
Schüssler, M. 1981, A&A, 94, L17
Skumanich, A. 1972, ApJ, 171, 565
Snodgrass, H. B. 1991, ApJ, 383, L85
Spiegel, E. A., & Zahn, J.-P. 1992, A&A, 265, 106
Takata, M., & Shibahashi, H. 1998, ApJ, 504, 1035
Talon, S., & Zahn, J.-P. 1998, A&A, 329, 315
Thompson, M. J., Toomre, J., Anderson, E., et al. 1996, Science, 272, 1300
Tomczyk, S., Schou, J., & Thompson, M. J. 1995, ApJ, 448, L57
Tripathy, S. C., & Antia, H. M. 1999, Sol. Phys., 186, 1
Ulrich, R. K. 1969, ApJ, 158, 427
Ulrich, R. K. 1970, ApJ, 162, 993

Unno, W., Osaki, Y., Ando, H., Saio, H., & Shibahashi, H. 1989, Nonradial Oscillations of Stars, 2nd ed., (University of Tokyo Press, Tokyo)
Ulrich, R. K., Boyden, J. E., Webster, L., Padilla, S. P., & Snodgrass, H. B. 1988, Sol. Phys., 117, 291
Vauclair, S. 2000, J. Astrophys. Astron., 21, 323
Vernazza, J. E., Avrett, E. H., & Loeser, R. 1981, ApJS, 45, 635
Vorontsov, S. V., Christensen-Dalsgaard, J., Schou, J., Strakhov, V. N., & Thompson, M. J. 2002, Science, 296, 101
Weiss, A., Flaskamp, M., & Tsytovich, V. N. 2001, A&A, 371, 1123
Wilson, P. R. 1987, Sol. Phys., 110, 59
Wittmann, A. D., Alge, E., & Bianda, M. 1993, Sol. Phys., 145, 205

The Active and Explosive Sun

A. Ambastha

Udaipur Solar Observatory, Physical Research Laboratory,
P. O. Box 198, Udaipur-313001, India
email: ambastha@prl.ernet.in

Abstract. The Sun's magnetic field and differential rotation give rise to much complexity in its structure and activity over a large range in both spatial and temporal scales. The most notable among these is the solar activity cycle of 11 years, or magnetic cycle of 22 years. On shorter time scales of a few seconds to several hours, spectacular explosive events occur in the solar atmosphere, such as, solar flares, prominence eruptions, and coronal mass ejections (CMEs). The explosive energy release takes place in the form of accelerated particles, bulk mass motion, and enhancement of radiation over the entire electromagnetic spectrum ranging from γ-rays to radio wavelengths. These solar transients are essentially the source of disturbance in the interplanetary medium, and also cause geomagnetic effects upon their encounter with the Earth. We present an account of the recent developments in our understanding of these phenomena using both space-borne, and ground-based observations.

1 Introduction

Solar activity has important relationships with the interplanetary weather, and the Earth's magnetosphere. The expanding solar wind, consisting of charged particles from the Sun, interacts with the Earth's magnetic field – pushing it towards the Earth on the sunlit side and stretching it at the night side. The resulting magnetosphere deflects the charged solar wind particles from entering the Earth (Fig. 1). Solar disturbances have been known to create electrical and magnetic disturbances on the Earth, and with the help of satellites it has become possible to track solar disturbances all the way from the Sun to its consequences at the Earth 4–5 days later. The level of activity on the Sun dramatically changes the average protective magnetic sheath around the Earth. The solar activity can disrupt communications and navigational equipment, damage satellites, and even cause power blackouts in high latitude locations. The magnetic cloud of plasma associated with solar storms can extend to a width of 50 million km when they reach the Earth. A visible manifestation of this solar-terrestrial relation is seen in the form of spectacular displays of polar lights in the upper atmosphere, *the aurora* (Latin for dawn, due to its resemblance to the predawn glow). The aurora is caused by high-energy charged particles streaming down along the Earth's polar magnetic field-lines. These solar particles collide with the atoms and molecules in the upper rarefied atmosphere. Various colours of the aurora are produced between 80–400 km altitude above the Earth's surface as the energy of solar particles is partly converted into visible light (Fig. 2a). Usually aurorae are sighted from locations near the polar latitudes, but when the Sun is more

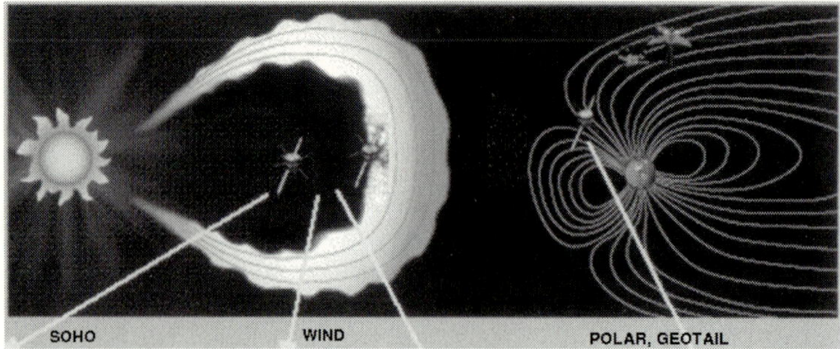

Fig. 1. The Sun-Earth connection: Solar activity, the interplanetary space, and the Earth's magnetosphere. The satellite SOHO (Solar and Heliospheric Observatory) detects the initial event on the Sun, WIND and SOHO carry out in-situ observations in the interplanetary medium, while POLAR, GEOTAIL and ground-based observatories study the magnetospheric and ground response

active, their observations have been reported from as far south as Florida in the USA. This is seen as an auroral oval in a picture of the northern terrestrial polar region taken from space (Fig. 2b).

The Sun's energy is generated by nuclear fusion reactions occurring in its central core, and is released at a steady rate. The Sun's interior is hidden under the opaque "photosphere", and is invisible even to powerful telescopes operating in the visible, X-ray or radio wavelengths. In the absence of a "direct" view, our knowledge of the Sun's interior was so far mostly based on the mathematical equations describing the physical processes operating within a star. Recent advancements in the field of helioseismology (or solar seismology), which is based on the accurate determination of frequencies of solar global oscillation modes, has been successful in providing an extremely revealing view of the Sun's anatomy all the way from its surface to the very central regions (Gough et al. 1996; Ambastha 1998; Antia, this volume). Figure 3 schematically shows the internal and outer structures of the Sun, along with some of the observable features associated with

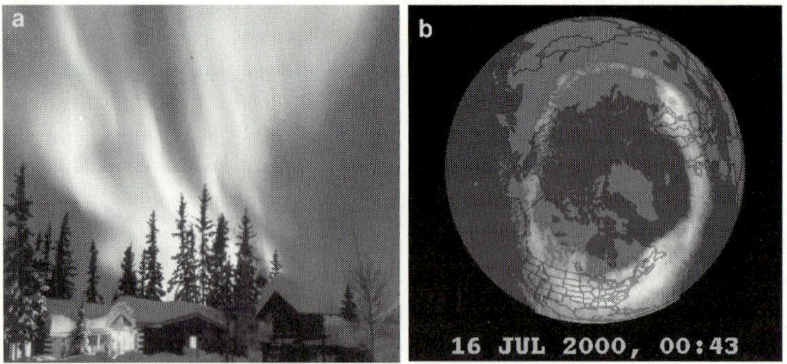

Fig. 2. Aurorae – The testimony to the solar-terrestrial connection: (a) A large auroral display. (b) An auroral oval encircling around the north pole as observed from space

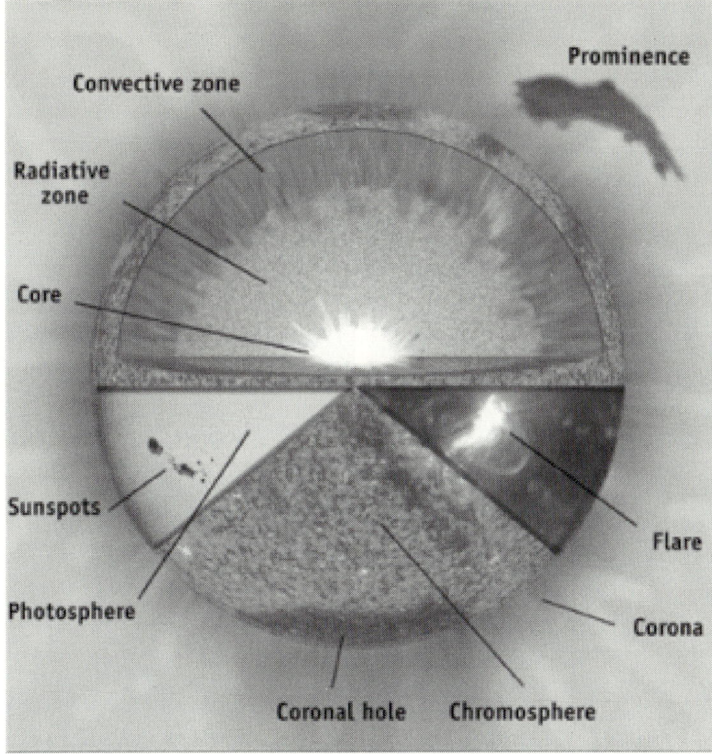

Fig. 3. Anatomy of the Sun – The structure and observable features in the outer layers of the solar atmosphere

the outer layers of the solar atmosphere, i.e., the photosphere, chromosphere, and corona. The energy generated in the central core propagates outward through the radiative and convective zones, and finally leaves the photosphere into the space. If only radiative diffusion of energy were operative, the radial flow of energy would have given rise to spatially uniform and steady release of radiation over the entire outer solar sphere. In reality, a variety of complex features are observed in the upper layers of the visible solar atmosphere, which depart from the spherically symmetric, and steady-state Sun.

The photosphere is a 100 km thick outer layer visible to the naked eye through broad-band solar filters, which is required to reduce the solar intensity considerably for safe viewing. A full disk photospheric image, i.e., *photoheliogram*, shows that the photospheric intensity falls off significantly toward the limb (Fig. 4a). This limb-darkening is due to the fact that we are seeing the higher photospheric layers of decreasing temperature as we look nearer the limb. Analysis of this effect provides a direct technique for determining the temperature variation in the photosphere with height.

Above the photosphere lies the hotter and more rarefied layer, *the chromosphere*, where the temperature increases to 20 000 K from the 5770 K at the

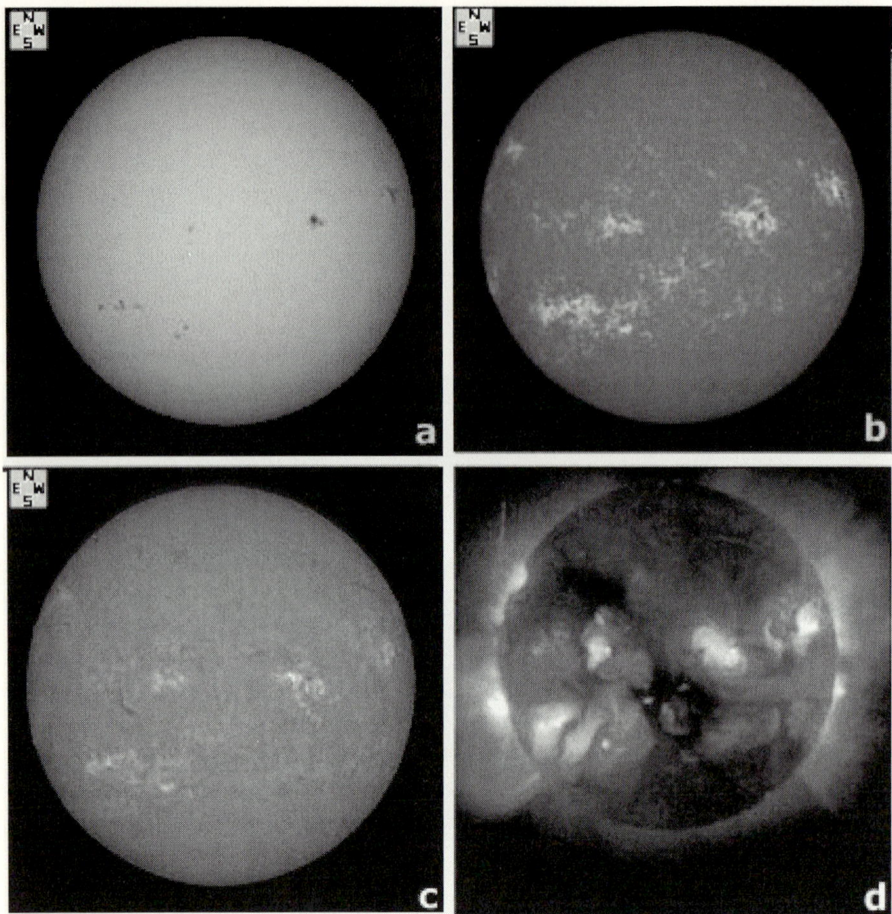

Fig. 4. Consecutively higher layers of the solar atmosphere as seen in different wavelengths on November 4, 2001: (a) The photosphere in white light showing dark sunspots. (b) The lower chromosphere in the Ca II K-line showing the sunspots surrounded by bright plages. (c) The mid-chromosphere in Hα showing bright active regions, and the dark, elongated filaments. (d) The corona seen in soft X-rays (SXR) by the Yohkoh satellite, showing the active regions marked by bright closed loop structures, interconnections between the active regions, and dark voids, called the coronal holes, prominently seen over the southern polar region

photosphere. The chromosphere is nearly transparent to the photospheric light, and is not visible to the naked eye except for a short time span just before (and after) a total solar eclipse. On other occasions, it can be seen using narrow-band filters centred at the Ca II 3934 Å K-line (Fig. 4b) or the Hα 6563 Å line (Fig. 4c), corresponding to the lower and middle chromosphere, respectively.

Going further above the chromosphere, there is a *transition region* with a steep gradient in density and temperature. This region is best seen in the UV spectrum of the Sun. Beyond the transition region is an extremely faint and hot corona which can be seen for the brief period of a total solar eclipse in white light, and in radio, EUV, and X-rays at other times (Fig. 4d). The relatively cool photosphere and chromosphere appear dark in these wavelengths. Thus, by changing the wavelength of light, successive layers of the Sun are revealed in the same way as various layers of an onion are peeled off. All these layers appear highly structured, and are observed to evolve at various scales. The large scale changes from the uniform, *quiet Sun*, are essentially brought about by the interaction of Sun's large-scale magnetic field with differential rotation and convection – the key factors that give rise to solar variability, or activity. A localised and transient release of energy in the form of explosive and eruptive phenomena also occurs over shorter time-scales due to the same interactions operating at smaller spatial dimensions.

2 The Signposts of Solar Activity

There are several signposts of solar activity observed at various layers of the solar atmosphere which indicate departures from a uniform and homogeneous Sun. Under exceptionally good atmospheric seeing conditions, a granular pattern covering the entire solar photosphere is seen even with a moderate aperture telescope (Fig. 5). *The granules* are bright isolated elements having a mean cell size of ≈ 1.76 seconds-of-arc (Roudier & Muller 1987), where 1 second-of-arc on the solar photosphere corresponds to ≈ 720 km. Dark, thin intergranular lanes separate these cells. The granules are essentially convective cells of upward-moving, hot parcels of gas. As they cool down, the granular material sinks back along the dark intergranular lanes. There is a central upflow of ~ 0.4 km s^{-1} surrounded by a horizontal outflow of ~ 0.25 km s^{-1}, and a downflow at the granular boundaries. The granules do not form fixed flow cells, but they continuously evolve – coalesce, expand, fragment, and explode – within a few minutes (Title et al. 1986). Their average lifetime is ~ 6 min. The process of evolution of granules is "non-stationary convection" in which the granules are heated by an underlying layer several times thicker than their diameter.

Apart from the granular structures, there are more difficult to observe active flow structures at larger scale, such as mesogranules ~ 10 seconds-of-arc size (Oda 1984), and supergranules ~ 40 seconds-of-arc size (Leighton, Noyes & Simon 1962). Supergranules have mean life-time of 36 hours (Worden & Simon 1976). These are best seen in measurements of the Doppler shift where light from material moving towards the observer is shifted to the blue, while light from material moving away is shifted to the red. These features cover the entire Sun, and continually evolving patterns are observed. Individual supergranules have flow speeds of about 0.5 km s^{-1}, i.e., 1800 km hr^{-1}. The fluid flows observed in supergranules carry magnetic field bundles to the edges of the cells where they produce the chromospheric network. The granules, mesogranules, and supergranules are generally interpreted as manifestation of convection, with an associated overshoot into the upper regions of the solar atmosphere.

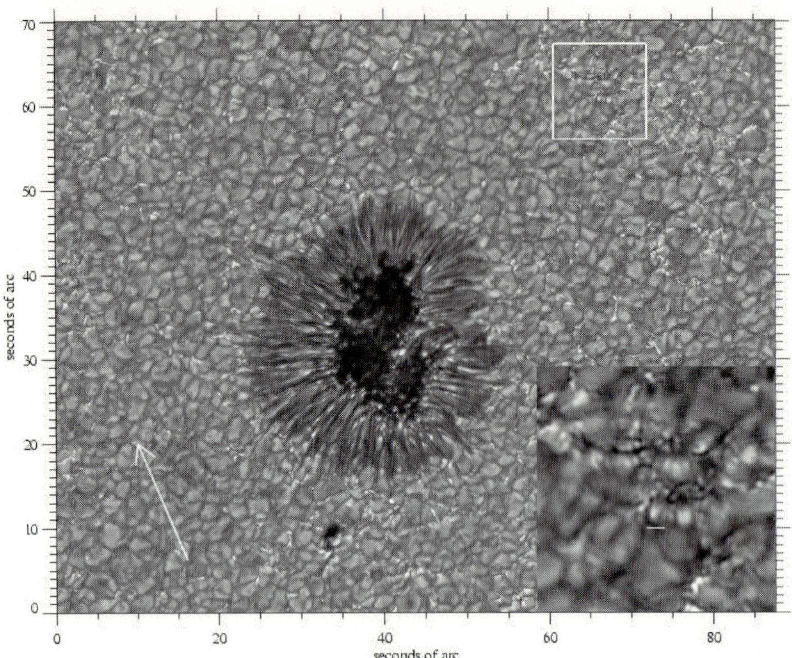

Fig. 5. A high resolution image of a sunspot in the sea of granules showing the fine radial penumbral features. This speckle-reconstructed G-band (4305 Å) image of NOAA 9407 was observed by the Dutch Open Telescope (DOT) on La Palma. The arrow marks the direction towards the disk centre, the dash in the inset in lower right (the enlargement of the marked granular area) measures 1 second-of-arc (adopted from Sütterlin 2001)

Small dark structures, such as the *pores*, and large sunspots are embedded in a sea of granules (Fig. 5). The sunspots are the most prominent and easiest to observe features in the photosphere. Large developed sunspots usually possess a dark central core, or *umbra*, surrounded by a lighter *penumbra*. High resolution observations show the penumbra to consist of radial fibril structures. Bright photospheric *faculae* are distributed around the sunspot, which are observed particularly more prominently when the sunspot is located near the Sun's limb, or edge.

Bright plages (the French word for beaches), or *chromospheric faculae*, are seen around the dark spots in the chromosphere. Plages are incandescent regions of gas with higher density. The sunspots and plages together define the spatial extent of solar active regions. These are localised centres of activity in the solar atmosphere. Compared with the quiet Sun regions, the active regions are conspicuous by enhanced emission over a broad spectral range, extending from soft X-ray to decimetric radio wavelengths. These active regions are associated with various forms of solar transients, which encompass a diverse range of phenomena.

A variety of other observable features exist in the chromosphere, such as fibril structures, spicules, the chromospheric network, dark elongated filaments on the disk, and cloud-like *prominences* hovering over the limb. *Spicules* are tiny jets, lasting a few minutes, and ejecting material outward into the corona at speeds of 20–30 km s^{-1}. The *chromospheric network* is a web-like pattern seen in $H\alpha$, and the Ca II K lines. These networks outline the supergranular cells, as seen in the photospheric velocity images, or Dopplergrams. The network is formed due to the presence of bundles of magnetic field-lines that are concentrated there by the fluid motions in the supergranules. The filaments are observed in the chromosphere, but not in the white light photospheric images. These are dense, cool clouds of material that are suspended above the solar surface and held in balance against gravity by loops of magnetic field. Both filaments and prominences are essentially the same phenomena except that the prominences appear bright against the dark sky-background. They can remain in a quiet or quiescent state for days or weeks. However, filaments can also erupt and rise off the Sun over a few minutes or hours as the magnetic loops supporting them slowly change, and the balance is lost.

3 Centres of Activity in the Solar Atmosphere – The Sunspots

Since the work of G. E. Hale, it is known that sunspots possess a strong magnetic field. Like the darker sunspots, the bright faculae are also areas of enhanced magnetic field strength, however, their field is concentrated in much smaller flux bundles than in the sunspots. One of the smallest magnetic structures seen in high resolution white light images are pores having sizes of a few granules, magnetic field strength of ≈ 1500 G, and life-time of ≈ 1 day. Dark sunspots usually form as the pores that grow in size. The sunspots are usually seen in groups and are essentially seats of strong magnetic fields that may reach up to 4000 G (Fig. 6). The pores and sunspots are observational evidences of magnetic flux concentration arising by convergent motions. Often the flux of one polarity is more concentrated than that of the other. In many cases, only a single, unipolar visible sunspot is seen surrounded by weak fluxes of the other polarity. The direction of the bipolar configuration, or the axis of the sunspot group, is oriented approximately east-west on the Sun, but the leading part is usually observed tilted toward the solar equator (*Joy's law*). Some indications exist that the extent of this tilt could be related with its activity level. Recent X-ray images show that remotely located sunspot groups, which appear to be isolated from each other at the photosphere may well have large scale interconnections through loop structures extending over the corona (Fig. 4d). Such connections exist between sunspot groups located in the same hemisphere, while trans-equatorial connections are also observed.

Sunspots occur in a low latitude belt between $\pm 40°$ around the solar equator. Their lifetimes have an enormous range from less than an hour to several months. The sizes range from close to the resolution limit of a telescope, i.e., about

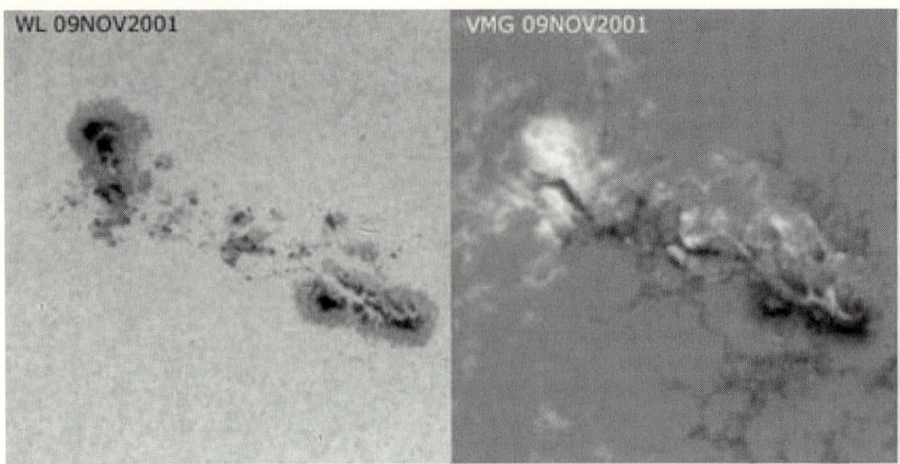

Fig. 6. A typical sunspot group, or active region. (a) White light image, and (b) the corresponding photospheric line-of-sight magnetogram taken on November 9, 2001 show the magnetic polarity distribution of the sunspots – dark is negative and bright is positive polarity

500 km to around 100 000 km. Even so, no more than $\sim 1\%$ of the Sun's visible hemisphere is covered by sunspots at a given time. They appear dark due to their relatively lower temperature (≈ 4000 K) than the surrounding (5770 K). The lower temperature of a sunspot is attributed to partial suppression of the convective thermal energy flux through the upper layers of the convection zone by the strong magnetic field. As the magnetic pressure in the sunspot umbra is a significant fraction of the total pressure, the gas density within the sunspot is lower and the gas is more transparent than the surrounding photosphere. Thus, we can see up to a greater depth in a sunspot, which leads to a depression of the umbra. This effect was discovered in 1769, and is called *Wilson effect*, which can be observed more clearly near the solar limb. Due to this effect, as a large sunspot makes a transit from the east-to-west limb of the Sun, there is a change of its symmetric shape. The lighter penumbra surrounding the central umbra of the sunspot is seen to be composed of many radial filaments along which the magnetic field is nearly horizontal. From the umbra, an outward flow of gas is observed which extends to approximately twice the penumbral radius. The maximum velocity of this flow is about 2 km s^{-1}. It is known as *Evershed effect* after its discovery by Evershed (1909) from the Kodaikanal Observatory, India. Larger sunspots show a *superpenumbra* extending far beyond the penumbra, best seen in the Hα line. Along the superpenumbral fibrils, an inverse Evershed flow of the order of ~ 20 km s^{-1} is observed. The Evershed effect is interpreted as a siphon flow along magnetic flux tubes joining two foot-points having different values of the gas pressure. A flow is driven from the high-pressure end to the low-pressure end.

The bright photospheric faculae and dark sunspots both represent sites of enhanced magnetic fields. Therefore, it is rather surprising that faculae (and chromospheric plages) appear bright while the sunspots are dark. The finding that magnetic flux tubes cause local depressions in the photospheric surface suggests a simple explanation of the excess brightness of photospheric faculae, and chromospheric plages. In both sunspot and facular flux tubes the energy balance is determined by the sum of heat transported by radiation and convection from below, and the net radiation into the flux tube from the side. These energy inputs are balanced against radiative losses into free space from the top of the plage or sunspot atmosphere. Magnetic field strength in the sunspot is $\sim 3000\,\mathrm{G}$, and the flux tubes have field strength up to $1500\,\mathrm{G}$. The width of flux bundles in a sunspot are thicker than the flux tubes in plages. In the sunspot's thicker flux tube, the inhibition of convection along the spot's axis is relatively much more important than any excess radiation into the umbra from the hot wall of the Wilson depression, and the result is a relatively cool atmosphere. In the thinner plage flux tube, the lateral radiation is relatively much more important, since it scales as the flux tube radius r, while the amount of convective inhibition scales as r^2. Flux tube models confirm that radiation from the hot convection zone into plage can result in an hotter atmosphere.

There are a large variety of sunspot groups observed on the Sun depending on their polarity, state of umbral and penumbral structures, and longitudinal extent. In 1938, Max Waldmeier introduced the *Zurich Classification System* for the sunspot groups. More recently, a classification scheme has been proposed by McIntosh (1990), which includes parameters such as shape, complexity of the largest spot in a group, compactness, the level of penumbral growth, etc. (Bhatnagar, this volume). These parameters have been found to be useful in predicting the level of activity of the sunspot groups. Several modifications have been subsequently introduced in order to address the limitation of the Zurich Classification (e.g., Beck et al. 1995).

The sunspot group is but a part of a much more comprehensive 3-dimensional entity known as an *active region*, which extends several tens of thousand kilometres into the solar atmosphere, and has a distinctive appearance in the chromosphere, transition region, and corona. Recent observations taken by the SOHO–MDI (Michelson Doppler Imager) instrument probes the hitherto invisible, deeper sub-photospheric structures of sunspots by *sounding* the solar interior (Fig. 7).

4 Sunspots and Solar Rotation

The position measurement of tracers such as, sunspots, faculae, and magnetic fluxes, made over a period of several days provides a simple and direct method of determination of the solar photospheric rotation. Other chromospheric tracers, such as, Hα filaments, plages, and structures in the green-line and white light corona have been used to obtain the chromospheric and coronal rotation rates. Doppler measurements are also used for determination of the rotation rate to an accuracy of about 1%. The Sun's rotation velocity of roughly 2 km s^{-1} near the

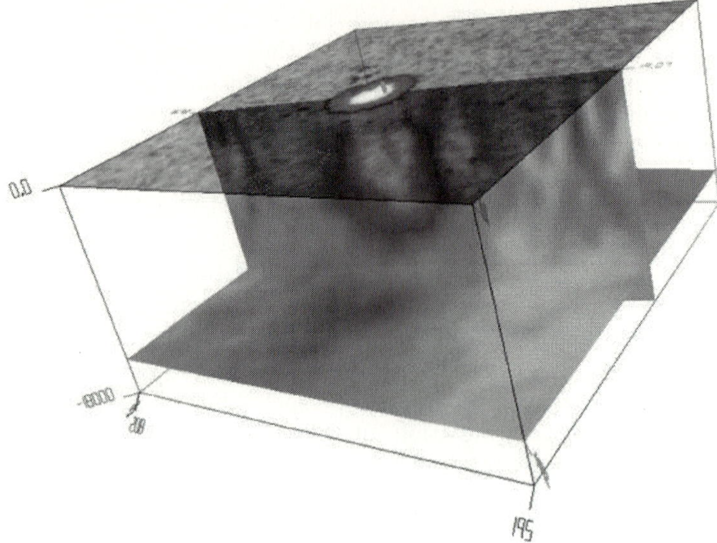

Fig. 7. The subsurface structure (sound speed) below a sunspot as derived from Doppler measurements by SOHO–MDI instrument using helioseismic technique. The surface intensity shows the sunspot with the central umbra surrounded by the somewhat filamentary penumbra. At the second plane, 24 000 km deep from the surface, faster sound speed areas are shown brighter, while slower sound speed are darker

equator carries structures such as sunspots across a telescope's field-of-view at a rate of about 10 seconds-of-arc per hour near the disk centre. The sidereal (i.e., with respect to fixed stars rather than to the revolving Earth) rotation rate at solar equator is found to be 2.84 μrad s^{-1}, which corresponds to a rotation period of approximately 25 days (Howard 1984). Detailed investigations have shown that the Sun does not rotate about its axis like a rigid object but rotates differentially; the rate of rotation decreases from the equator towards the poles. The east-west transit of sunspots lying within the sunspot latitude zone $\pm 16°$ takes $\sim$ 27.2753 days as seen from Earth (i.e., the synodic rotation period). The photospheric (sidereal) rotation rate is given by $\Omega = 14.05 - 1.492 \sin^2 \theta - 2.606 \sin^4 \theta$ degrees per day, where θ is the latitude (Snodgrass 1984). The effect of this differential rotation is that the rotation period at latitudes $\pm 60°$ is around 4 days longer than that at the equator. There is also evidence that the rotation rates of the north and south solar hemispheres can be different. It is also expected that there is a radial gradient of the solar rotation, which might in principle be obtained from Fraunhofer lines formed at different heights in the solar atmosphere. In addition to the other evolutionary processes, the differential rotation causes considerable change in the observable features on solar surface.

The spherical surface of the solar atmosphere can be mapped on to a flat synoptic chart during the course of one solar rotation. This chart uses the heliographic latitudes $[-90°, +90°]$ plotted against the heliographic longitudes $[0°, 360°]$ for the mapping of the observed features, such as, the sunspots, facu-

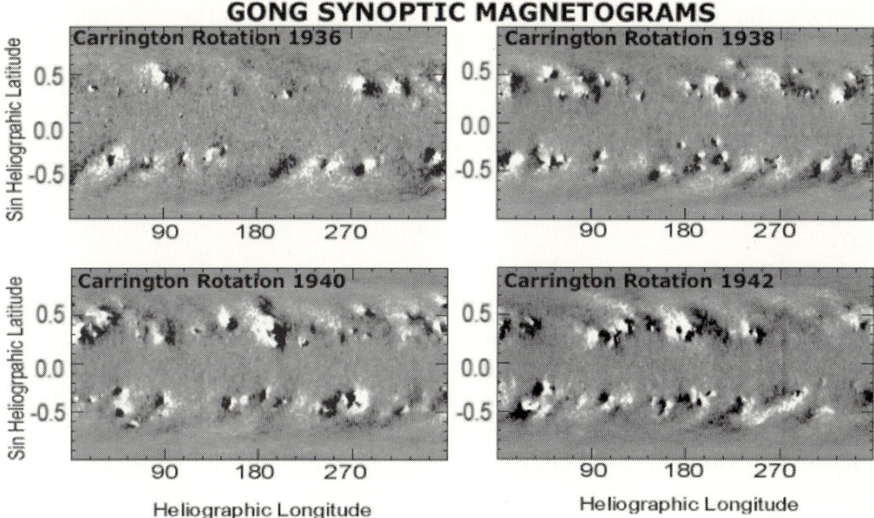

Fig. 8. Evolution of magnetic fluxes associated with active regions shown over a period of six solar rotations, i.e., from Carrington Rotation Number 1936 to 1942. The data were obtained from the instrument operated by the Global Oscillations Network Group (GONG) Project (Harvey et al. 1996)

lae, filaments, magnetic flux, etc. Daily images of the Sun are used to produce synoptic maps that represent the entire surface of the Sun in various ways. Each daily observation is remapped into longitude measured from the central meridian and both latitude and its sine. These daily remapped images are weighted, shifted to the appropriate Carrington longitude and then merged with data from other days to form representations of features on the solar surface. In case of the magnetic synoptic chart, the line-of-sight magnetic field measurements are converted to approximate flux density before the remapping phase, by assuming that the fields are vertical. This is a bad assumption in strong active regions but is a fairly good assumption for weak active regions and network structures. Such synoptic charts give an overview of the structures observed during the course of a synodic solar rotation without taking into account the temporal evolution. Figure 8 shows a sequence of synoptic charts of magnetic fluxes. Considerable magnetic activity on the solar photosphere, in the form of decay and birth of active regions, is seen at several locations during this period of a few Carrington rotations.

5 The 11 Year Solar Activity Cycle

Solar activity is best measured by a quantitative index, the *sunspot number*, related to the number of sunspot groups and individual sunspots present on the Sun on a given day. Rudolf Wolf introduced in 1848 a simple and globally used *Wolf number* of sunspots, or Relative Sunspot Number defined as $R = k(10g+f)$, where g is the number of spot groups, f is the number of all the individual spots in these groups, and k is a reduction factor representing the atmospheric conditions, efficiency of the telescope and the observer (see Bhatnagar, this volume). Wolf used $k = 1$ for his refractor with an 8-cm aperture and focal-length of 110 cm. For his hand-held telescope used by him while travelling, Wolf adopted $k = 1.5$. For example, one would get $R = 11$ for a single spot on the Sun. The largest value since 1749 that has been recorded was $R = 355$ on December 24–25, 1957.

A more recent trend in solar physics is to use the radio flux at 11 cm wavelength as a measure of solar activity, which is a better reflection of the magnetic fields on the Sun. Another quantitative parameter linked to the magnetic field strength is the sunspot area. For the period 1874–1938, the annual mean of the sunspot area number A_G from Greenwich, and the Wolf number R_Z from Zurich were shown to be linked by the relation $A_G = 16.7 R_Z$. Similarly, a simple relation between the magnetic flux density B_m (in Gauss) in the centre of a *stable* spot with an area A_s (in millionths of the visible solar hemisphere) is given by:

$$B_m = \frac{3700 A_s}{A_s + 66} \, . \tag{1}$$

The temporal activity of the Sun is shown remarkably well by the daily change in the number of sunspots. The monthly mean relative sunspot numbers (and similarly, the area), when plotted over several years and decades, reveal a pattern of periodic rise and fall of solar activity (Fig. 9). The average length of the activity cycle between the maxima is ≈ 11.1 years as reported by Wolf in 1853. But, it has been found to vary in individual cases between 8 and 15 years. The rise from minimum to maximum (average 4.8 years) takes less time than the fall (average 6.2 years). The numbering of cycles starts at zero from 1749; the cycle beginning in 1986 is number 22, and the on-going cycle is 23, for which the ascending phase began in September 1996.

5.1 Mapping Sunspot Positions During Activity Cycles: The Butterfly Diagram

The spatial coordinates of sunspots in the visible solar hemisphere can be plotted against the time axis in the position maps to show the temporal development and motion of sunspot groups on the Sun. An interesting "butterfly diagram" results when the sunspot latitude positions are thus plotted, which shows the zone movement of sunspot appearance (Fig. 10). Each Carrington rotation is averaged in longitude at each value of sine latitude and the result is placed in one column of the butterfly map. White light photoheliograms taken over a few days are sufficient to indicate that sunspots are not uniformly distributed over the Sun, but they occur in certain patterns in two latitude zones, called activity zones. These zones are located nearly symmetrically around the solar equator. With the progress of the solar cycle, a migration of the heliographic latitude of sunspot zones towards lower latitudes is observed. This migration is known

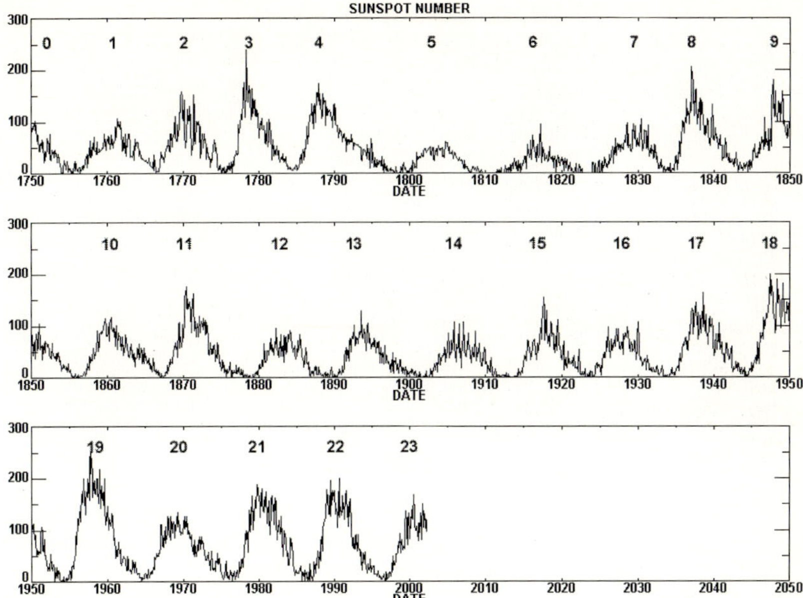

Fig. 9. The 11 year sunspot activity cycles starting from Cycle 0 around 1750 to Cycle 23 having the peak activity around 2000–01. Note the variation in the amplitude of the cycles, particularly the large amplitude of the cycle 19

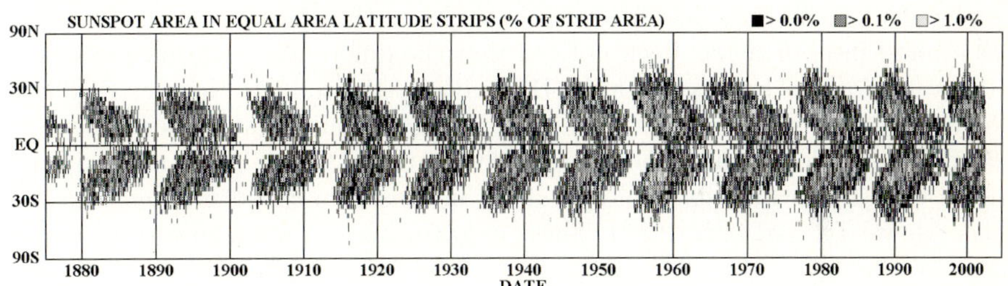

Fig. 10. A Butterfly Diagram obtained using the sunspot areas

as *Spörer's law.* Shortly before the sunspot minimum phase, sunspot groups corresponding to the next cycle appear at high latitudes $\pm(30\text{–}40^\circ)$. As the cycle progresses, these sunspot groups appear closer towards the equator until just after the next minimum when they disappear in its vicinity. The latitude distribution of sunspots is a good method to determine the time of sunspot minimum for a given cycle. In fact, around this time sunspots belonging to the old cycle (located close to the equator), and those of the new cycle (spots appearing at high latitudes) are found to coexist.

Full disk solar magnetograms have become available since 1975, which provide magnetic flux data for sunspots, and also show the regions of comparatively weak magnetic fluxes. A magnetic butterfly diagram, similar to that for sunspot

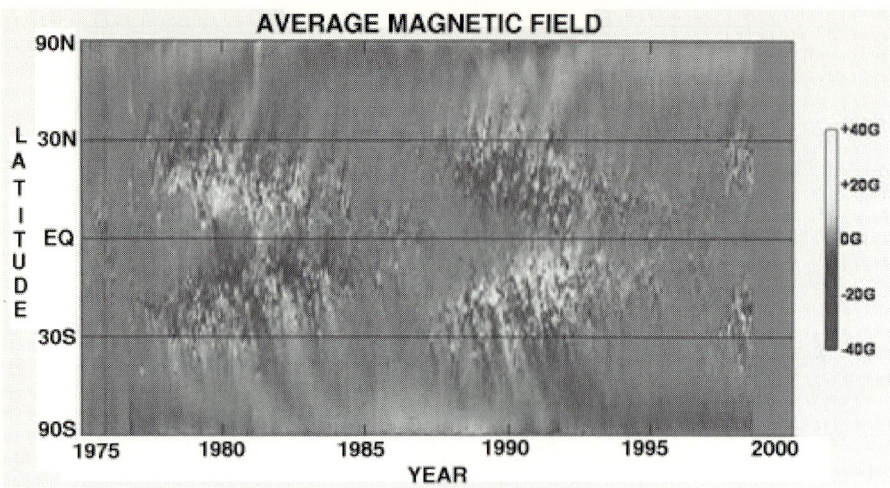

Fig. 11. A butterfly diagram obtained from line-of-sight magnetograms shows the poleward and equatorward motions of solar photospheric magnetic fluxes corresponding to the activity cycles 21 and 22 (adopted from Lites 2000)

positions, results when flux-positions are plotted with time. From the magnetic butterfly diagram, in addition to the equatorward motion of sunspot-zones observed from the sunspot butterfly diagram, poleward magnetic flux motions also become evident at high latitudes (Fig. 11). From the observations of the solar surface and interior it has been noticed that the differential rotation profile exhibits variations with the solar cycle (Howard & LaBonte 1980; Howe et al. 2000). The existence of slow and fast zonal bands of rotation has also been shown by surface observations, which show these bands migrate from high to low latitudes during the 11-year activity cycle, resulting in a pattern called *torsional oscillations* (Howard & LaBonte 1980). These migrating bands seem to correlate with the migrating magnetic activity bands shown by the butterfly diagram. However, the relation between the zonal shear flow and activity bands is not understood. Antia & Basu (2000) have studied the temporal variations of the rotation rate in the solar interior using Global Oscillation Network Group (GONG) data obtained during 1995–1999. They find alternating latitudinal bands of faster and slower rotation that move toward the equator with time, observed as torsional oscillations at the surface. This flow pattern appears to persist to a depth of at least $0.1R_\odot$ and its magnitude is well correlated with the solar activity indices.

As seen from full disk magnetograms, magnetic polarities of the leading (right) and the trailing (left) sunspots are different between the Sun's northern and southern hemispheres. Furthermore, all the spots' polarities assume the opposite sign from one 11-year cycle to the next, which is accompanied by a reversal of the Sun's general magnetic field (Fig. 12). This is known as *Hale's polarity law*. Thus, the period of solar magnetic cycle is $\sim$ 22 years, or two sunspot activity cycles, during which the magnetic polarity of solar poles goes

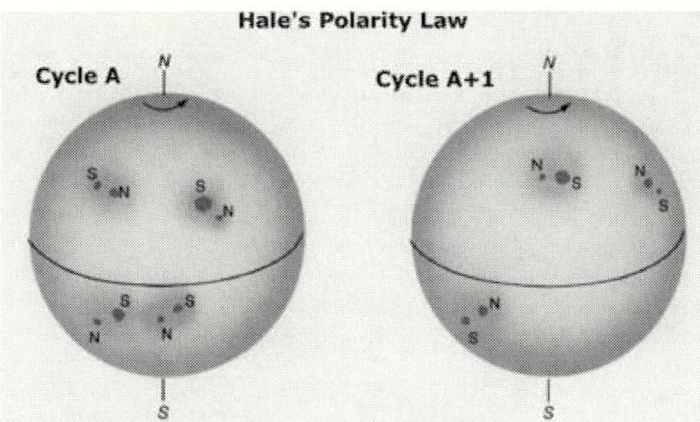

Fig. 12. Hale's polarity law: Switching of polarities during two activity cycles. The northern (N) and southern (S) poles are marked on the solar rotation axis. The N–S marked on the sunspots indicate the magnetic field polarities, which are shown to switch from one cycle to another and are always opposite in the northern and southern hemispheres

through one cycle. The reversal of magnetic polarity takes place at the poles around the maximum of the sunspot activity. It is also significant that the mean poloidal field is antisymmetric with respect to the equatorial plane, as is the case with the mean toroidal field.

5.2 The Changing Face of the Sun over the Solar Cycle

The Sun goes through considerable amount of change in its outer layers as observed at all wavelengths, from the minimum phase of an activity cycle to the maximum. During the minimum phase, the Sun's disk is nearly uniform with little or no sunspots, and no enhanced regions of magnetic fluxes or X-ray emission. As the cycle goes through the ascending phase and nears the maximum, large active regions begin to appear. Also, the frequency of occurrence of explosive phenomena, such as flares starts to increase. Figure 13 shows the progress of the current cycle 23 and increase of activity from 1996 to 2002. As the cycle approaches the minimum phase, the number of active regions will start waning, and the solar disk will become featureless once again. Although the number of flares also reduces during the minimum phase, sporadic flare events of large magnitude are known to occur. Such flares observed in the minimum phase are associated with large erupting filaments, and not with active regions. The flare energy is essentially derived from the magnetic energy stored in highly twisted filaments, which gets released during the eruption process. Recent observations from space show that large coronal mass ejections (CMEs) also occur irrespective of the Sun's marching toward the minimum phase.

5.3 Solar Activity Cycles of Long Periods

The sunspot observations before 1749 are few and far between, but it appears that there was a period of very little sunspot activity for a 70-year period between

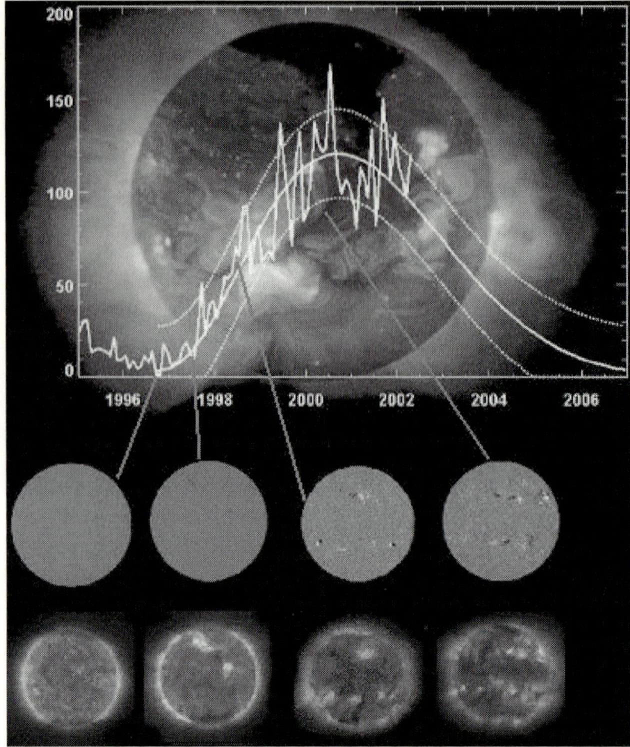

Fig. 13. The Sun's 11-year solar cycle as reflected by the number of sunspots recorded to date and as projected (dotted line) for Cycle 23. Selected SOHO EIT 195 Å and SOHO MDI magnetogram images show the rising level of solar activity as the cycle progresses

1645–1715, which is called the *Maunder Minimum* (Eddy 1983). It is interesting to note that the time of Maunder minimum coincided with an unusually cold period in the northern hemisphere, known as the *Little Ice Age*. From the sunspot activity cycles, we can easily notice that the amplitude of the cycles vary, and it does so non-randomly; a series of larger maxima appears to follow a series of lower ones. For example, after a low 14th cycle during 1901–1913, the amplitude of maxima gradually increased to its highest value in the 19th cycle during 1954–1964 (Fig. 9). This indicates that *long period sunspot cycles* are also operative, such as the 80-year cycle reported by Gleissberg (1952). The existence of *long cycles* shows that the Sun does *remember* how active it was in the previous cycles, and there is a link between the 11-year cycles.

Accurate forecasting of sunspot activity has proved to be difficult. This difficulty essentially indicates that not just one but several long cycles of varying lengths are perhaps operating (Wilson 1992; Letfus 1994). These cycles could be discovered using the spectral analysis technique if long periods of reliable data existed. Unfortunately, reliable Wolf numbers have only been available since 1750.

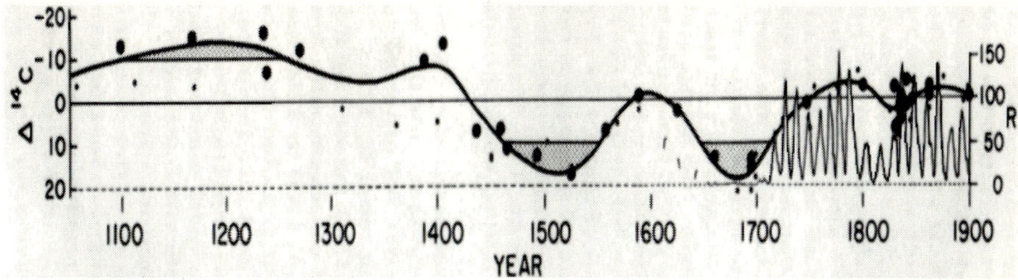

Fig. 14. Solar activity cycles as inferred from ^{14}C dating of tree rings for the period 1050–1900. The periods of large deviations are shaded, which correspond to the Grand Maximum (1100–1250), Spörer Minimum (1460–1550), and Maunder Minimum (1645–1715). The Wolf sunspot number is also shown after 1610 (e.g., Eddy 1983)

Perhaps the most important evidence for long term solar activity fluctuations has come from other means, notably the ^{14}C dating technique applied on the records frozen in old tree rings to track the history of solar activity backwards to several thousand years. Apart from giving further evidence of the Maunder Minimum during 1645–1715, the tree ring data provide indications of other periods of anomalous solar activity, such as, the Spörer Minimum during 1450–1550, and the abnormally high activity period of 1100–1250, termed as the "Grand Maximum"(Fig. 14).

It is evident that the Sun does not maintain a *clock-work* of regular sunspot activity or magnetic cycle. It spends long periods in a relatively quiescent mode, such as, the Maunder or Spörer minima, and also in abnormally high activity state, such as the Grand Maximum. Ultimately, a successful theory of solar dynamo should explain why the activity cycle is so irregular, and how it regenerates itself after having turned itself off.

5.4 Babcock's Model of the Solar Activity and the Magnetic Cycle

Several heuristic models have been proposed to explain the solar activity and magnetic cycles, with some based on mathematical support. Each of these models has offered some insight into possible cyclic processes, but none has provided an explanation to all the available data. These models can be classified as (a) relaxation models (e.g., Babcock 1961), (b) forced oscillator models (e.g., Bracewell 1988), or (c) dynamo wave models (e.g., Parker 1955; Krause & Radler 1980). Both (a) and (c) are essentially based on a solar dynamo theory (see also Venkatakrishnan, this volume).

According to the relaxation model put forward by Horace Babcock, let us assume that in the first stage, which coincides with an epoch of low activity, the magnetic field lines are essentially polar, i.e., they run primarily from south to north as an 11-year cycle begins (Fig. 15). The latitudinal differential rotation of the Sun begins to stretch the field lines beneath the faster-spinning equatorial region. Thus, the initial poloidal (north-south) field is essentially changed into a toroidal (east-west) configuration, in which the lines of force are near-circles

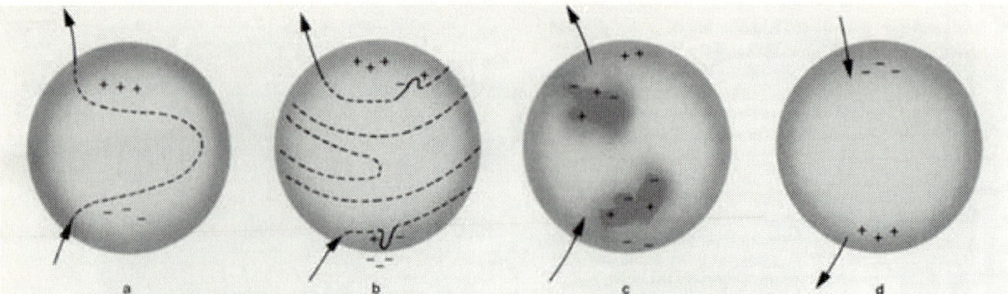

Fig. 15. Babcock's kinematic model of Solar Cycle

around the solar axis. In time the stretching wraps the lines several times around the Sun as in (b), causing them to intertwine and intensify. Ultimately they are driven to the surface as magnetic loops, by local convection or magnetic buoyancy. Each rising loop spawns an active region once it breaks into the surface shown in (c), creating bipolar active-region groups that obey Hale's polarity laws. As the cycle nears its end in (d), the leading regions drift toward the equator, where their opposite polarities mix and cancel; the trailing regions drift poleward where their polarity cancels and replaces the existing fields of opposite polarity. The entire sequence then repeats, except that all polarities have been reversed. Thus a complete magnetic cycle takes 22 years to complete. This model still provides the best conceptual framework for our understanding of the main features of the Sun's magnetic cycle, namely, the generation of active regions, the Hale's polarity laws, Spörer's law, and the 22-year solar magnetic polarity oscillation. However, this model is mainly kinematic and based on the observed properties of photospheric fields, such as differential rotation with latitude, eruption of active regions with tilted dipole axes, and subsequent separation of p- and f-polarities as the active regions evolve. The dynamics behind these phenomena is not well understood.

The relationship between the poloidal and toroidal components of the solar mean field in a cyclic manner prompts us to ask the more general question of the origin of solar magnetism. In recent years, dynamo models have progressed beyond purely kinematic solutions where $\boldsymbol{v}$ is assumed, to fully dynamical solutions of the induction equation along with the coupled mass, momentum, and energy relations for the solar plasma:

$$\frac{\partial \boldsymbol{B}}{\partial t} = \boldsymbol{\nabla} \times (\boldsymbol{v} \times \boldsymbol{B}) - \frac{c^2}{4\pi\sigma} \boldsymbol{\nabla} \times (\boldsymbol{\nabla} \times \boldsymbol{B}) , \qquad (2)$$

where σ is called the electrical conductivity. The first term in the right hand side of the induction equation represents the inducing effect of the motion of the material upon the magnetic field. The second term corresponds to the *Ohmic* dissipation of the field arising due to the finite electrical resistance. Using the induction equation, these models compute a representation of the velocity fields of large scale solar convection, of the Sun's differential rotation profile, and of the magnetic fields that are generated by these motions. Several workers have studied the generation of solar magnetic field by induction due to moving conductors, as a self-excited dynamo (Cowling 1934; Bullard & Gellman 1954; Yoshimura 1975). Comparison of the results with recent observations raises important questions about the processes invoked in the Babcock's model. In the Babcock model, the magnetic field is intensified by both latitudinal and radial gradients in the Sun's angular rotation rate. The latitudinal gradients produce a toroidal field component from the poloidal field. The radial gradients provide further twist to the field-lines through a roller-bearing effect. Steenbeck and Krause (1969), and Stix (1976, 1991) have numerically obtained some aspects of the sunspot activity cycle and butterfly diagram (see Venkatakrishnan, this volume).

6 Explosive, Eruptive Phenomena on the Sun

Apart from the variation in the solar activity over global scales and on long time scales, local and short-lived explosive phenomena also occur in the solar atmosphere. The energy released from these explosive events lead to bulk mass motion, accelerated particles, and enhancement of radiation over the entire electromagnetic spectrum, ranging from γ-rays to radio wavelengths. Solar flares, filament/prominence eruptions, and coronal mass ejections are some of the most spectacular transient and explosive phenomena displayed by the violent Sun particularly when it is at, or near the maximum phase of the solar activity cycle.

6.1 Solar Flares

The most striking explosive form of solar activity are solar flares. A flare is essentially a sudden catastrophic release of energy appearing as enhancement of electromagnetic radiation over a very wide range, and as mass, particle, wave and shock wave motions. Most flares occur in active regions in the neighbourhood of sunspots; being more frequent when the active region is in a rapid development stage. Because of the association with active regions, the frequency of flares also follows the 11-year sunspot cycle but with some deviations. However, flares may also occur when an active region has decayed, or lost all its sunspots. As much as 10^{32} ergs of energy may be released in a large flare in a matter of a few minutes to hours. About 25% of this energy can appear in the visible wavelengths. Considering such a flare lasting an hour, and covering an area of 3×10^9 km^2, the rate of energy release in the flare can be estimated as 2×10^{11} W km^{-2}. This is nearly 300 times lower than the energy emitted by the photosphere, which is 6×10^{13} W km^{-2}. Therefore, normally the flares are not visible in the photosphere except in some exceptionally high energy and impulsive events of so called *white light* flares. In fact, it is remarkable that a solar flare was first discovered by Carrington and Hodgson in 1859, through the observation of a white light flare seen as sudden brightening in the photospheric layer. It is not only the amount of energy released in flares, but also the suddenness of its release which make the flares so spectacular. Few astronomical phenomena are as rapid as solar flares. Other very time-dependent phenomena are X-ray

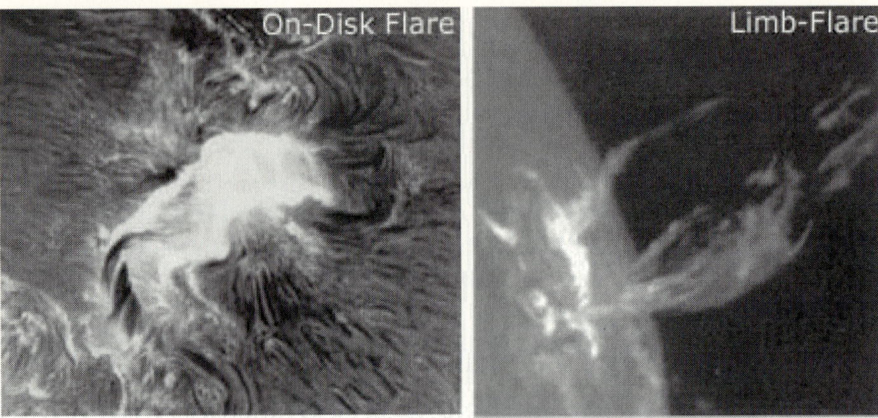

Fig. 16. (a) Optical part of a large flare as seen on the disk in Hα. (b) The vertical extent of a flare-associated material ejection, and chromospheric bright ribbons shown in a flare located near the Sun's limb

emissions associated with certain binary star systems, cosmic γ-ray bursts, and flares in other stars.

Flares are most widely and traditionally observed at ground-based observatories as sudden chromospheric brightening, using narrow passband optical filters centred at the Hα-line at 6563 Å (Fig. 16). In many cases, flares tend to occur repeatedly at the same location, and display similar spatial structures during the course of evolution of an active region. Such flares are traditionally called *homologous*. A part of the stored energy is released in the flare, after which the conditions are restored back due to favourable magnetic evolution in the neighbourhood, and the next homologous flare ensues. This process continues till most of the stored energy has been released, and the active region has reached a relaxed state. Some major flares are known to trigger secondary flares in the neighbouring regions, and also in remote active regions, which may be magnetically connected. These flares are called *sympathetic* as they occur in response to the primary flare event elsewhere.

On the solar limb, a flare appears as a bright mound and then develops rapidly in size, associated with the ejection of material in the form of spray, or surges (Fig. 16b). Surges are ascending, then descending motions of material along an almost straight path, with upward velocities of 50–200 km s^{-1}, reaching heights of about 100 000 km. In many active regions, repetitive occurrence of surges is observed during their initial phase of birth on the solar surface. On the other hand, sprays are more explosive events in which fragments of material are ejected out at velocities of up to 2000 km s^{-1}, greater than the solar escape velocity. As the spray material rises upward, it is accelerated and gradually disappears from the Hα passband. However, the ejected material continues to move outward, and can be followed in other wavelengths. Flares were first erroneously considered to be a chromospheric phenomenon. With the development of observational techniques in other wavelengths, notably radio UV, EUV, X-rays, it is now realised

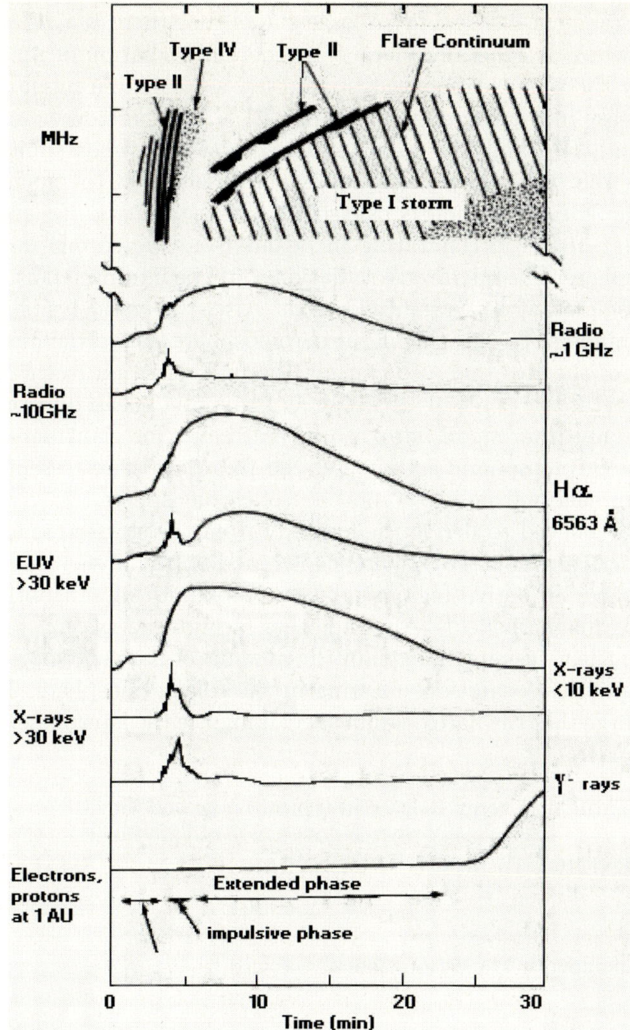

Fig. 17. A schematic representation of the different phases of a typical solar flare as observed in electromagnetic radiation and particle motion. The top part shows various types of radio bursts associated with the flares (adopted from Dulk, McLean & Nelson 1985; see also Kane 1974)

that the chromospheric Hα brightening is just one facet of a much more complex 3-dimensional process of energy release in a flare; the chromospheric flare being only the proverbial tip of the iceberg.

Most of the optically *invisible* flare emission comes from the hot coronal plasma, and not from the chromosphere. They are recorded in spectral regions ranging from the radio wavelengths of $\sim 3\,\mathrm{km}$ (i.e., frequency $\sim 100\,\mathrm{kHz}$), to the extreme ultra-violet (EUV), and X-ray, i.e., $< 0.06\,\text{Å}$ ($> 200\,\mathrm{keV}$) (Fig. 17).

The spectral range of flare energy release shows that there is a *precursor* phase of the flare, which is generally marked by thermal radiation of up to 10^7 K. It is followed by the *impulsive* phase lasting a few seconds to a minute, consisting of energy release in a burst of γ-ray, X-ray, EUV, and microwave emission. On the other hand, Hα, and soft X-ray profiles show somewhat similar extended phase during the flare. From X-ray emission observed in the solar flares, one infers that the X-ray emission is caused by bremsstrahlung of electrons in the energy range 1–10 keV precipitating in beams from the corona down into the denser atmosphere. The microwave radiation could be interpreted as synchrotron radiation associated with the same electrons. After the impulsive phase, thermal radiation dominates. The heating of the chromosphere leads to an increasing level of excitation of the Hα, and also causing thermal radiation in the form of soft X-rays. Energy in the range of 10^{28}–10^{33} ergs may be released in times ranging from a few minutes for small flares, to several hours for giant flares (Table 1). The observed chromospheric activity related to a flare arises essentially due to the transfer of energy from the corona to the chromosphere.

Flares are believed to derive their energy from the stressed magnetic structures of the active region. This inference is based on the observational fact that most flares occur in active regions, in close proximity of sunspots, which are seats of strong magnetic fields. Also, no other form of energy, viz., gravitational, thermal, or nuclear fusion can explain the amount of flare energy released, considering the physical conditions existing in the solar atmosphere. The released energy is so large that the flares significantly influence the interplanetary medium

Table 1. Energy Release in typical Large and Small Flares

FORM OF ENERGY RELEASE	LARGE FLARE (Ergs)	SMALL FLARE (Ergs)
Hydrodynamic flow (I. P. ejection, shock)	4×10^{32}	-
RADIATIONS:		
SXR, UV	5×10^{31}	$< 10^{29}$
Optical (continuum)	3×10^{31}	-
Hα emission	3×10^{30}	10^{26}
HXR	5×10^{26}	10^{24}
γ-ray	2×10^{25}	-
Radio Emission	2×10^{24}	$< 10^{21}$
ACCELERATED PARTICLES:		
Electrons	3×10^{31}	10^{27}
Protons	3×10^{31}	-
DURATION:	minutes to hours	up to several minutes

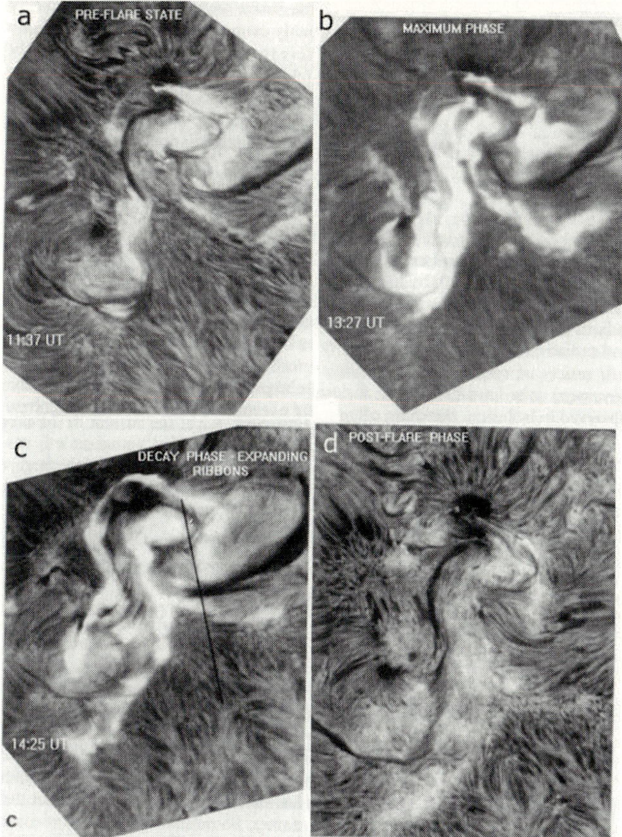

Fig. 18. Spatial and temporal evolution of a large flare seen in the chromospheric absorption line Hα 6563 Å (adopted from Bruzek 1979)

and terrestrial atmosphere. The extreme diversity of flares, flare-associated phenomena, and the ongoing search for the catastrophic trigger mechanism make the flare-research an important challenge in solar physics.

In many energetic flares, elongated bright ribbons are seen in Hα (two ribbon flares). Usually these ribbons occur on either sides of a dark active filament, which traces the magnetic polarity inversion line, or the *neutral line*. All flares seen in Hα are also seen as soft X-ray brightenings or bursts. Larger flares give rise to hard X-rays (HXR), microwaves (MW), γ-rays and accelerated particles. Flares occur in the active bipolar magnetic structures seen at the photosphere as sunspots. The field-lines joining these sunspots generally close at the corona. As the Hα emission is observed at both sides of the neutral line, the flare energy release involves such closed magnetic structure, or *loops*.

The rapid rearrangement of the magnetic field configuration in an active region, where a large flare occurred, is shown typically by a sequence of Hα images (Fig. 18). During the first stage of the pre-flare state, a dark filament running

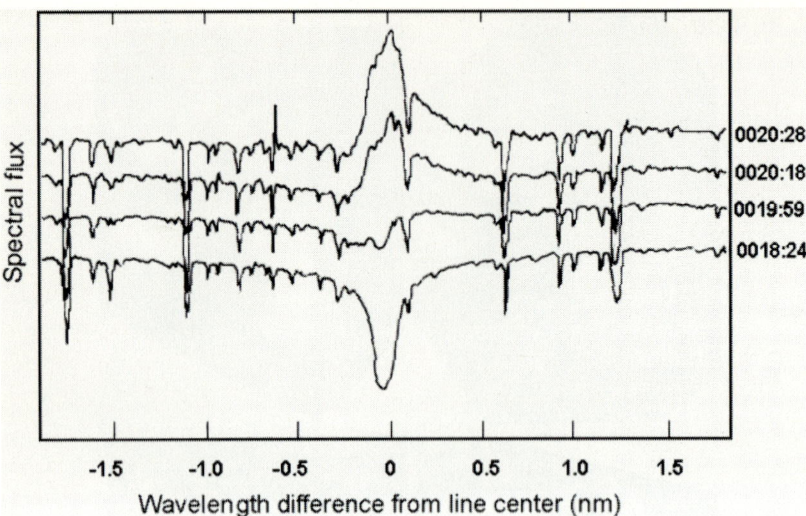

Fig. 19. Successive Hα-line profiles at a location in a flare on June 20, 1982 (time indicated to the right of each curve). The spectral line has initially an absorption profile, but it develops a strong emission core. The wavelength scale is measured from the centre of the Hα-line (after Ichimoto & Kurokawa 1984)

along the magnetic inversion line becomes activated, and starts rising. A few minutes later, the magnetic energy stored in the filament's structure is released, and appears as bright ribbons of a flare at either sides of the magnetic inversion line in the maximum phase (b). These bright ribbons show a rapid expansion often accompanied by a strong increase in brightness, called flash phase. During the decay phase of the flare, the flare ribbons gradually decay in intensity, post-flare loops begin forming across the location of the flare, and material of the erupted filament starts settling down (c). Later, a reformed filament showing the post-flare magnetic field configuration results (d). This reformation implies that after the flare, the initial magnetic structure is largely restored in this particular event, however, this restoration may not occur every time.

The Fraunhofer line-profiles also remarkably show the dynamics of the flare activity. In fact, many strong lines are reversed from absorption to emission, and in case of a large flare, the central intensity may be more than double compared to the continuum. As the flare ribbons brighten, the absorption lines begin to change to emission indicating the tremendous heating of the material involved in the flare process (Fig. 19). These observations can be used to infer the velocity and temperatures in the flaring-sites.

The tendency of an active region to flare is generally dependent on the magnetic complexity of the region. Solar vector magnetic field measurements in the photosphere have been used to derive conditions favouring the flare occurrence. One of the methods is to infer how the magnetic field lines are *sheared* across the magnetic inversion lines, however, not all flares can be explained by just one parameter. There are a large variety of observational features associated with

Table 2. Flare Classification in Hα, Radio and X-Rays

Hα Class	Area ($10^{-6} A_{\odot}$)	5000 MHz Flux (sfu)	SXR Class (0.1–0.8 nm Flux in Watt m^{-2})
S	< 200	5	C2 (2×10^{-6})
1	200–500	30	M3 (3×10^{-5})
2	500–1200	300	X1 (1×10^{-4})
3	1200–2400	3000	X5 (5×10^{-4})
4	> 2400	30000	X9 (9×10^{-4})

flares, such as, large and complex sunspots, changing spot-areas (Ambastha & Bhatnagar 1988), complex magnetic fluxes, magnetic field gradients, new emerging fluxes (Vorpahl 1973; Choudhary, Ambastha & Ai 1998), flux cancellations (Wang & Shi 1993; Mathew & Ambastha 2000), rapid spot motions and their collisions, magnetic shear (or stress) (Hagyard et al. 1984), and abnormal polarities such as δ-spots.

6.2 Flare Classification

Flares occur in a large range of spatial and temporal scales and energies. A simple scheme for classifying the Hα flare importance is used by the National Oceanic and Atmospheric Administration, USA (NOAA) based on both flare area (corrected for foreshortening), and brightness as measured in the Hα line (Table 2). The optical flares are classified as sub-flare, Class 1, Class 2, etc., based on the area of flare brightening observed in the line-centre of Hα. Another parameter signifying the brightness level of the flare is generally suffixed to the area class. The brightness levels are generally divided as faint (F), normal (N), or bright (B). For example, a bright flare of *importance* 1 in Hα will be denoted as Class 1B flare. The corresponding typical radio fluxes are also listed based on the amount of flux measured in solar flux unit (s.f.u.) at 5000 MHz frequency. Here $1\,\text{s.f.u.} = 10^4\,\text{Jy} = 10^{-22}$ Watts m^{-2} Hz^{-1}. A more quantitative classification is given by the integrated X-ray flux associated with flares, based on soft X-ray data from monitoring satellites, called Geostationary Operational Environmental Satellites (GOES). Table 2 gives typical X-ray fluxes corresponding to various classes of optical flares, which are only given as approximate measures. It should be noted that there is no general correlation between a large Hα flare with the similar large X-ray class. There are instances when little optical brightening was observed, while the X-ray instruments reported a major X-class flare. Similarly, a large class 2B flare may be associated only with a modest M-class X-ray event. The onset, maximum phase, end times, and peak emission are measured by GOES and are made available by NOAA in the Solar Geophysical Data Reports. A network of ground-based solar observatories around the world also monitors solar flares in Hα and provides reports to the World Data Centre located at Boulder (USA).

Reports and observational data on solar activity and space weather is readily available to the users through the internet. For example, the Royal Greenwich Observatory (RGO) compiled sunspot observations from a small network of observatories to produce a dataset of daily observations starting in May of 1874. The observatory concluded this dataset in 1976 after the US Air Force (USAF) started compiling data from its own Solar Optical Observing Network (SOON). This work was continued with the help of the US National Oceanic and Atmospheric Administration (NOAA) with much of the same information being compiled through to the present. The entire sunspot dataset is available from 1874–2002 at *http://science.msfc.nasa.gov/ssl/pad/solar/greenwch.htm.* The current sunspot number, the status of solar activity, and space weather condition can be obtained at *http://spaceweather.com.* Current full disk solar images, and active regions in various wavelengths are available in an excellent web-site made available by the Big Bear Solar Observatory (BBSO) at *http://www.bbso.njit.edu/arm/latest/.* In addition, one could refer to the web-site of the Solar Data Analysis Center (SDAC) of NASA Goddard Space Flight Center at *http://umbra.nascom.nasa.gov/*, and the Space Environment Center (SEC) site at *http://www.sec.noaa.gov/today.html* for current solar images. For a day of interest, one can obtain a report about solar and geophysical activity, solar active regions, space weather alerts, etc., from the Space Environment Center (SEC) archives at *http://www.sec.noaa.gov/majordomo_archive.cgi/.* Some other web-sites of interest are the Solar Geophysical Data publication available at *http://sgd.ngdc.noaa.gov/sgd/jsp/solarindex.jsp*, and *http://www.sec.noaa.gov/Data/solar.html.* Similarly, a image archive is available for obtaining solar images for any day of interest from 1998 to the current date at *http://www.sec.noaa.gov/solar_images/index.html.*

6.3 The Standard Flare Model and the Main Phases of Flares

During their evolution, flares show considerable temporal and spatial dynamic activity over all wavelengths. From a large number of flare observations, it emerges that most of the flares exhibit three main phases, as follows (e.g., Stix 1991):

The preflare state of buildup of energy and flare trigger: Magnetic energy is stored in the corona due either to motions of photospheric foot-points, or to a current carrying field below the photosphere. A cool, dense filament is formed, which is suspended by a magnetic field. The field evolves slowly through equilibrium states, finally reaching a non-equilibrium which causes the field to rise. The field erupts outward into the interplanetary space, ejecting chromospheric and coronal mass.

The impulsive phase: Reconnection of the magnetic field provides the plasma heating and particle acceleration. Large Hα and X-ray post-flare loops form around the magnetic polarity reversal or neutral line. Apart from Hα, this phase

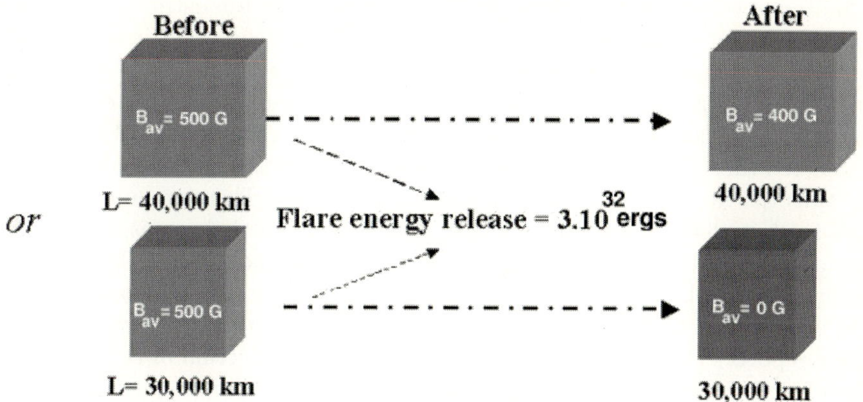

Fig. 20. Schematic diagram showing change in average magnetic field corresponding to the energy release from a large flare in the solar atmosphere

is evident in hard X-rays, radio, microwave emission, UV, EUV, acceleration of particles, and radio bursts.

The gradual and post-flare phase: During this phase, a slow increase in soft X-ray emission occurs, which is caused by filling of loops and arches. The duration of this phase ranges from a few minutes to several hours.

6.4 Fundamental Questions About Flares

There are several major problems related to the flares awaiting answers. Some of these are: (a) Where and how is the flare energy stored? There are no direct observable phenomena showing the precise location. (b) Why is the stored energy released? Only fragmentary data exist on this aspect. (c) Where is the flare energy released? Preliminary indications exist which point toward the coronal regions. (d) What happens to the energy after release? Enormous data exist on this aspect, which essentially indicates the after-effect of the flare in various wavelengths.

As flares mostly occur in active regions, and involve magnetic structures, it is believed that the source of flare energy is magnetic. The magnetic energy contained in a volume of solar atmosphere, i.e., $W = [B^2/8\pi]L^3$, when released in the form of a flare is somehow converted into heat, bulk kinetic energy, particle acceleration, and radiation. For example, let us consider the magnetic energy contained in a cubical volume of solar atmosphere, with a side length of $L =$ 40 000 km, and an average magnetic field of $B_{\rm av} = 500\,$G. After the release of energy equivalent to a typical large flare, i.e., 3×10^{32} ergs, the average field in this volume would decrease to $B_{\rm av} = 400\,$G (or to $B_{\rm av} = 0\,$G, for a cube having sides of $L = 30\,000$ km with the same initial $B_{\rm av}$) (Fig. 20).

The above-mentioned scenario is an oversimplified one, particularly due to the fact that most of the magnetic energy annihilation perhaps takes place in a limited volume in the coronal medium, and not over the entire volume corresponding to an active region's dimensions. Unfortunately, so far it has not been possible to quantitatively measure the coronal magnetic field due to the lack of suitable coronal spectral lines. As a result, accurate and direct detection of changes at the coronal heights has not been possible. The magnetic field-lines are connected through the various layers of the solar atmosphere, and coronal fields are thus tied to the photospheric magnetic structures. Therefore, one would expect to observe a measurable variation at the photospheric layer associated with any changes in the primary flare site located at coronal heights. The resulting changes at the photosphere are expected to be only minor, which require reliable and accurate measurements of magnetic fields free from instrumental artifacts. Some reports exist in the literature of claiming the detection of changes in the line-of-sight component, B_L, of the magnetic field corresponding to major flares. However, these claims remain ambiguous as there are several observational difficulties affecting the results. Some quantitative estimates of changes have been reported in parameters such as magnetic shear, electric current density etc., which are derived from observed transverse component B_t of the magnetic fields (Ambastha, Hagyard & West 1993). However, changes of a diverse variety have been found for different classes of flares, therefore the results need to be confirmed further. More recently, a sudden decrease in the magnetic energy was found in a flare region, using high cadence photospheric magnetograms from SOHO/MDI (Kosovichev & Zharkova 1999). This may perhaps be a direct evidence of magnetic energy release in solar flares. However, it is generally felt that the accuracy and sensitivity of magnetic field measurements is required to be improved further. In addition, it is needed to develop techniques for measuring fields at the higher levels of the solar atmosphere, i.e., in the chromosphere and corona.

6.5 Potential and Force-Free Magnetic Fields

Magnetic energy required for flares is generally believed to be stored in active regions when stressing the coronal magnetic field configuration to non-potential states. This could happen by sub-photospheric flux motions acting on a simple potential field structure, which is in the lowest energy state. Since these motions ($v \sim 10^5$ cm s^{-1}) are much smaller than the Alfvén velocity ($v_A \sim B/(4\pi\rho)^{1/2} = 2 \times 10^9$ cm s^{-1}), the equation of motion in the static approximation is given by (e.g., Priest 1982):

$$\rho \frac{\mathrm{d}\boldsymbol{v}}{\mathrm{d}t} = \frac{1}{c}(\boldsymbol{J} \times \boldsymbol{B}) - \boldsymbol{\nabla} p + \rho \boldsymbol{g} = 0 \,, \tag{3}$$

where, p, $\boldsymbol{g}$, ρ, $\boldsymbol{v}$, and $\boldsymbol{J}$ are the pressure, gravitational acceleration, density, velocity, and the electric current density, respectively. For the solar atmosphere, both pressure and gravitational terms are small, therefore, the magnetic force $\boldsymbol{J} \times \boldsymbol{B} = \frac{c}{4\pi}(\boldsymbol{\nabla} \times \boldsymbol{B}) \times \boldsymbol{B} = 0$, which implies zero Lorentz force, i.e., a force-free configuration of magnetic fields. This condition is satisfied when: (i) $\boldsymbol{B} = 0$

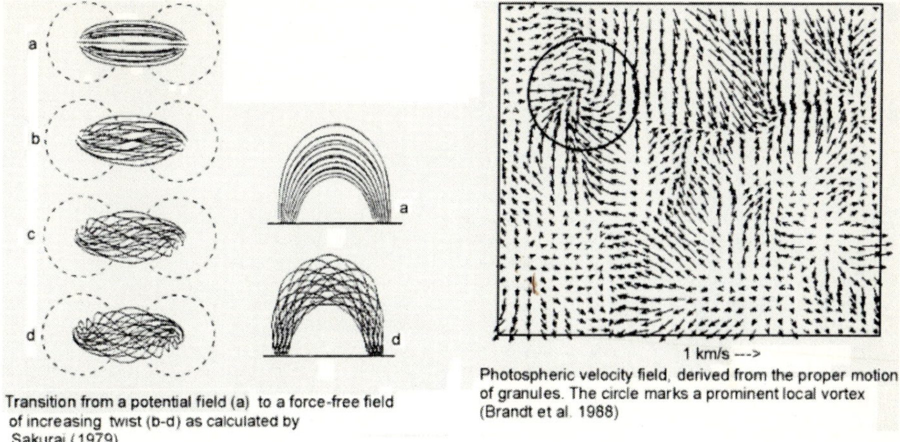

Fig. 21. Transition from a potential to a non-potential magnetic structure

(i.e., a trivial solution), or (ii) $\nabla \times \boldsymbol{B} = 0$, (i.e., $\boldsymbol{B} = \nabla\phi$, which corresponds to potential field structures), or (iii) $\nabla \times \boldsymbol{B} = \alpha \boldsymbol{B}$, where $\alpha = \alpha(\boldsymbol{r})$. On applying the divergence operator on (iii), one gets $\boldsymbol{B} \cdot \nabla\alpha = 0$, i.e., α is constant along the field lines. The condition (iii) corresponds to stressed magnetic fields with non-zero electric current $\boldsymbol{J}$ parallel to $\boldsymbol{B}$ everywhere, so that the flow of charged particles does not cross the field, thereby avoiding a large $\boldsymbol{J} \times \boldsymbol{B}$ force. The equation (2) of induction in the hydrodynamic approximation may be written as:

$$\frac{\partial}{\partial t}\left(\frac{|\boldsymbol{B}|^2}{8\pi}\right) = \frac{1}{4\pi}\nabla \cdot [(\boldsymbol{v} \times \boldsymbol{B}) \times \boldsymbol{B}] - \frac{|\boldsymbol{J}|^2}{\alpha} , \tag{4}$$

where we have used $\nabla \times \boldsymbol{B} = \frac{4\pi}{c}\boldsymbol{J}$, and the force-free condition, $\boldsymbol{J} \times \boldsymbol{B} = 0$. The rate of variation of magnetic energy, δM, is, therefore, given by:

$$\delta M = \frac{\partial}{\partial t}\int_V \frac{|\boldsymbol{B}|^2}{8\pi}\, dV . \tag{5}$$

From this relation, it is evident that photospheric motions and currents may increase the magnetic energy in an active region. The observed photospheric flux tube motions could easily lead to significant departures from the lowest energy state, which corresponds to the potential field configuration. As a result, the active regions could store adequate free magnetic energy, which may eventually be released in flares (Ambastha & Bhatnagar 1988).

The main effect of the electric currents in a force-free field is to introduce twist, or "helicity" into the field structures. The helical structure of a force-free field as an example is shown in Fig. 21(a–d), adopted from Sakurai (1979). He calculated the transition from a potential field to a force-free, non-potential field generated by an increased twist due to the motion of the foot-points of the magnetic loops. Such photospheric velocities derived from observed proper motion of granules show a local vortex, corresponding to the twisting motion at the

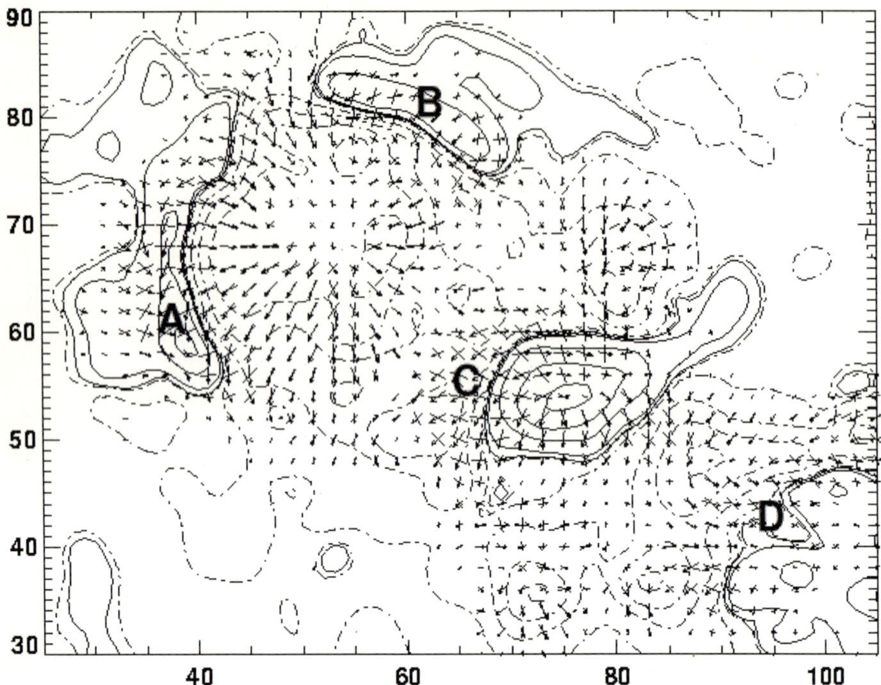

Fig. 22. A photospheric vector magnetogram of the solar active region NOAA 6555 obtained by the NASA–MSFC magnetograph on March 23, 1991. The composite vector magnetic map shows the line-of-sight component of the magnetic field, B_L, as the continuous (*dash-dotted*) contours, corresponding to the positive (*negative*) polarity distribution in the active region, at ± [10, 100, 500, 1000, 1500] G levels. The overlaid line-segments with arrow-heads show the magnitude and azimuthal angles of the calculated potential transverse fields ${B_t}^{\text{pot}}$, while those without arrow-heads show the observed transverse fields ${B_t}^{\text{obs}}$, plotted in the range [200, 1000] G. The sites of strongly non-potential fields are conspicuous by the locations of large departure of the observed transverse fields from the potential fields. These sites are usually expected to be associated with flares

photosphere as seen in the right hand panel of Fig. 21 (Brandt et al. 1988). The vector magnetograph observations of solar active regions usually show locations of such highly non-potential structures within an active region, which are produced by the photospheric motions. Generally, flare-ribbons have been observed to form near these locations, where the directions of the observed transverse component of the magnetic field ${B_t}^{\text{obs}}$ shows large departures from the directions of the potential transverse, ${B_t}^{\text{pot}}$ field. Figure 22 shows a composite vector magnetogram of a large active region NOAA 6555 obtained by the solar facility operated at the NASA–Marshall Space Flight Center, showing the sites of large non-potentialities marked as "A", "B", "C" and "D". NOAA 6555 was called to be a superactive region during its disk transit of March 17–31, 1991, as it produced a very large number of flares including several of the X-class. Signifi-

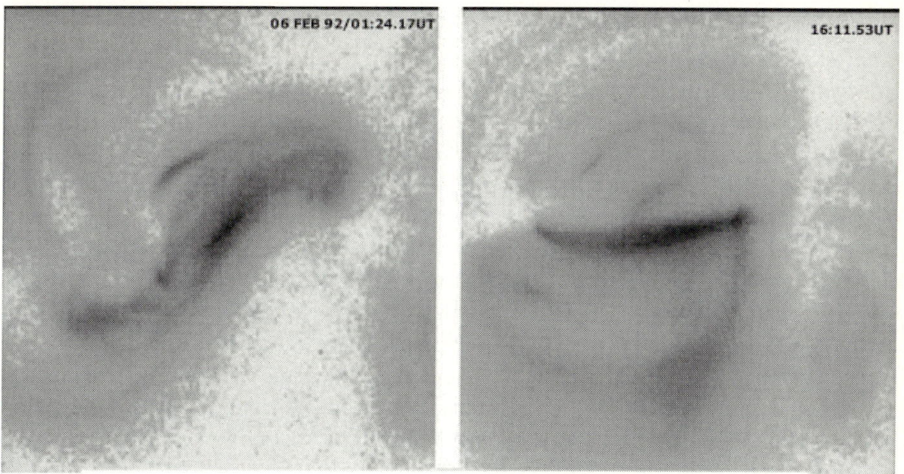

Fig. 23. Observational evidence of the simplification of strong non-potential configuration to a potential configuration after a large flare of February 6, 1992, as inferred from Yohkoh X-ray observation (adopted from Shimizu 1996)

cantly, the magnetogram shows a large flux imbalance in NOAA 6555, as evident from the large areas of negative polarity contours compared to that of positive polarity contours. In order that the condition $\nabla \cdot \boldsymbol{B} = 0$ is obeyed, the large flux imbalance would imply that the active region was magnetically connected with other remotely located active regions. Another important observation was that the location "C" having the largest magnetic stress in NOAA 6555 was not associated with its X-class flares, but the site "A" of rapid evolution associated with a δ-spot. Therefore, it may be inferred that apart from magnetic non-potentiality, additional conditions are required to produce major flares, such as, rapid temporal changes (Ambastha, Hagyard & West 1993), new flux emergence (Choudhary, Ambastha & Ai 1998), and the development of strong field gradients, etc. The observed photospheric motions during the course of evolution of NOAA 6555 produced adequate conditions in the overlying coronal altitude, such as the appearance of a magnetic null ($\boldsymbol{B} = 0$) prior to its major X-class flares (Fontenla et al. 1995).

It is expected that the extent of magnetic non-potentiality should relax towards a lower energy state corresponding to the potential fields subsequent to a flare, after the excess or "free" energy available in the active region is released. There are some direct observations which indicate such a relaxation of the non-potential configuration after a large flare, for example, an event observed by Yohkoh in X-rays (Fig. 23), and evolution of Hα arcades observed during a large duration flare (Choudhary, Gary & Ambastha 1999). There are several reports in the literature about the detection of changes in the magnetic shear measured at the photosphere, associated with large flares. Contrary to the expectations, some reports indicate an increase in the photospheric shear (Wang et al. 1994), while, there are also reports of a decrease in shear followed by an increase (Ambastha,

Hagyard & West 1993). It is expected that magnetic shear should decrease after a major flare, therefore, the reports of shear increase are somewhat perplexing. However, it should be noted that the measurements are available only at the photospheric layer, and not at the "primary" site of the flare, located in the corona. Furthermore, it is important to be rather careful about interpreting the magnetic field measurements even at the photosphere as several observational effects could introduce errors in the magnetic field estimations. These include the error arising from the effect of the flare on the spectral line-profile itself, which is used for magnetic field measurement. The cross-talk between the much stronger circularly polarised light (used to derive the line-of-sight component of the field, B_L) into the weaker linearly polarised light (used for deriving the transverse field B_t) could also affect the measurement. Therefore, the results on shear changes derived from the photospheric measurements of the magnetic field remain rather ambiguous at present, until the instrumental accuracies are improved to the desired extent.

6.6 Solar Coronal Plasma Conditions and Magnetic Reconnection

In order to understand the requirements for the release of magnetic energy, let us consider the plasma conditions in the solar coronal plasma with a temperature $T = 3 \times 10^6$ K. In this hot plasma, we find that the electrical resistivity is extremely small, i.e.,

$$\eta = \sigma^{-1} = 1.5 \times 10^{-7} T^{-3/2} \approx 3 \times 10^{-17} \text{ e.s.u.} , \tag{6}$$

where, σ is the electrical conductivity. The magnetic Reynolds number for a magnetised plasma is defined as the ratio of two time scales, i.e., the time scale of Ohmic diffusion ($\tau_d = \frac{4\pi L^2}{c^2\eta}$), and the "advective time scale" ($\tau_{\text{adv}} = \frac{L}{v}$), i.e.,

$$R_m \equiv \frac{\tau_d}{\tau_{\text{adv}}} \equiv \frac{4\pi v L}{c^2 \eta} \sim \frac{\boldsymbol{\nabla} \times (\boldsymbol{v} \times \boldsymbol{B})}{\frac{c^2\eta}{4\pi} \boldsymbol{\nabla} \times (\boldsymbol{\nabla} \times \boldsymbol{B})} , \tag{7}$$

It may be noted that R_m can be expressed as the ratio of the induction term $\boldsymbol{\nabla} \times (\boldsymbol{v} \times \boldsymbol{B})$ over the Ohmic dissipation term $\frac{c^2\eta}{4\pi} \boldsymbol{\nabla} \times (\eta \boldsymbol{\nabla} \times \boldsymbol{B})$ of the induction equation (2) by replacing the vector quantities by their magnitudes, and the spatial derivative $[\boldsymbol{\nabla}\times]$ by $1/L$, where L is a characteristic length scale. For the solar plasma, R_m is very large. For example, at the chromosphere, let us assume $T \approx 10^4$ K, characteristic velocities v $\sim 10^6$ cm s^{-1}, and characteristic length scales of $L = 10^9$ cm. We find that the magnetic Reynolds number $R_m = 10^8 >> 1$. In the lower corona, T and v are both about two orders of magnitude larger, therefore, $R_m \approx 10^{13}$. This implies that the solar plasma has very "high conductivity", or that $\tau_d >> \tau_{\text{adv}}$. Therefore, the field-lines are essentially frozen-in the highly conducting solar plasma, i.e., the radiating plasma is a good tracer of the magnetic field lines in the solar atmosphere. The UV, EUV, and X-ray coronal images obtained by space-borne instruments faithfully display the magnetic structures associated with active regions and other solar features.

The large value of R_m makes it very difficult to release energy stored in the stressed magnetic fields, as such an energy release process proceeds over the diffusion time scale. Thus, the corresponding time scale for conversion of magnetic energy to heat is given by:

$$\tau_d \sim \frac{B}{\partial B/\partial t} \sim \frac{4\pi L^2}{c^2} = 5 \times 10^{14}\,\mathrm{s} = 10^7 \text{ years!!} \tag{8}$$

This is evidently a much longer time than the typical time of 10^3 s for observed flares, requiring that either the characteristic length-scale L is very small, i.e., $L \sim 1\,\mathrm{km}$, or η is small, in which case, we need to invoke turbulence in the coronal plasma. Such a situation could be met near neutral sheets, where the concept of classical electrical resistivity may not hold. Reconnection with rapid diffusion of magnetic fields could occur in a variety of ways in such situations (e.g., Sturrock et al. 1986).

6.7 A Flare Model as Inferred from Recent Observations

Masuda et al. (1995) found hard X-ray sources well above soft X-ray loops in several compact loop flares observed near the solar limb, in addition to double-foot point sources at the peak time of the impulsive phase. This was believed to be the evidence for magnetic reconnection at the loop-top coronal site, where the primary flare energy process took place. If the reconnection hypothesis is correct, a hot plasma or plasmoid ejection is expected to be associated with these flares. Using several limb flares observed by Yohkoh, it was found that the flares were associated with X-ray plasma ejections high above the soft X-ray loop, and the velocity of ejections was found in the range 50–400 km s^{-1}. The result supports the hypothesis of magnetic reconnection for the class of impulsive compact loop flares. Recent soft X-ray observations have helped in constructing temperature maps of coronal regions associated with flares, which show that the hottest temperature domains coincide with coronal HXR emission (Fig. 24a). An analysis of a coronal HXR flare was made using a precise timing of the HXR variability detected by the Compton Observatory. The time-of-flight localisation of the acceleration site is found to be consistent with the above-the-loop-top location of the HXR source (Fig. 24b). The geometry synthesised from the observations supports the flare model by Shibata et al. (1995). A reconnection site in the corona above the SXR source is supposed to drive a rapid flow, which impinges on the denser material in the magnetic loop, creates the HXR emissions, and the high temperature areas (Fig. 25).

6.8 Solar Quakes Produced by Large Flares

The analysis of data from the SOHO/Michelson Doppler Imager (MDI) showed that a large X-class flare of July 1996 produced a significant circular wave packet emanating from the flare site in NOAA active region 7978 (Kosovichev & Zharkova 1998). This is the first clear acoustic signature of a large X2.6/1B solar

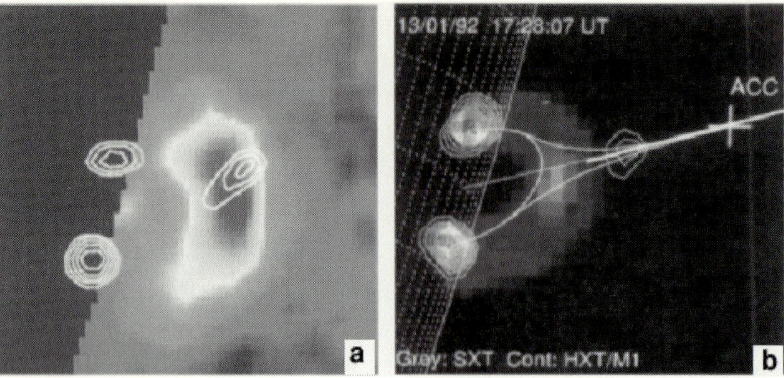

Fig. 24. (a) Temperature map generated from soft X-ray images. Superimposed coronal HXR contours coincide with the hottest temperature domain $\sim 20 \times 10^6$ K. (b) An analysis of the coronal HXR flare using the timing of HXR variability detected by the Compton Observatory. Time-of-flight localisation of the acceleration site, "+", is consistent with the above-the-loop location of the HXR source observed by Yohkoh HXT (adopted from Shimizu 1996)

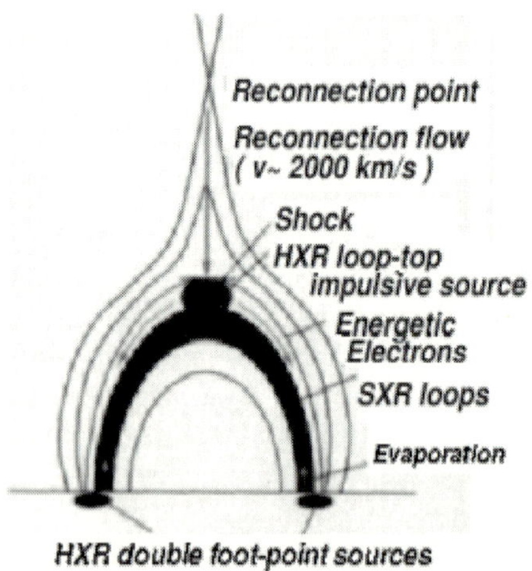

Fig. 25. The geometry synthesised from the observations supports a unified flare model proposed by Shibata et al. (1995). A reconnection site in the corona above the SXR source drives a rapid flow, which impinges on the denser material in the magnetic loop and creates both hard X-ray and high temperatures (adopted from Shimizu 1996)

flare, which was the only flare of moderate size to occur at the time of Sun's activity minimum in 1996. It was found that the flare-generated solar quake contained about 40 000 times the energy released in the great earthquake that devastated San Francisco in 1906. The quake produced what appears to be large ripples spreading over the Sun's surface, which are like surface waves on a pond produced by a stone. In an hour, the waves travelled a distance ten times that of the Earth's diameter. The waves accelerated from 10 km s^{-1} to 115 km s^{-1} as they travelled outward and disappeared into the photosphere (Fig. 26). For

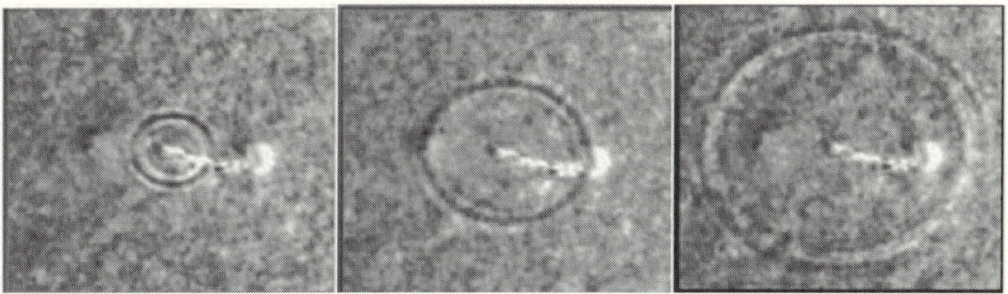

Fig. 26. Sequence of velocity images obtained by SOHO–MDI, showing the expanding solar quake associated with a large X-class flare

another X-class flare, rapidly propagating magnetic waves were also found by Kosovichev & Zharkova (1999).

6.9 Eruptive Prominences and Filaments

Prominences belong to the chromosphere according to their state of ionisation, and are best seen in chromospheric lines such as Hα or Ca II K. Several pinkish clouds extending to 50 000 km or more above solar limb can often be seen during a total solar eclipse. On the disk, they appear as dark, thin and rather long filaments (Fig. 27a), and bright cloud-like structures on the limb as seen against the sky background (Fig. 27b). Typical values for temperature, thickness, height, and length are 7000 K, 5000 km, 50 000 km, and 200 000 km, respectively. They are found mainly in two groups: *polar crown* filaments which are seen in high latitude poleward locations, and the other group of filaments observed in active mid-latitudes between active regions. The polar crown filaments appear a few years after a sunspot maximum, and then show a poleward migration, reaching the pole around the time of the succeeding maximum. Some filaments also develop within active regions, but these are more active, having shorter life times depending on the state of evolution within the active region. High speed plasma motions of about 5 km s^{-1} have been observed even in quiescent prominences. Since the prominences protrude far above the average chromosphere, they are essentially cooler chromospheric matter surrounded by 100 times hotter, coronal matter at 10^6 K. The lateral pressure equilibrium demands that the prominence density be 100 times the coronal density. The *quiescent* prominences remain stable for weeks or even months. The long life-time of quiescent prominences indicates a remarkable stability of their MHD equilibrium.

The quiescent filaments lie along magnetic polarity inversion lines, i.e., $B_L = 0$, separating large areas of weak fields of opposite polarity. Direct measurements of the prominence magnetic field using the Zeeman and Hanle effects indicate fields of 5–10 G in quiescent prominences. The Hanle effect is observed in lines such as Na I D_3 arising in the tenuous prominence plasma by resonance scattering of photospheric light. This scattering produces a linear polarisation of the scattered radiation, parallel to the limb. The Zeeman splitting of the atomic levels by the prominence magnetic field can cause a partial depolarisation of the

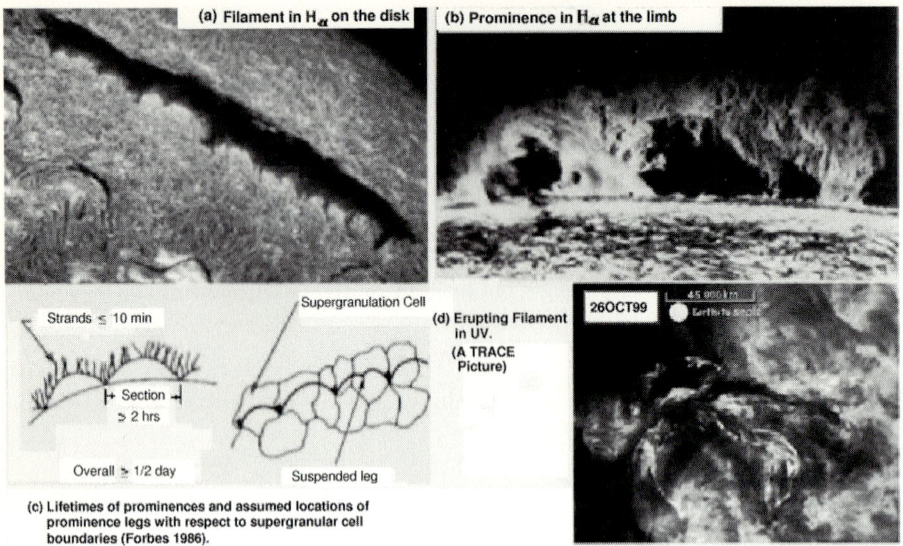

Fig. 27. (a) A large dense chromospheric filament seen on the disk. (b) A prominence as seen on the limb against the sky background. (c) Schematic diagram showing the connection of prominence legs with the chromosphere. (d) An erupting filament of October 26, 1999, observed by TRACE satellite. For scale, Earth's size is indicated in this frame

line radiation and also a rotation of its linear polarisation plane. The degree of residual measured polarisation left after this Hanle depolarisation effect can be related to the magnetic field intensity and direction using models of the line radiative transfer in the prominence plasma.

The total mass in a few large prominences is comparable to that in the whole coronal volume, and the prominence plasma seems to be draining downward at a rate sufficient to deplete the corona within a few hours. The rapid draining of prominence plasma suggests that a mass supply from the chromosphere below is required. These observational facts imply a substantial and continual material circulation between the corona and chromosphere, which may have an important role in the corona's mass balance. The issue of prominence stability, i.e., the dense prominence material's support against gravity was addressed originally by Kippenhahn & Schlüter (1957) (K–S), according to which a horizontal magnetic field is bent downward by the prominence mass (Fig. 28a,b). A Lorentz force acting to balance gravity is generated by a current running transverse to the field lines, along the filament axis. The magnetic field associated with this current alters the original magnetic field to the bent shape. The filament is supported by the tension in the bent field lines, whose foot-points are anchored into the photospheric plasma. A prediction of this model is that the magnetic vector should thread the filament in the direction joining the two magnetic polarities observed in the either side of the filament.

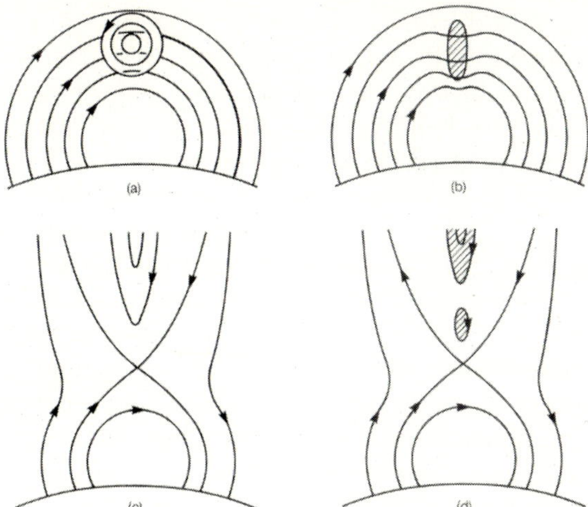

Fig. 28. Schematic diagrams illustrating the magnetic field-line geometries for filament support against gravity, suggested by (a,b) Kippenhahn and Schlüter (K–S model), and (c,d) Kuperus and Raadu (K–R model). In K–S model, an electric current flows perpendicularly to the magnetic field lines of an arch-like structure, the associated field lines of which add vectorially to the field of the arch to give a "sagging" loop geometry. In the K–R model, material condenses in a current sheet, with isolated knots forming which sink down until supported by the field below

Another model (Fig. 28c,d) was put forward by Kuperus & Raadu (1974) (K–R), which associated the filaments with material condensing within a current sheet. The support against gravity of the dense filament plasma is provided by the vertical gradient in magnetic field-line tension. As seen in this model, the closed field-lines thread the prominence both in the direction expected from straight connections between polarities observed to either side of the prominence, and also in the opposite direction. The high latitude quiescent prominences, such as polar crown filaments exhibit the magnetic field structure consistent with the K–R model. The lower prominences having heights below 3×10^4 km, which occur in active latitudes, are consistent with the K–S model. More detailed models proposed by Eric Priest and colleagues explain the formation and support of prominence by thermal instability at the tops of several loops in an arcade structure, into which material from the chromosphere is "siphoned" (Steele & Priest 1990; Demoulin & Priest 1990; see also the articles in Priest 1989). An objection to these models is that, although the quiescent prominences are observed to lie along the polarity reversal lines, the field-lines do not run perpendicular to the length of the prominence, but almost parallel to it. In fact the magnetic geometry could be in general helical as seen in many prominences, particularly well seen during their eruption. It is clear that the theoretical models do not match with the observations, and one major problem remains in the accurate measurements of magnetic fields in and around the prominences.

Prominences are observed to be connected to the chromosphere at the supergranulation cell boundaries by legs or foot-points (Fig. 27c). As this connection is gradually lost due to the evolution at lower layers, the magneto-static balance of the prominence is disturbed. A static balance between the magnetic pressure $B^2/8\pi$ and weight per unit area of an overlying prominence in quasi-static equilibrium may last for days or weeks. When the static balance is lost, it turns into explosive eruptive prominence (Fig. 27d). On the disk, the eruption of filaments in Hα appears as its sudden disappearance, and the phenomenon is known as *disparition brusque*. The eruption of filaments is accompanied by less spectacular events, for example, a brightening of the underlying chromosphere, and a slow increase and decrease of soft X-ray emission. However, many of the large scale coronal mass ejections that are detected by space-borne instruments appear to be outward moving remnants of quiescent prominences. In many cases, prominence eruptions appear to be spontaneous, as no other event, such as a flare, seem to have occurred in the neighbourhood. They perhaps occur just because of the undergoing slow evolution of the fields in their surroundings or due to a new flux emergence close by. In some events, a large flare releases a shock wave, called *Moreton wave*, which could disturb the stability of the filament and cause its eventual eruption.

Helicity is a commonly observed feature associated with eruptive prominences, both on microscopic as well as on macroscopic scales. In some cases, two or more tubes of matter are seen helically intertwined in a rope-like structure (Srivastava, Ambastha & Bhatnagar 1991). In addition, a chromospheric flare brightening and coronal ejection may result as a consequence. These are essentially large quiescent prominences that became unstable due to a destabilising evolution of magnetic fields in their neighbourhood, and erupt outward at speeds of the order of a few hundred km s^{-1} (Fig. 29). Eventually they disappear, although there are also cases when filaments reformed within a few hours to days. Figure 30 shows eruptive filaments observed on the solar disk, and on the limb as a large spray of material seen in Hα.

Multi-wavelength observations contribute significantly to the understanding of erupting filaments, as in the case of flares. For example, a spectacular eruptive prominence observation was made at three wavelengths: at 6563 Å Hα from the Norikura coronagraph, in SXR by Yohkoh, and at 17 GHz microwave by the Nobeyama Radio-Heliograph (Fig. 31). These data taken together strikingly confirm the large scale magnetic reconnection picture developed for coronal eruptions.

6.10 Coronal Mass Ejections (CMEs) – Large Scale Eruptions of Magnetic Clouds from the Sun

During a total solar eclipse, the structure of the outer rarefied solar atmosphere, the corona, is seen for a brief period of the totality (e.g., Billings 1966). Large scale streamers extending to several solar radii can be seen at these events (Fig. 32). The corona on the disk is revealed in broad-band soft X-ray, UV, and EUV pictures taken by space-borne instruments which show bright active

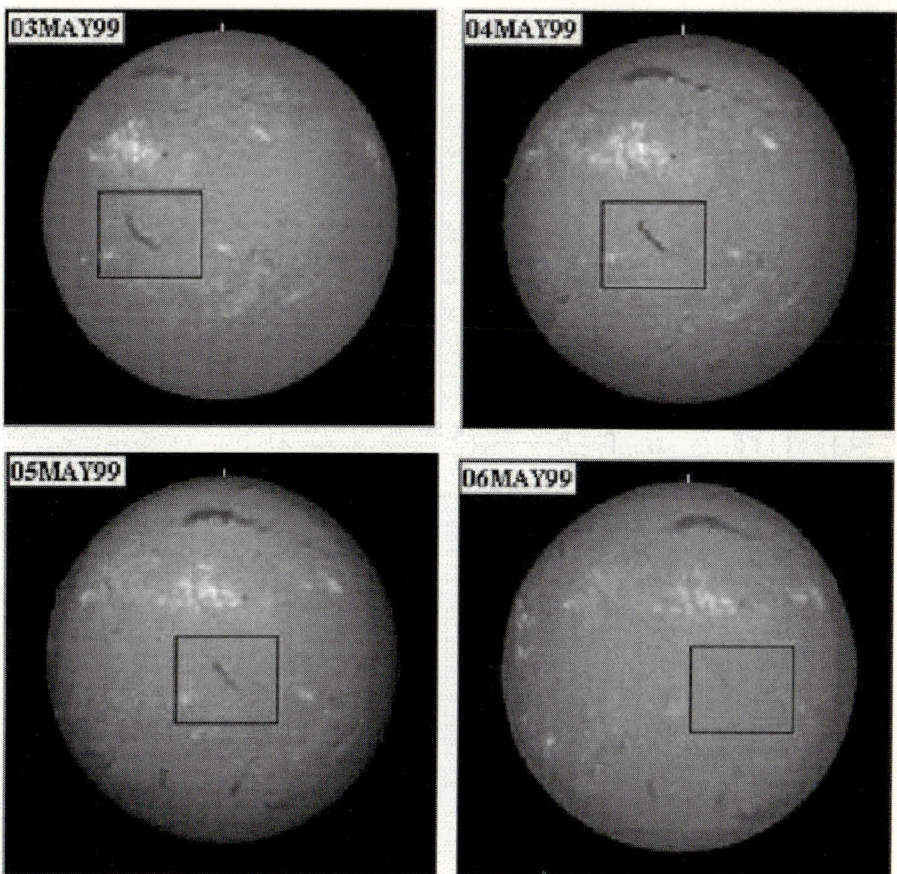

Fig. 29. Time-evolution of chromospheric filaments – A disparition brusque event, or erupting filament as seen in full disk Hα images tracked during the period May 3–6, 1999 (USO pictures)

regions, and loops (e.g., Phillips 1992). They also show the dark, elongated regions extending generally in the north-south direction, in which the emission is about 2–3 times lower than the surrounding regions (Fig. 33). These are called *coronal holes*, where the magnetic field configuration generally has open structures. Rapid motions and brightness changes in the white light corona have been noticed using space-borne coronagraphs. The space-borne instruments allow the observations of these eruptive transients up to much greater distances from the Sun than ground-based ones (Tousey 1973). Large-scale coronal mass ejections (CMEs), were first seen in 1973 by OSO-7 (Orbiting Solar Observatory-7), and later by Skylab mission (1973–74), and now by SOHO (Fig. 34). The CMEs are essentially large scale magnetic structures expelled from the Sun due to MHD processes involving interaction between plasma and magnetic field in closed magnetic field regions, such as, active regions, and filaments/prominences.

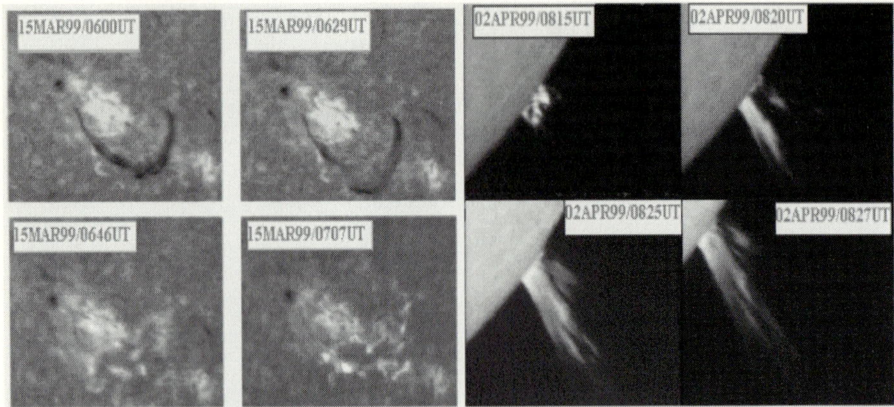

Fig. 30. On-disk, and limb view of eruptive prominences in Hα (USO pictures)

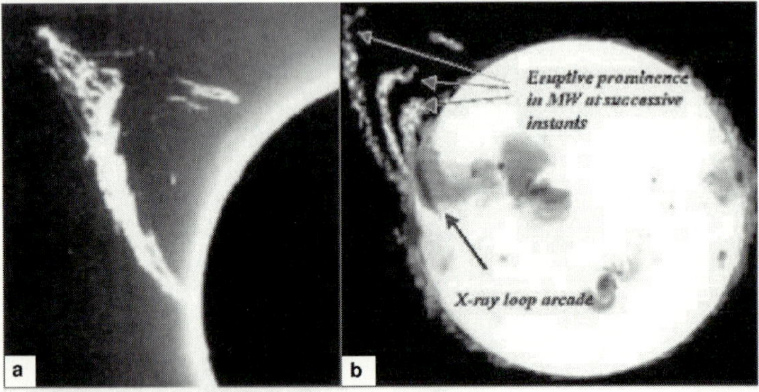

Fig. 31. A prominence eruption observed at multi-frequency. (a) Hα 6563 Å image obtained from Norikura Coronagraph, (b) A composite of Yohkoh SXR and 17 GHz microwave images obtained from Nobeyama Radio-Heliograph (adopted from Shimizu 1996)

Flux emergence near such pre-existing closed field structures has been considered as a possible trigger of CMEs (Feynman & Martin 1995; Wang & Sheeley 1999). CMEs are important solar phenomena, as they influence the physical conditions of the interplanetary medium. CMEs in the solar wind were detected as magnetic clouds or ejecta, from the data obtained by the *Interplanetary Magnetic Platform* (IMP), Helios and Voyager spacecrafts (Burlaga et al. 1981). Using the interplanetary scintillation (IPS) observations, CMEs have been detected over the entire Sun-Earth distance (Manoharan et al. 1995). Yohkoh observations have shown that CME events often belong to a class of soft X-ray events, termed as long-duration events or LDE's, which are often related to prominence eruptions. Srivastava et al. (1998) have studied the solar origins of intense geomagnetic storms and the role of CMEs.

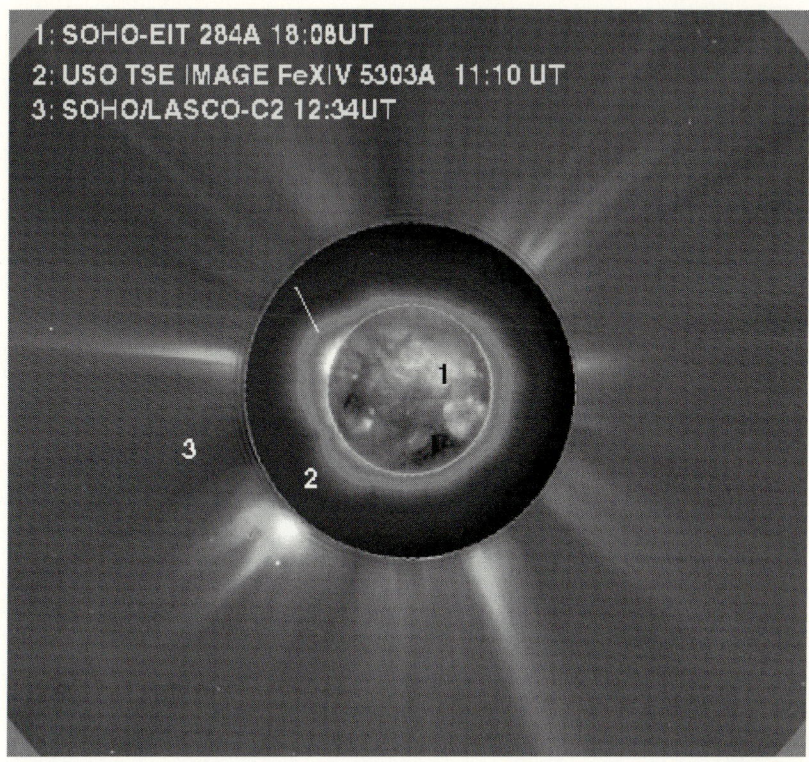

Fig. 32. A composite picture of the total solar eclipse of June 21, 2001 made using the inner coronal image taken in Fe XIV 5303 Å by the Udaipur Solar Observatory (USO) team from Lusaka (Zambia), SOHO–EIT 284 Å on-disk coronal image, and SOHO/LASCO-C2 white light large scale corona. The large scale LASCO image shows the streamers

It is still rather unclear as to how the CMEs are initiated, although the recent spacecraft observations have provided extensive information about the source regions, and the initial phase of CMEs. There are CMEs associated with flares, and also those without any perceptible chromospheric association. A large number of CMEs are associated with eruptive prominences, which occur throughout the activity cycle, and not just during solar maximum phase. Formation and eruption of prominences is a central issue of CME initiation, and there are reports that consider the eruption of prominences as the cause of CMEs (Wu et al. 2000). However, there is a complex relationship between the filament, coronal cavity, and the frontal structure before and during the eruption.

The speeds of ejection vary widely, with a maximum of > 1000 km s^{-1} for the most energetic events, but down to about 10 km s^{-1} for CMEs with no associated activity, generally seen near a solar activity minimum period. Some 10^{13} kg of mass are ejected so that the kinetic energy may be 5×10^{31} ergs. To this should be added the magnetic energy and the enthalpy, yielding a total that

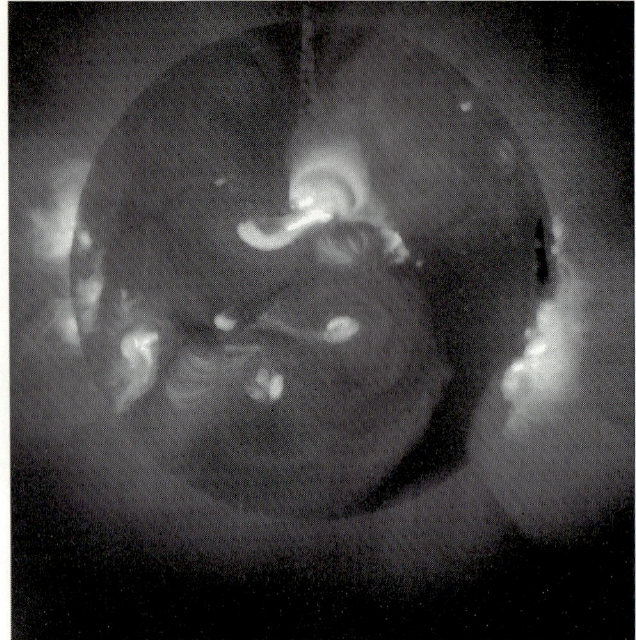

Fig. 33. A full field soft X-ray image taken by Yohkoh SXT telescope on December 14, 2001 showing the corona on-disk and around the limb. This picture was one of the last images taken by the Yohkoh spacecraft

may exceed 10^{32} ergs, which make them comparable to or more energetic than large flares.

The three dimensional form of CMEs is debatable, i.e., it is generally not clear whether the leading bright rims are bubbles or loops? Halo CMEs expanding symmetrically around the Sun provide the clue about the 3-D nature of CMEs (Howard et al. 1982). These are the class of CMEs, which are ejected earthwards, and are of particular importance to space weather related to geomagnetic storms (Gosling et al. 1991). Many halo CMEs are associated with eruption of Hα filaments occurring near the disk centre, therefore Hα patrols hold promise to forecast the onset of halo CMEs. Additional information from EUV, X-ray or Hα observations are required to distinguish between the halo CMEs moving towards the Earth, and away from Earth. The halo CMEs are now being routinely observed by the sensitive SOHO/LASCO (Large Angle and Spectrometric Coronagraph) coronagraph.

Both thermal and non-thermal signatures associated with CMEs can be detected at radio wavelengths. Thermal emission depends on the temperature, density and magnetic field of the region, and also on the frequency, while, the non-thermal emission depends on the density and energy of the non-thermal electrons. Therefore, radio techniques can be used to identify the mechanisms operating to produce the thermal or non-thermal component of radio emission.

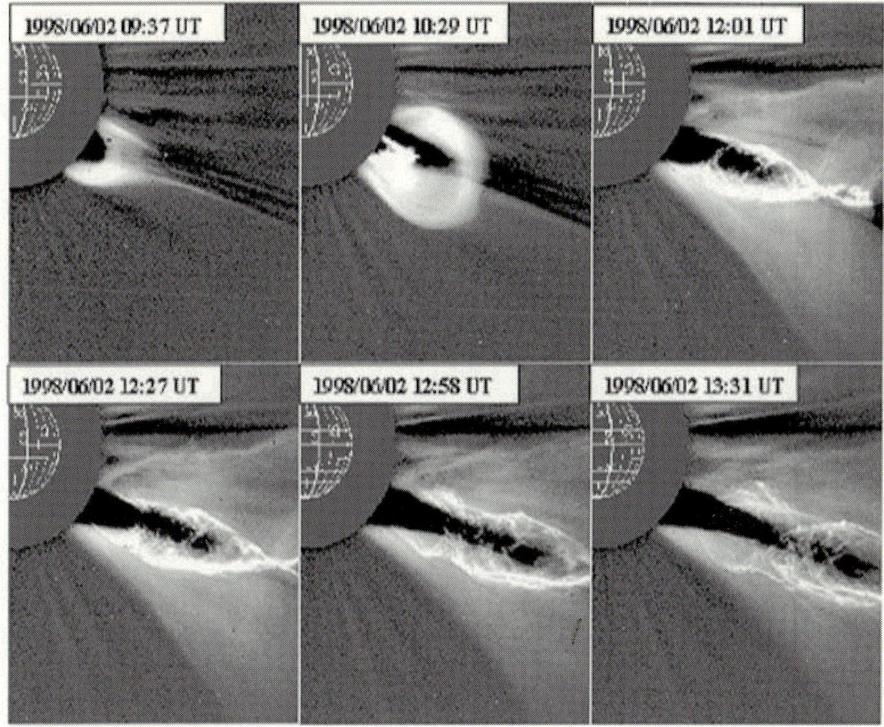

Fig. 34. A large CME observed by SOHO on June 2, 1998 originating on the SW solar limb (adopted from Srivastava & Schwenn 2000)

There has been much speculation in the modelling of CMEs. In one picture, CMEs are due to ropes of magnetic field that rise by magnetic buoyancy, the gas being less dense on the inside of the rope than outside. Another view is that CMEs result from an untwisting of the field lines. The possibility that flare-associated CMEs are propelled by a pressure pulse due to the flare is perhaps unlikely in view of the fact that the CMEs appear to precede the flare. A CME happens because of a large-scale departure from equilibrium when the magnetic and gravitational forces no longer balance the coronal gas expansion, and lifts off with constant speed, either slow or fast. Accordingly, the CME is not the *result* of a flare or erupting prominence, but all three may be due to the loss of equilibrium. It is clear that multi-wavelength, multi-instrument data are required to construct a complete picture of the CME phenomenon.

UV observations of coronal plasma phenomena has been provided with unprecedented resolution by the Transition Region and Coronal Explorer (TRACE) satellite. It was launched on April 2, 1998 for joint observations with SOHO (Solar and Heliospheric Observatory) during the maximum phase of the current solar cycle. It has a capability of a 1 arc-sec resolution, and takes observations in various wavelengths with a few second time resolution. TRACE observations have pointed towards a corona comprised of thin loops that are naturally dynamic and continuously evolving. These very thin loops are heated on a time span of minutes to tens of minutes, after which the heating stops or changes significantly. The heating appears to occur primarily in the lowest 10 000 to 20 000 km of the magnetic field lines in the coronal segments. There is strong evidence suggesting that the lower altitude heating is intermittent on time scales of a minute or less, suggesting that the loops are driven from somewhere near the loop footpoints. Yohkoh, SOHO, and TRACE have considerably advanced our knowledge about flares, eruptive prominences, and CMEs during the past decade, and are expected to continue providing further information for the understanding of these important phenomena.

References

Ambastha, A. 1998, Resonance, 3, 18
Ambastha, A., & Bhatnagar, A. 1988, J. Astrophys. Astron., 9,137
Ambastha, A., Hagyard, M. J., & West, E. A. 1993, Sol. Phys., 148, 277
Antia, H. M., & Basu, S. 2000, ApJ, 541, 442
Babcock, H. W. 1961, ApJ, 133, 572
Beck, R., Hilbrecht, H., Reinsch, K. & Volker, P. 1995, Solar Astronomy Handbook (Willmann-Bell Inc., Richmond)
Billings, D. E. 1966, A Guide to Solar Corona (Academic Press, New York)
Bracewell, R. N. 1988, Sol. Phys., 117, 261
Brandt, P. N., Scharmer, G. B., Ferguson, S., Shine, R. A., Tarbell, T. D., & Title, A. M. 1988, Nature, 335, 238
Bruzek, A. 1979, Sol. Phys., 61, 35
Bullard, E., & Gellman, H. 1954, Phil. Trans. R. Soc. London, A247, 213
Burlaga, L., Sittler, E., Mariani, F., & Schwenn, R. 1981, J. Geophys. Res., 86, 6673
Choudhary, D. P., Ambastha, A., & Ai, G. 1998, Sol. Phys., 179, 133
Choudhary, D. P., Gary, G. A., & Ambastha, A. 1999, in XIX NSO/SP Workshop on High Resolution Solar Physics, eds. T. R. Rimmele, K. S. Balasubramaniam, & R. R. Radick, Astron. Soc. Pacific Conf. series 183, 523
Cowling, T. G. 1934, MNRAS, 94, 39
Demoulin, P., & Priest, E. R. 1990 in Dynamics of Quiescent Prominences, IAU Coll. 117, eds. V. Ruzdjak & E. Tandberg-Hanssen (Springer-Verlag, New York), 269
Dulk, G. A., McLean, D. J., & Nelson, G. J. 1985, in Solar Radiophysics: Studies of emission from the Sun at metre wavelengths, eds. D. J. McLean & N. R. Labrum (Cambridge University Press, Cambridge), 53
Eddy, J. A. 1983, Sol. Phys., 89, 195
Evershed, J. 1909, MNRAS, 69, 454
Feynman, J., & Martin, S. F. 1995, J. Geophys. Res., 100, 3355
Fontenla, J. M., Ambastha, A., Kalman, B., & Csepura, Gy. 1995, ApJ, 440, 894
Gleissberg, W. 1952, Die Häufigkeit der Sonnenflecken (Akademie-Verlag, Berlin)
Gosling, J. T., McComas, D. J., Phillips, J. L., & Bame, S. J. 1991, J. Geophys. Res., 96, 7831
Gough, D. O., Leibacher, J. W., Scherrer, P. H., & Toomre, J. 1996, Science, 272, 1281
Hagyard, M. J., Smith, J. B., Jr., Teuber, D., & West, E. A. 1984, Sol. Phys., 91, 115
Harvey, J. W., Hill, F., Hubbard R. P., et al. 1996, Science, 272, 1284
Howard, R. 1984, ARA&A, 22, 131

Howard, R., & LaBonte, B. J. 1980, ApJ, 239, L33
Howard, R. A., Michels, D. J., Sheeley, N. R., Jr., & Kooman, M. J. 1982, ApJ, 263, L101
Howe, R., Christensen-Dalsgaard, J., Hill, F., Komm, R. W., Larsen, R. M., Schou, J., Thompson, M. J., & Toomre, J. 2000, ApJ, 533, L163
Ichimoto, K., & Kurokawa, H. 1984, Sol. Phys., 93, 105
Kane, S. R. 1974, in Coronal Disturbances, I. A. U. Symposium no. 57, ed. G. A. Newkirk (Reidel, Dordrecht, Boston), 105
Kippenhahn, R., & Schlüter, A. 1957, Z. Astrophys., 43, 36
Krause, F., & Radler, K. -H. 1980, Mean-field Magnetohydrodynamics and Dynamo Theory (Akademie-Verlag, Berlin)
Kosovichev, A. G., & Zharkova, V. V. 1998, Nature, 393, 317
Kosovichev, A. G., & Zharkova, V. V. 1999, Sol. Phys., 190, 459
Kuperus, M., & Raadu, M. A. 1974, A&A, 31, 189
Leighton, R. B., Noyes, R. W., & Simon, G. W. 1962, ApJ, 135, 474
Letfus, V. 1994, Sol. Phys., 149, 405
Lites, B. 2000, Rev. Geophys., 38, 1
Manoharan, P. K., Ananthakrishnan, S., Dryer, M., Detman, T. R., Leinbach, H., Kojima, M., Watanabe, T., & Kahn, J. 1995, Sol. Phys., 156, 377
Masuda, S., Kosugi, T., Hara, H., Sakao, T., Shibata, K., & Tsuneta, S. 1995, PASJ, 47, 677
Mathew, S. K., & Ambastha, A. 2000, Sol. Phys., 197, 75
McIntosh, P. S. 1990, Sol. Phys., 125, 251
Oda, N. 1984, Sol. Phys., 93, 243
Parker, E. N. 1955, ApJ, 122, 293
Phillips, K. J. H. 1992, Guide to the Sun (Cambridge University Press, Cambridge)
Priest, E. R. 1982, Solar Magnetohydrodynamics (D. Reidel, Dordrecht)
Priest, E. R. 1989, Dynamics and Structure of Quiescent Solar Prominences (Kluwer Academic Publishers, Dordrecht)
Roudier, Th., & Muller, R. 1987, Sol. Phys., 107, 11
Sakurai, T. 1979, PASJ, 31, 209
Shibata, K., Masuda, S., Shimojo, M., Hara, H., Yokoyama, T., Tsuneta, S., Kosugi, T., & Ogawara, Y. 1995, ApJ, 451, L83
Shimizu, T. (ed.) 1996, Yohkoh Views the Sun – The First Five Years, Institute of Space and Astronautical Science, National Astronomical Observatory, Japan.
Snodgrass, H. B. 1984, Sol. Phys., 94, 13
Srivastava, N., & Schwenn, R. 2000, in The Outer Heliosphere: Beyond the Planets, eds. K. Scherer, H. Fichtner, & E. Marsch (Copernicus Gesellschaft E. V., Katlenberg-Lindau), 367
Srivastava, N., Ambastha, A., & Bhatnagar, A. 1991, Sol. Phys., 133, 339
Srivastava, N., Gonzalez, W. D., Gonzalez, A. L. C., & Masuda, S. 1998, Sol. Phys., 183, 419
Steele, C. D., & Priest, E. R. 1990, in Dynamics of Quiescent Prominences, IAU Coll. 117, eds. V. Ruzdjak & E. Tandberg-Hanssen (Springer-Verlag, New York), 275
Steenbeck, M., & Krause, F. 1969, Astron. Nachr., 291, 49
Stix, M. 1976, in Basic Mechanisms of Solar Activity, IAU Symp. 71, eds. V. Bumba, & J. Kleczek (D. Reidel, Dordrecht), 367
Stix, M. 1991, The Sun: An Introduction (Springer-Verlag, Berlin)
Sturrock, P. A., Holzer, T. E., Mihalas, D. M., & Ulrich, R. K. (eds.) 1986, Physics of the Sun, Vol.II, (D. Reidel, Dordrecht).

Sütterlin, P. 2001, A&A, 374, L21
Title, A. M., Tarbell, T. D., Simon, G. W., & the SOUP Team 1986, Adv. Space Res., 6, 253
Tousey, R. 1973, Space Res., 13, 713
Vorpahl, J. A. 1973, Sol. Phys., 28, 115
Wang, H., Ewell, M. W., Jr., Zirin, H., & Ai, G. 1994, ApJ, 424, 436
Wang, J., & Shi, Z. 1993, Sol. Phys., 143, 119
Wang, Y. -M., & Sheeley, N. R., Jr. 1999, ApJ, 510, L157
Wilson, R. M. 1992, Sol. Phys., 140, 181
Worden, S. P., & Simon, G. W. 1976, Sol. Phys., 46, 73
Wu, S. T., Guo, W. P., Plunkett, S. P., Schmieder, B., & Simnett, G. 2000, J. Atmosph. Solar-Terrestrial Phys., 62, 1489
Yoshimura, H. 1975, ApJ, 201, 740

Magnetic Flux Tubes and Activity on the Sun

S.S. Hasan

Indian Institute of Astrophysics, Koramangala, Bangalore-560034, India
email: hasan@iiap.ernet.in

Abstract. Activity on the Sun is associated with magnetic fields, involving a complex interaction between the field and plasma. In this review I focus on three fundamental aspects of magnetic activity: (a) the generation, storage and emergence of magnetic fields from the solar interior; (b) the nature of the surface magnetic fields, especially in the form of small-scale flux tubes; and (c) dynamical processes in flux tubes and heating of the magnetic chromosphere.

1 Introduction

Activity on the Sun and on other stars is closely related to the magnetic field, involving a complex interaction between the field and plasma. Magnetic activity occurs at all levels in the solar atmosphere. In the photosphere, it gives rise to sunspots, whose distribution with time exhibits a 11-year cycle for reasons which are still not fully understood. Bright areas called *faculae* near spots and the overlying *plages* associated with enhanced chromospheric heating are another manifestation of magnetic activity. In the corona, prominences or filaments which are cool and dense structures, can become active and erupt. Another example of magnetic activity are solar flares, which represent a sudden (on a time scale of about 1 hour) release of large amounts of energy (10^{29}–10^{32} erg) in the form of radiation and fast particles. Coronal mass ejections (CMEs), which represent the ejection of coronal matter into space, are yet another manifestation of activity.

As already stated, the magnetic field is responsible for the above forms of activity. However, the above-mentioned phenomena are not entirely distinct, but in fact represent a response of the solar plasma to changes in the underlying magnetic field topology and dynamics. Active regions on the Sun are believed to form due to the emergence of bipolar magnetic flux from the convection zone. This field is responsible for the rich diversity of processes in the solar atmosphere. Magnetic flux upon emerging in the photosphere, gets concentrated into discrete structures which range from sunspots at the largest scale down to tiny *faculae*. Any progress in understanding the different forms of activity must necessarily focus on the physics of magnetic flux tubes at different locations, ranging from the deep convection zone to the upper atmosphere.

With the above paradigm in mind, let us first consider certain fundamental questions, which need to be addressed to understand the basic nature of magnetic activity:

- How is the magnetic field generated, maintained and dispersed?
- What are its properties such as structure, strength, geometry?
- What are the dynamical processes associated with magnetic fields?
- What role do magnetic fields play in energy transport?

In this review, we shall examine some aspects of the above questions. Section 2 is devoted to a discussion of the origin of the magnetic field in the deep convection zone and its subsequent transport to the photosphere. In Sect. 3, we examine the nature of photospheric magnetic structures as well as mechanisms that can intensify the field to the observed strengths (typically in the kilogauss range). Section 4 deals with dynamical processes associated with magnetic flux tubes and their role in heating the chromosphere and corona. Finally in Sect. 5 we present a summary and the broad conclusions of our review.

2 Magnetic Fields in the Solar Interior and Flux Emergence

It is now generally believed that the solar magnetic field is generated through dynamo action at the base of the convection zone by the interplay of differential rotation, helical convective motions and meridional circulation.

2.1 Solar Dynamo

The idea that a dynamo is responsible for generating the solar magnetic field was proposed long ago by Larmor (1919) and further developed by Cowling (1933), Parker (1955), Steenbeck, Krause & Rädler (1966) and others. The essential point is that a magnetic field is maintained by currents induced in a plasma by its motion across field lines. This motion with velocity $\boldsymbol{v}$ across a field $\boldsymbol{B}$ leads to an induced electric field $\boldsymbol{v} \times \boldsymbol{B}/c$ which generates an electric current by Ohm's law $\boldsymbol{j} = \sigma(\boldsymbol{E} + \boldsymbol{v} \times \boldsymbol{B}/c)$, where $\boldsymbol{E}$ is the electric field, c the light speed and σ is the electrical conductivity. From Ampère's law, $\boldsymbol{\nabla} \times \boldsymbol{B} = 4\pi \boldsymbol{j}/c$ this current produces a magnetic field and hence an electric field from the induction equation $c\boldsymbol{\nabla} \times \boldsymbol{E} = -\partial \boldsymbol{B}/\partial t$. The combined effect of the current and magnetic field creates a Lorentz force $\boldsymbol{j} \times \boldsymbol{B}/c$ on the plasma which opposes the force that drives the motions. The above equations can be combined to yield the following equation in terms of $\boldsymbol{B}$, (named induction equation as well or conservation law of the magnetic field) which is the starting point for the dynamo theory (e.g., Priest 1982):

$$\frac{\partial \boldsymbol{B}}{\partial t} = \boldsymbol{\nabla} \times (\boldsymbol{v} \times \boldsymbol{B}) + \eta \nabla^2 \boldsymbol{B} \,, \quad (1)$$

where $\eta = c^2/4\pi\sigma$ is the magnetic diffusivity. In addition, we need to solve the MHD equations to self-consistently take into account the back-reaction of the flow on the field. Often this is neglected in the interest of mathematical tractability and (1) alone is solved – such models are referred to as kinematic dynamos.

Qualitatively, the dynamo involves the generation of a toroidal component of the magnetic field (in the azimuthal direction) from a poloidal (meridional) magnetic field due to differential rotation (ω effect). The next step is the regeneration of the poloidal field from the toroidal field. A workable model for this was proposed by Parker (1955) by incorporating the effects of helical motions arising from Coriolis forces. This is the so called α-effect, which can be estimated more formally using mean-field magnetohydrodynamics (Steenbeck, Krause & Rädler 1966, see also Stix 1989). Such models were able, by a suitable adjustment of parameters, to reproduce many aspects of the solar cycle, including the well known "butterfly diagram", that depicts the sunspot distribution with time over a solar cycle (Ambastha, this volume). However, in order to accomplish this, the angular velocity of the Sun needs to increase with depth (i.e., $d\omega/dr < 0$) by up to 40%, which is inconsistent with the results of helioseismology (Antia, this volume). There are other problems associated with the need to store toroidal magnetic flux for a significant fraction of a sunspot cycle (to overcome the difficulty of magnetic buoyancy). These and other questions are still a matter of investigation. More details on the dynamo mechanism can be found in the review by Venkatakrishnan (this volume).

2.2 Seat of the Dynamo

Observations show that the magnetic field in active regions (AR for short) is generally bipolar with the field essentially directed parallel to the equator with opposite orientations in both the hemispheres (Hale's polarity law). This suggests that the dominant component of the Sun's magnetic field is toroidal. The emerging bipolar loops in AR indicate that these toroidal fields are formed in the sub-surface layers. Based on estimates of the total azimuthal magnetic flux appearing during an 11-year period, Parker (1979) derived a lower limit for the toroidal field strength of 10^2 G located at a depth of at least 10^5 km in order to remain submerged for several years. More refined estimates (see the review by Schmitt 1993 and references therein) reveal that due to the combined effects of magnetic buoyancy, convective instabilities and fragmentation mechanisms, magnetic flux stored at the convection zone base would escape to the surface on a time scale typically of a month. This is clearly too short a time for the dynamo to be effective.

Another argument against the dynamo being located within the convection zone is based on the results from helioseismology which have shown that the solar rotation in the convection zone does not have a significant radial gradient, other than in a shear layer near the base of the convection zone called the *tachocline* (Spiegel & Zahn 1992). This gradient in angular velocity is required for the efficient operation of the dynamo.

The above difficulties can be circumvented if the location of the (toroidal) magnetic field and the seat of the dynamo is just below the base of the convection zone in a thin sub-adiabatic layer also referred to as "the convective overshoot layer" (e.g., Spiegel & Weiss 1980). In this layer, with a thickness of around 10^4 km (van Ballegooijen 1982), fields in the range 10^4–10^5 G can be stored over sufficiently long periods for the dynamo action to occur before their escape due to buoyancy. That the field is probably located in the overshoot layer is also suggested by the fact that if the field were stored below in the radiative zone, which is stably stratified, it is unlikely that it would ever be able to break out. Yet there are magnetically active stars (dMe stars) where the convection zone reaches all the way to the star's centre, which would not have room for such an overshoot layer.

2.3 Flux Emergence

Once the field has been amplified by the dynamo, it needs to be released into the convection zone by some mechanism, where it can be transported to the surface by magnetic buoyancy (Parker 1955).

In order to understand magnetic buoyancy, let us consider an isolated horizontal flux tube in pressure equilibrium with its non-magnetic surroundings, so that

$$p_i + \frac{B^2}{8\pi} = p_e \ , \tag{2}$$

where p_i and p_e are the internal and external gas pressures respectively and B denotes the uniform field strength in the flux tube. If the internal and external temperatures are equal so that $T_i = T_e$ (thermal equilibrium), then since $p_i < p_e$, the gas in the tube is less dense than its surrounding ($\rho_i < \rho_e$), implying that the tube will rise under the influence of gravity.

The "release" of magnetic flux can occur through an instability in the upper part of the layer where the field is stored, and where $\mathrm{d}B/\mathrm{d}z < 0$ (Parker 1955; Gilman 1970). Here z (outward vertically directed) is the height. The instability bends the tubes allowing material to flow down from the apex along the sides (Fig. 1), thereby increasing the effect of buoyancy owing to the depletion of material at the top. This instability is in fact similar to the hydromagnetic Rayleigh–Taylor instability in a plasma and has been generalised to include rotation (e.g., Spruit & van Ballegooijen 1982; van Ballegooijen 1983; Ferriz-Mas & Schüssler 1993, 1995). Fields thus get transported to the photosphere where they emerge as bipolar magnetic loops which form active regions (see Fig. 1).

The rise of a magnetic flux tube from the base of the convection zone to the surface has been extensively studied using numerical simulations (e.g., Moreno-Insertis 1986; Choudhuri & Gilman 1987; Fan, Fisher & DeLuca 1993; Schüssler et al. 1994; Caligari, Moreno-Insertis & Schüssler 1995; see the review by Fisher et al. 2000 and references therein). The rise of a flux tube is influenced predominantly by the interplay between the following forces: buoyancy, aerodynamic drag and the Coriolis force. If the field strength is below the local dynamic equipartition value B_{eq}, (which can be calculated by equating the dynamic pressure of the flow $\frac{1}{2}\rho v^2$, where ρ is the density and v the flow speed, with the magnetic pressure $B_{\mathrm{eq}}^2/8\pi$), the inertial forces associated with rotation would force the tube to rise keeping a constant distance from the rotation axis instead of rising roughly in the radial direction. Other arguments suggest that for rising

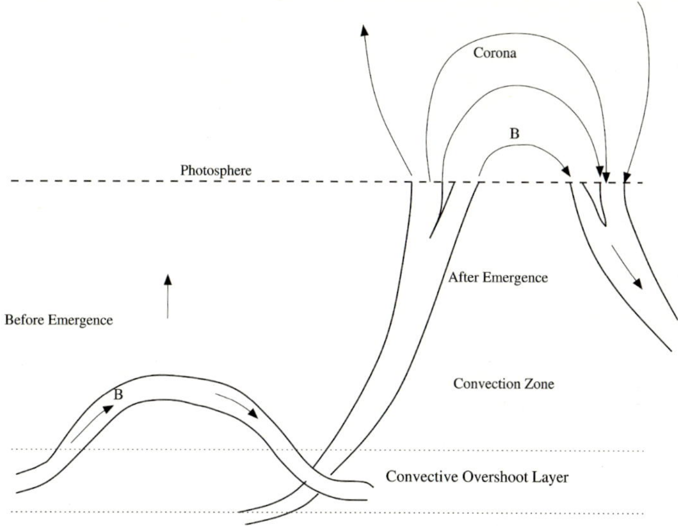

Fig. 1. Schematic diagram of the likely sub-photospheric magnetic structure of an emerging active region (after Fisher et al. 2000)

tubes with field strengths $B \leq B_{\rm eq}$ ($B_{\rm eq} \sim 10^4$ G at the convection zone base) the local gas pressure inside the tube rises to the ambient value during the ascent and the tube will "explode". These studies conclude that fields greater than $\sim 10^5$ G are necessarily required to overcome the above difficulties.

The Coriolis force, along with the requirement that the magnetic field at the convection zone base be greater than 10^5 G has been able to account for the fact that flux tubes emerge within the activity belts. Furthermore, it was shown (e.g., D'Silva & Choudhuri 1993; Fan, Fisher & McClymont 1994; Schüssler et al. 1994; Caligari, Moreno-Insertis & Schüssler 1995) that such tubes can account for Joy's law, according to which the average orientation of bipolar active regions is tilted slightly away from the azimuthal direction – the tilt angle is roughly proportional to latitude.

A heuristic explanation of Joy's law has been offered by Fan, Fisher & McClymont (1994) who consider a cartoon model, shown in Fig. 2, which depicts the different forces acting on a rising flux tube. The left panel in Fig. 2 shows a flux tube moving upwards under the combined influence of the buoyancy force $\boldsymbol{F}_{\rm B}$ (directed upwards) and the aerodynamic drag $\boldsymbol{F}_{\rm D}$ (directed downwards). Due to the Coriolis force, the tube is twisted into a backward 'S' shape (in the northern hemisphere) when viewed from above (right panel in Fig. 2). By balancing the Coriolis force with the magnetic tension, they estimated that the tilt angle α is given by $\alpha \approx \Phi^{1/4} \sin\theta$ where Φ is the magnetic flux in the tube and θ is the latitude (see Fig. 2). This estimate appears to be fairly robust as was shown by Fisher, Fan & Howard (1995) by testing it against the tilt angles of a large sample of spot groups.

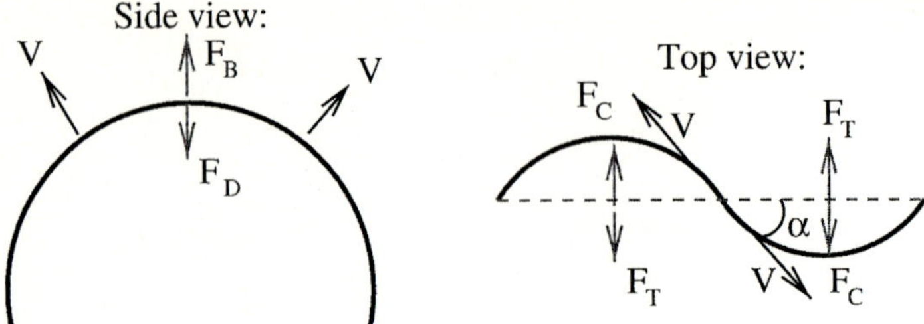

Fig. 2. Illustration of the how the Coriolis force influences a rising flux tube emerging in the convection zone. Side view: The upward motion depends on a balance of the forces due to magnetic buoyancy F_B and the aerodynamic drag F_D. Top view: The Coriolis force twists the rising tube loop into a 'S' shape. The magnetic tension force tends to straighten the tube. A balance between the two leads to the prediction that $\alpha \sim \Phi^{1/4} \sin\theta$ (from Fisher et al. 2000)

Most of the simulations for the transport of magnetic flux tubes have relied on the so-called "thin flux tube approximation" (Roberts & Webb 1978) which idealises a tube as an infinitely thin structure surrounded by a field-free medium. By neglecting variations along the radial direction, the motion of the tube moving as a single entity through the ambient medium, with which it is constantly in pressure balance, can be essentially reduced to a 1-D problem (Spruit 1981). However, since the pressure decreases outwards in the convection zone, the internal pressure and field strength must decrease in order to maintain pressure balance (see (2)). The decrease in field strength $\boldsymbol{B}$ implies that the radius must increase (to conserve magnetic flux $\Phi_{\mathrm{M}} \sim Br^2 = \mathrm{const.}$) so that as the tube rises, the thin flux tube approximation will eventually break down. This typically occurs when the top of the loops are about 10 000 km below the surface.

Recently, calculations have been carried out by taking into account the finite thickness of a tube (Emonet & Moreno-Insertis 1998; Fan, Zweibel & Lantz 1998; see also the review by Fisher et al. 2000 and references therein). The main findings of these simulations is that a twist in the magnetic field counteracts the tendency of tubes to fragment as they approach the top of the convection zone. The next step would be to have full 3-D simulations which model a tube as it ascends from the convective overshoot layer.

3 Nature of the Surface Magnetic Field

In the previous section we showed that the origin of the surface magnetic field lies in a dynamo located at the base of the convection zone. Let us now turn our attention to the field in the photosphere and above. It is well known that the solar magnetic field at the surface primarily consists of discrete elements, varying in size from 50 000 km for the largest sunspots down to 100–300 km for intense flux tubes, with field strengths in the range 1500–2000 G (e.g., Stenflo 1994; see the review by Solanki 1993, and references therein). More extended structures such as *plages* and the chromospheric network consist of assemblies of intense flux tubes. *Plages* are the chromospheric manifestations of *faculae*, photospheric field elements that appear as bright structures. The network essentially consists of a pattern that defines the boundaries of large convective cells called supergranules. Magnetic flux tubes typically fill about 30% of the total surface area in a plage. In this study we focus on the small-scale magnetic field and exclude sunspots, mainly because limitations of space prevent us from doing adequate justice to this topic. The reader is referred to several excellent monographs on this subject (e.g., Thomas & Weiss 1992; Bogdan 2000).

Magnetic flux is injected into the photosphere from the interior in the form of bipolar concentrations. They emerge as coherent flux tubes in the photosphere on a wide range of scales. The source function $n(A)$ of the frequency of emerging bipolar regions as a function of area A is a smooth monotonically decreasing function of A (e.g., Zwaan & Harvey 1994). How does the distribution of flux-tube size continue to flux tubes with smaller diameters below the resolution limit? Although it is not possible to determine the number of individual flux tubes in a plage, one can attempt a qualitative estimate. Let us compare the area in active region plages (including the region between flux tubes) with that occupied by sunspots. The ratio of the plage to sunspot area changes from 12 at cycle maximum to 25 at minimum (Chapman, Cookson & Dobias 1997), which implies that the surface area of plages is much larger than than of sunspots. Taking into account the magnetic elements in the network of the quiet Sun, increases the ratio even further. Since 10–20% of the surface area is typically covered by fields in a plage, the amount of magnetic flux carried by plages and sunspots is comparable at maximum (Solanki 1999).

Upon emergence as large coherent entities (the largest one being sunspots), flux tubes fragment into the enhanced network. Active regions survive in this *plage* state for several days depending upon their size (Harvey 1993; Schrijver & Harvey 1994). Harvey finds a typical time given by $t_{\rm AR} \approx 15(\Phi/10^{21}\,{\rm Mx})$ days, where Φ is the total magnetic flux. Once active regions begin their decay, the flux escapes from them into the quiet photosphere, where it is passively moved by the supergranular flows to the boundaries and into the network, where it cancels and replaces the old flux. The emerging flux is at about the dynamic equipartition field strength $B_{\rm eq}$. For a typical flow speed $v \approx 2$ km s^{-1}, the dynamic equipartition field $B_{\rm eq} \approx 400\,$G. Consequently, some local process must be present which not only resists the tendency of the field elements to fragment, but also intensifies the magnetic field to the observed values in the kG range.

3.1 Formation of Intense Flux Tubes in the Photosphere

It is is generally believed that the formation of intense flux tubes occurs through a combination of two mechanisms: namely *flux expulsion* (Parker 1963) and *convective collapse* (Parker 1978).

Flux Expulsion

Flux expulsion is the interaction of convection with a vertical magnetic field, leading to the expulsion of the field from the interior of a convection cell to its boundaries, which can be quantitatively studied using the induction equation given by (1). Since the electrical conductivity of the solar plasma is very high, one should in principal be able to generate very high field strengths. This process has been extensively studied numerically (e.g., Weiss 1966, 1981; Galloway, Proctor & Weiss 1978; Nordlund 1983; Stein & Nordlund 2000). These calculations have shown that the convective flows concentrate the magnetic field in filament channels located in regions between the edges of the convective cells, which are also the sites of cool downflowing material.

However, when one includes the back-reaction of the field on the flow, it appears unlikely that the maximum field strengths that can be achieved by this process can exceed the dynamical equipartition value of several hundred Gauss. This suggests that some other mechanism is needed to intensify the field further to strengths observed in the magnetic network. A possible way of achieving this is through *convective collapse.*

Convective Collapse

Convective collapse is an instability driven by the superadiabatic temperature gradient just below the photosphere, which was first suggested by Parker (1978) as a mechanism responsible for the formation of intense flux tubes on the Sun. In order to understand this process, let us consider a "thin" vertical flux tube of cylindrical cross-section on the solar surface extending through the photosphere and into the convection zone of the Sun. By "thin" we mean that all physical quantities in the tube at a particular height (z) are essentially constant in the radial direction. Let us idealise the tube as a tapered cylinder (see Fig. 3) whose cross-section increases with height (in view of the decrease of confining external pressure) and focus on a gas element in the tube at a height z with temperature T_i, density ρ_i and pressure p_i. Initially we assume that the tube is in hydrostatic equilibrium, so that the vertical force F_z acting on this element is:

$$F_z = -g\rho_i - \frac{\mathrm{d}p_i}{\mathrm{d}z} = 0 \,, \tag{3}$$

where g is the acceleration due to gravity (acting in the downward direction). Equation (3) expresses a balance between the buoyancy and pressure forces. We now displace this gas element downwards by a distance $\Delta z < 0$. At the new location z', let the temperature, density and pressure be denoted by T_i',ρ_i' and p_i' respectively. The force acting on the fluid element at the displaced position is given by:

$$F_z' = -g\rho_i' - \frac{\mathrm{d}p_i'}{\mathrm{d}z} \,. \tag{4}$$

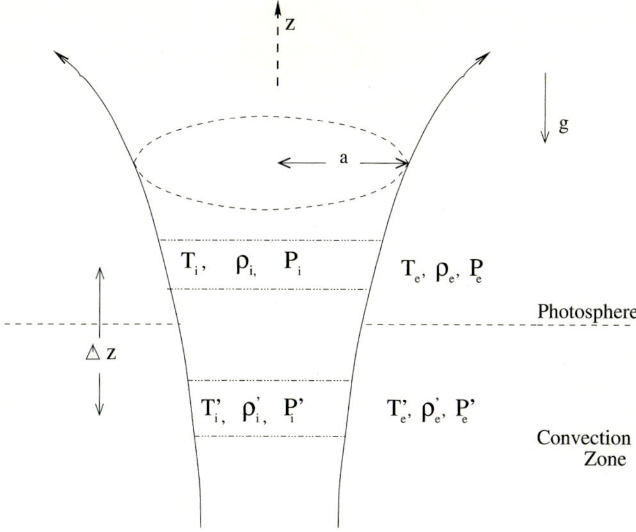

Fig. 3. Physical explanation for the occurrence of a convective instability in a vertical magnetic flux tube extending vertically through the photosphere and convection zone. The atmosphere inside and outside the tube is initially in hydrostatic equilibrium and the internal and external temperatures are equal at each geometric level. Consider a small downwards adiabatic displacement Δz of a fluid element. Assuming that this element remains in pressure equilibrium with the ambient medium, it will experience a negative buoyancy if the temperature gradient inside the tube is greater than the adiabatic value. For more details see the discussion in the text

Writing $F_z' = F_z + \Delta F_z$ along with similar expressions for ρ_i' and p_i', where ΔF_z, the Lagrangian perturbation in the force F_z on the fluid element, is given by:

$$\Delta F_z = -g\Delta\rho_i - \frac{d\Delta p_i}{dz} \, . \tag{5}$$

For an adiabatic displacement,

$$\frac{\Delta\rho_i}{\rho_i} = \frac{1}{\gamma\chi_\rho}\frac{\Delta p_i}{p_i} \, , \tag{6}$$

where γ is the ratio of specific heats and χ_ρ is defined by (12). Let us assume that at all instants the gas inside the tube is in pressure balance with its surroundings and that $T_i = T_e$, where T_e is the external temperature at the same height. Using (2), we find:

$$\Delta p_i = \frac{\beta}{\beta+1}\Delta p_e = \frac{\beta}{\beta+1}\frac{\mathrm{d}p_e}{\mathrm{d}z}\Delta z \, , \tag{7}$$

where $\beta = 8\pi p_i/B^2$. Substituting (6) and (7) in (5), we find:

$$\begin{aligned} \Delta F_z &= -\frac{\beta}{\beta+1}\rho_e g\left(\frac{1}{\gamma\chi_\rho}\frac{\mathrm{d}\ln p_e}{\mathrm{d}z} + \frac{1}{\rho_e g}\frac{\mathrm{d}^2 p_e}{\mathrm{d}z^2}\right)\Delta z \, , \\ &= -\frac{\beta}{\beta+1}\rho_e\omega_{\mathrm{BV}}^2\Delta z \, , \end{aligned} \tag{8}$$

assuming that the external gas is in hydrostatic equilibrium at all heights. In (8), $\omega_{\rm BV}^2$ denotes the Brunt-Väisälä frequency given by:

$$\omega_{\rm BV}^2 = -g\left[\frac{{\rm d}\ln\rho}{{\rm d}z} - \frac{1}{\gamma\chi_\rho}\frac{{\rm d}\ln p}{{\rm d}z}\right], \tag{9}$$

$$= -g\frac{\chi_T}{\chi_\rho}\left(\left|\frac{{\rm d}\ln T}{{\rm d}z}\right| - \left|\frac{{\rm d}\ln T}{{\rm d}z}\right|_{\rm ad}\right), \tag{10}$$

$$= -\gamma\frac{g^2}{c_S^2}\,\chi_T\left(\nabla - \nabla_{\rm ad}\right), \tag{11}$$

where $\nabla \equiv {\rm d}\ln T/{\rm d}\ln p$, $c_S = \sqrt{\gamma\chi_\rho\, p/\rho}$ is the sound speed and for convenience we have dropped the subscript 'e' on all quantities appearing in the right hand side of the expression for $\omega_{\rm BV}^2$. The quantities χ_ρ and χ_T (which incorporate the effects of ionisation) are defined by:

$$\chi_\rho = \left(\frac{\partial\ln p}{\partial\ln\rho}\right)_T \quad \text{and} \quad \chi_T = \left(\frac{\partial\ln p}{\partial\ln T}\right)_\rho . \tag{12}$$

The second term inside the brackets in (10) denotes the temperature gradient if the stratification were adiabatic. Thus, $\omega_{\rm BV}^2$ is a measure of the superadiabaticity of the fluid. If $\omega_{\rm BV}^2 < 0$, we find from (8), that the displaced fluid element experiences a downward force, which implies that the configuration is unstable. Thus, the condition for instability is:

$$\left|\frac{d\ln T}{dz}\right| > \left|\frac{d\ln T}{dz}\right|_{\rm ad}, \tag{13}$$

or alternatively if $\nabla > \nabla_{\rm ad}$ which is the well-known Schwarzschild criterion for convective instability. We see that as the field strength increases (i.e., as β decreases) the force on the displaced element decreases, which implies that the magnetic field has an inhibiting effect on the instability.

Physically, the above instability occurs because a gas element (for a superadiabatic stratification) that is displaced downwards finds itself in an environment where it is cooler and denser than its immediate surroundings and therefore experiences a negative buoyancy. The instability has the effect of evacuating the upper portion of the tube (just below the photosphere which is superadiabatic), leading to a local reduction in pressure. Consequently, the tube collapses to a configuration with a higher field strength to maintain pressure balance with the ambient medium. The enhanced field resists the tendency of the tube to collapse. A careful analysis based on a linear stability calculation, reveals that *convective collapse* occurs if $\beta > \beta_c$, where β_c denotes the value for marginal stability (Webb & Roberts 1978; Spruit & Zweibel 1979; Unno & Ando 1979). For the solar stratification, Spruit & Zweibel (1979) found $\beta_c = 1.83$, which corresponds to a surface value of the magnetic field strength of about 1300 G, whereas Rajaguru & Hasan (2000) obtained $\beta_c = 1.64$ using a slightly different solar model.

Detailed nonlinear calculations have been carried out for *thin* flux tubes (Hasan 1983, 1984, 1985; Venkatakrishnan 1983, 1985; Takeuchi 1993, 1995) and for a 2-D flux sheet (Grossmann-Doerth, Schüssler & Steiner 1998) in order to follow the time dependent evolution of the instability. For closed boundary conditions at both ends of the tube (consistent with the linear results with Spruit & Zweibel 1979), Hasan (1984) found that a tube with a surface field of about 700 G (corresponding to $\beta \approx 3$) undergoes collapse to a state in which the field has an average strength of about 1250 G. For an adiabatic flow, the fluid in the tube is unable to get rid of its momentum and in the final state it exhibits oscillations. On the other hand, it was argued by Takeuchi (1993) that if one adopts an open boundary condition at the bottom, the fluid in the tube eventually settles down to hydrostatic equilibrium. It should be noted that field intensification occurs mainly in the surface layers, i.e., locally in a layer ~ 100 km below the photospheric surface. Calculations (Hasan 1984, 1985) suggest that the field does not increase much above the marginal stability limit of $\beta \approx 1.8$, whereas observations suggest a lower value ($\beta \approx 0.3$) (e.g., Solanki 1993 and references therein). In the author's opinion, no convincing calculation has yet demonstrated how such strong fields can be generated. A possible clue to this problem might lie in the fact that intense flux tubes occur in regions associated with cool downflowing plumes of gas. Preliminary calculations have been carried out by Hasan & van Ballegooijen (1998) in which the inclusion of Reynolds stresses and the cool material just outside the tube have been incorporated in an equilibrium model for a flux tube. We find that these effects are likely to play an important role in producing intense fields with high field strength.

Let us now consider the consequences of including lateral heat exchange between the flux tube and the external medium. It can be shown (Hasan 1986) that the effect of horizontal radiative transport is to counteract the instability, which is most efficient when the gas within the tube is thermally insulated from its surroundings. However, in reality this insulation is reduced due to the leakage of heat into the tube from the ambient medium. Since the time scale for radiative heat exchange decreases with the tube radius, the critical value of β for the onset of instability (β_c) increases with decreasing photospheric radius (Hasan 1986). The situation becomes more complicated if one also incorporates vertical radiative transport, which has a destabilising effect and tends to enhance the convective collapse due to a cooling associated with radiative losses in the vertical direction (Rajaguru & Hasan 2000). Using a refined treatment of radiative transfer Rajaguru & Hasan (2000) derived a relation between the critical photospheric radius a_0 and the field strength $B_{\rm ph}$, which demarcates the separation between convectively unstable and stable tubes. Figure 4 depicts the variation (solid curve) of $B_{\rm ph}$ as a function of the photospheric radius $a_{\rm ph}$ (the magnetic flux $B\pi a_{\rm ph}^2$ is shown on the upper scale). For comparison, we also show data from observations: the squares are from Solanki et al. (1996); the light shaded region from Lin (1995) and Lin & Rimmele (1999); and the dark shaded region is from Sánchez Almeida & Lites (2000). All points above the solid curve denote stable configurations whereas those below are unstable. A flux tube that is

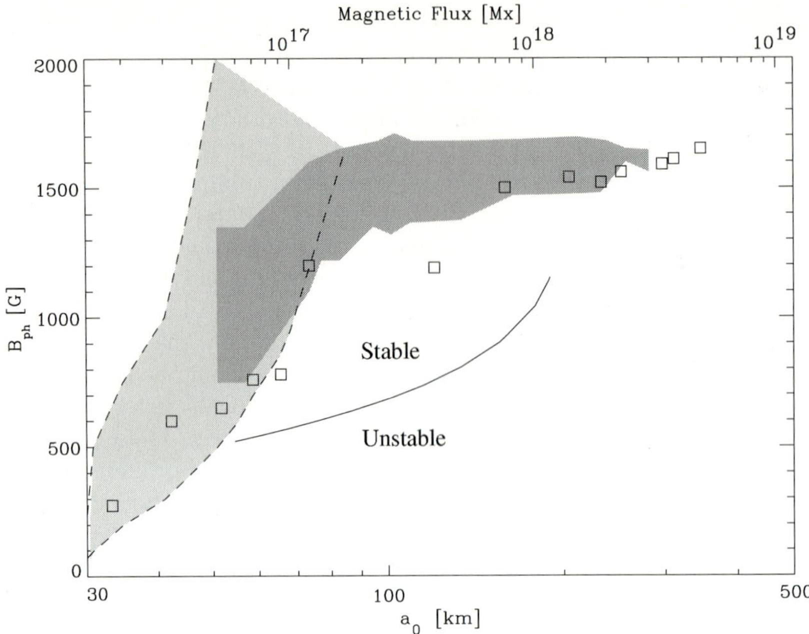

Fig. 4. Relationship between the critical photospheric radius a_0 and photospheric field strength B_{ph} (solid line) which demarcates the separation between convectively stable and unstable tubes. The top scale denotes the magnetic flux. Superposed are data from observations: squares are from Solanki et al. (1996); the light shaded region from Lin (1995) and Lin & Rimmele (1999); and the dark shaded region is from Sánchez Almeida & Lites (2000). All points above the *solid* curve denote stable configurations whereas those below are unstable. A flux tube that is unstable will trace out a vertical path (since magnetic flux is conserved for an ideal plasma) in this diagram

unstable (region below the solid curve) will trace out a vertical path (since magnetic flux is conserved for an ideal plasma) in this diagram. Once it crosses the solid curve, the collapse will cease and the tube will be stable. We find that the maximum field strength for which the tube is unstable is 1160 G, corresponding to $\beta_c = 2.45$ and a critical radius of $a_0 = 190$ km, which translates to a critical flux $\Phi_c = 1.31 \times 10^{18}$ Mx. Tubes with a magnetic flux greater than Φ_c are invariably in a collapsed state with a field above 1160 G. Figure 4 suggests a division of observed tubes into two groups with: (a) flux concentrations with flux above Φ_c which are associated with kilogauss strong-field network elements in which their field strength is weakly dependent on their flux content (Solanki et al. 1996); and (b) elements with flux lower than Φ_c in which the field has a significant variation with flux. Such flux tubes can be identified with weak to moderate fields, which may possibly occur in the internetwork region, that is the interior of the supergranulation cells.

The above discussion has focused on the relation between the critical field strength and radius for stability. Let us now consider the situation for $\beta < \beta_c$.

In this case, it can be shown that the tube exhibits overstability (Hasan 1985, 1986; Rajaguru & Hasan 2000). Physically, overstability involves a delicate balance among the following three processes: (a) buoyancy, which drives the convective instability in the presence of a superadiabatic temperature gradient, (b) a restoring force associated with the magnetic and gas pressure, and (c) a dissipative mechanism such as radiative damping. When the magnetic field is strong enough to counteract the instability, the tube exhibits undamped oscillations in the adiabatic limit. However, in the presence of lateral radiative transport, which depends on the flux tube radius, the driving force due to buoyancy is reduced in such a way that the net restoring force is greater during the return to equilibrium than it is during the departure away from equilibrium. Consequently, during each oscillation cycle, energy is extracted from the radiation field and converted into mechanical motions. However, for field strengths that are typical in the solar network ($\beta \approx 0.3$), the time scale for damping in the vertical direction becomes sufficiently small so as to counteract the overstability which is driven by lateral heat exchange. Rajaguru & Hasan (2000) found that such flux tubes with radius greater than about 170 km are no longer overstable but subject to radiative damping.

3.2 The Thermal Structure of Intense Flux Tubes

In the previous section we have seen that the formation of small-scale flux tubes is strongly influenced by radiative energy transport. This effect is particularly important in the photospheric layers where the radiative heat exchange time scale becomes comparable or even less than the time scale for dynamical processes. Let us now turn to the thermodynamics of intense flux tubes, which requires a detailed treatment of the energy balance and energy transport mechanisms. The earliest models of tubes were provided by Spruit (1976) who incorporated radiation and convection in a crude way by solving a heat transport equation. This work demonstrated that the radiation field provides a tight link between the thermal structure of the tube and that of the ambient medium. Further developments in models came through the use of a more refined treatment of radiative transfer (e.g., Ferrari et al. 1985; Kalkofen et al. 1986; Steiner & Stenflo 1989; Fabiani Bendicho, Kneer & Trujillo Bueno 1992; Pizzo, MacGregor & Kunasz 1993a,b; Fawzy et al. 2002). The inclusion of a more general energy transport equation with a self-consistent treatment of convective, radiative and mechanical energy transport was carried out by Hasan (1988) and Hasan & Kalkofen (1994). In addition to the above studies that treat a static tube, there are several time-dependent studies, such as those carried out for a thin flux tube by Hasan (1984, 1985, 1991) and Takeuchi (1999); for flux sheaths in 2-D by Deinzer et al. (1984), Knölker, Schüssler & Weisshaar (1988), Grossmann-Doerth et al. (1989), Steiner et al. (1994, 1996), and for fields in 3-D by Nordlund & Stein (1989, 1990).

The temperature structure of a flux tube is governed by the interplay of vertical and horizontal energy transport. In the photosphere, this is due to radiative energy transport, but in the sub-photosphere and below, convective transport also needs to be taken into account. Furthermore, we need to go beyond the thin flux tube approximation in the photosphere, and self-consistently treat the flux tube and ambient medium in the treatment of radiative transfer. Indeed, such an approach was taken by Hasan, Kalkofen & Steiner (1999) in which a static equilibrium model with multi-dimensional radiative transport was developed.

The main feature of the improved model is that the flux tube and ambient medium are treated self-consistently, through which the effect of the tube on the surrounding atmosphere can be clearly discerned. Indeed, as our calculations confirm, there is a thermal boundary layer at the tube-external medium interface, with an extension comparable to the horizontal scale of the tube in the optically shallow regions. This can have important physical consequences for the thermodynamic structure of flux tubes.

Let us consider a vertical magnetic flux tube of circular cross section and radius a embedded in a nonmagnetised atmosphere. We adopt a cylindrical coordinate system and assume rotational symmetry about the tube axis. For simplicity we use the thin flux tube approximation to treat the magnetostatic equation. We assume that the pressure and magnetic field are specified and use the energy transport and radiative transfer equations to determine the thermal structure of the tube. We use β to parameterise our models.

We first construct model atmospheres for the ambient medium and the flux tube using the method described in Hasan & Kalkofen (1994). Briefly, the external atmosphere is generated by solving the equations of hydrostatic and energy equilibrium, assuming a plane-parallel medium. Flux tube models are then constructed by solving the magnetostatic equations for a thin tube along with the radiative transfer equation (for a grey atmosphere with 8 polar angles). The energy equation also incorporates convective transport, which we model using a mixing-length formalism, with an additional parameter characterising the partial inhibition of convection in the flux tube. We fix the value of the magnetic field using the thin-flux tube approximation. However, the temperature structure is determined by solving the energy transport equation, given by:

$$\nabla\cdot\boldsymbol{F}_{\mathrm{R}} = 4\pi\kappa(S - J) = -\nabla\cdot\boldsymbol{F}_{\mathrm{C}} \, , \tag{14}$$

where $\boldsymbol{F}_{\mathrm{R}}$ and $\boldsymbol{F}_{\mathrm{C}}$ denote the radiative and convective flux, S is the frequency-integrated source function and J is the mean radiation intensity. In the photosphere and in the shallow layers of the sub-photosphere, the convective flux can be neglected, which implies that the atmosphere is in radiative equilibrium. The mean radiation intensity is determined by solving the following radiative transfer equation:

$$(\boldsymbol{n}\cdot\nabla)I = \kappa(S - I) \, , \tag{15}$$

where $I = I(\boldsymbol{r}, \boldsymbol{n})$ denotes the frequency-integrated specific intensity of the radiation field and κ is the Rosseland mean opacity per unit distance. The solution of (15) needs to be carried out over all space so as to include the flux tube and the surrounding atmosphere. If the opacity and source function are known, then (15) can be formally solved to calculate I, and hence determine the frequency-

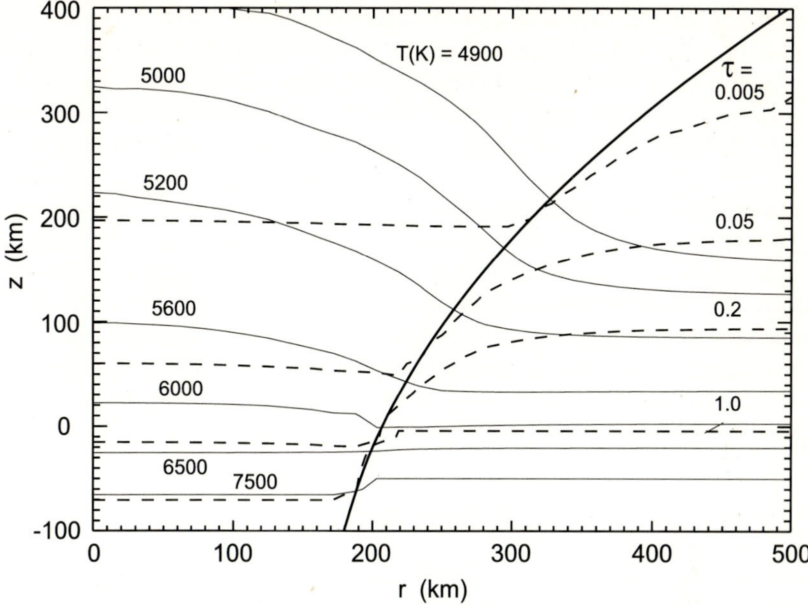

Fig. 5. Isotherms (*solid* lines) and lines of constant optical depth (*dashed* lines) in a flux tube and the surrounding medium for $a_0 = 200\,\mathrm{km}$ and $\beta_0 = 0.5$. The values of the contours are given above each curve (after Hasan, Kalkofen & Steiner 1999). The heavy *solid* line denotes the boundary of the tube

integrated mean intensity J and radiative flux $\boldsymbol{F}_R$ as follows:

$$J = \frac{1}{4\pi} \oint I(\boldsymbol{r}, \boldsymbol{n})\, \mathrm{d}\Omega\,, \tag{16}$$

and

$$\boldsymbol{F}_R = 4\pi \oint I(\boldsymbol{r}, \boldsymbol{n})\boldsymbol{n}\, \mathrm{d}\Omega\,, \tag{17}$$

where the integration is over solid angle. We assume local thermodynamic equilibrium (LTE) and equate S to the frequency-integrated Planck function, so that $S = \sigma T^4/\pi$. We consider layers where convection can be neglected, so that we can assume radiative equilibrium. We include the presence of convection indirectly by choosing upward intensities at the lower boundary which are compatible with convective energy transport. For a specified temperature structure, the equation of hydrostatic equilibrium can be integrated (using the perfect gas law) to determine the pressure and density at each height in the atmosphere. These values in turn can be used to determine the opacity.

Figure 5 shows the isotherms (thin solid curves) in a flux tube of radius 200 km at $z = 0$ corresponding to a value of $\beta = 0.5$. The dashed curves denote lines of constant optical depth. The heavy solid lines denotes the boundary of the flux tube. In the photosphere, we find that the temperature in the flux tube is higher than that in the ambient medium; as the height increases the isotherms are shifted upwards by larger distances. On the other hand, in the subphotosphere, the isotherms dip downwards, due to the reduced energy transport owing to inhibition of convection in these layers. Close to $z = 0$, the isotherms are essentially flat, reflecting the fact that in these layers horizontal radiative transport is very efficient. The regions of influence of the flux tube on the ambient medium increases with height (owing to the reduced density and hence increased photon mean free path).

2-D radiative transfer effects are probably the reason why small flux tubes appear as bright features in photospheric radiation. However, for thicker tubes, the horizontal optical thickness insulates the interior more from the ambient medium, and the cooling due to inhibition of convection may dominate. Consequently, they appear as dark structures (pores or sunspots). The transition between bright and dark structures occurs for a tube diameter of about 600 km (Knölker & Schüssler 1988).

4 Dynamical Processes and Heating of the Magnetic Chromosphere

The solar chromosphere plays an important role as the lower boundary of the heliosphere, and therefore for radiation that affects the ionisation state of the upper terrestrial atmosphere as well as for the origin of the solar wind and the generation of coronal mass ejections. It is thus of great interest to understand the chromosphere and, in particular, the state of the gas in its upper layers. In the quiet chromosphere we distinguish the magnetic network on the boundary of supergranulation cells, where strong magnetic fields are organised in magnetic flux tubes, and internetwork regions in the cell interior, where magnetic fields are weak and dynamically unimportant.

Ground-based observations of the Ca II H and K lines, which are formed in the low chromosphere, show similar emission from network and internetwork regions. While instantaneous bright points from the internetwork may outshine network bright points (see Fig. 1 of Lites, Rutten & Kalkofen 1993, hereafter LRK93), the long-time average intensity shows total calcium emission from the network to be more important (see Fig. 1 of von Uexküll & Kneer 1995). In addition to the higher intensity of the network bright points, their period is longer, $\sim$ 7 minutes (LRK93; Curdt & Heinzel 1998), and the time variation of their intensity profile much less peaked.

Space-based observations of UV spectral lines and continua provide important constraints on the structure and dynamics of the chromosphere and chromosphere–corona transition region. Observations with SUMER have shown that the UV lines are always in emission, consistent with semi-empirical models in which the temperature in the chromosphere increases with height at all times (Vernazza, Avrett & Loeser 1981). Internetwork regions show large-scale coherent oscillations with length scales of 3–7 Mm and periods between 120 and 200 s in spectral lines of neutral and singly ionised species, and sometimes also in lines from higher ionisation states (Carlsson, Judge & Wilhelm 1997;

Wikstøl et al. 2000; McIntosh et al. 2001). These oscillations have also been seen with TRACE (Rutten, de Pontieu & Lites 1999; Judge, Tarbell & Wilhelm 2001). These observations show that there are upward propagating waves in the non-magnetic chromosphere that occasionally drive oscillations in the overlying transition region. Network regions are brighter than internetwork regions, and show strong oscillatory power only at lower frequencies (Judge, Carlsson & Wilhelm 1997). Transition region lines from the network show persistent redshifts and the line widths indicate the presence of subsonic, unresolved non-thermal Doppler motions of several kilometres per second (Dere & Mason 1993; Peter 2000, 2001). Furthermore, there is a strong correlation between high intensity and redshift (Hansteen, Betta & Carlsson 2000). Curdt & Heinzel (1998) found evidence for upward propagating waves within the network (also see Heinzel & Curdt 1999). However, the wave modes responsible for these oscillations have not yet been identified.

The phenomena in the magnetic network and in the non-magnetic cell interior show superficial similarity. Yet the physical processes occurring in the two media are most likely different. It is therefore instructive to compare our understanding of the phenomena in the network with that in the cell interior. The steady radiative emission of the non-magnetic chromosphere is well described by the empirical models of Vernazza, Avrett & Loeser (1981) and Fontenla, Avrett & Loeser (1993). This implies that the non-magnetic chromosphere is continually heated, perhaps by ubiquitous weak shocks. In addition there are stronger, more intermittent shocks that are responsible for the internetwork bright points seen in the Ca II H and K lines (Carlsson & Stein 1995, 1997). According to the empirical models by Vernazza, Avrett & Loeser (1981) and Fontenla, Avrett & Loeser (1993), the temperature structure of the magnetic chromosphere is very similar to that of the non-magnetic chromosphere, suggesting that the heating mechanisms in the two media may be similar. However, the statistics of H and K line asymmetries and the periods of oscillations in the magnetic network are significantly different from those of calcium bright points in the cell interior. These differences may find an explanation in the wave modes and the mechanisms of excitation of oscillations in the two media.

In this overview, we shall focus on dynamical processes occurring in the network and their role in heating the magnetic chromosphere. A complete model must explain the nature and period of the oscillations observed in the network as well as its heating. Furthermore, the model must be compatible with observations.

4.1 Longitudinal and Transverse Waves in Flux Tubes

The magnetic field in the network can be idealised in terms of isolated vertical flux tubes in the photosphere which fan out with height. It is well known that flux tubes support a variety of wave modes. The detailed behaviour of these modes for thin flux tubes has been extensively studied (for a recent review see Roberts & Ulmschneider 1998). The modes that we shall be concerned with are the sausage or longitudinal mode (Defouw 1976; Roberts & Webb 1978) and the

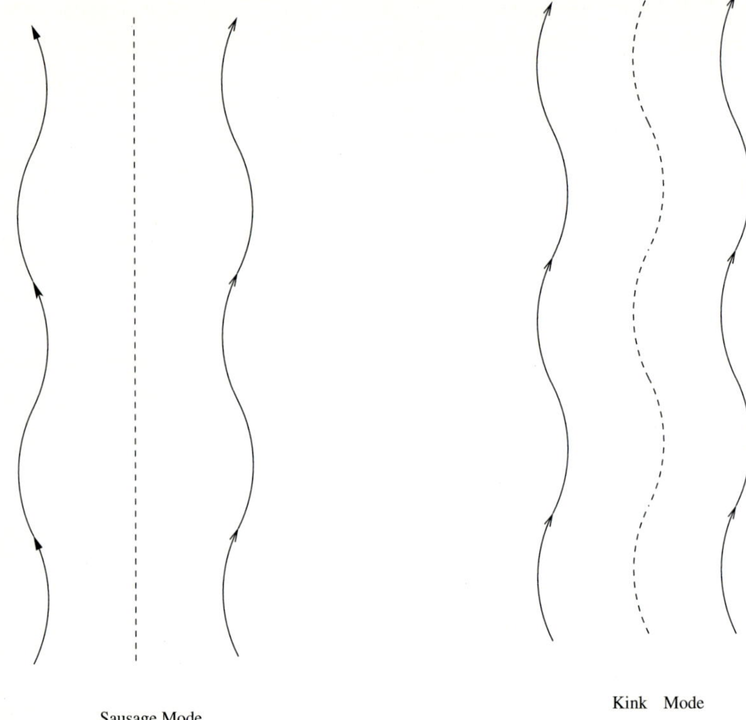

Fig. 6. Form of the perturbation associated with sausage (left panel) and kink (right panel) waves in a magnetic flux tube

kink or transverse mode (Ryutov & Ryutova 1976; Parker 1979; Spruit 1982). Figure 6 schematically shows the general form of the perturbations of the flux tube in the above modes.

The earliest studies on MHD wave excitation were based on extensions of the Lighthill (1952) mechanism (Osterbrock 1961; Musielak & Rosner 1987; Collins 1989, 1992). More recently, Musielak, Rosner & Ulmschneider (1989), Musielak et al. (1995), Huang, Musielak & Ulmschneider (1995) and Ulmschneider & Musielak (1998) examined the generation of longitudinal and transverse waves in a flux tube through turbulent motions in the convection zone. An alternative scenario motivated by the observations of Muller & Roudier (1992) and Muller et al. (1994) suggests that transverse waves can be generated through the impulse imparted by granules to magnetic flux tubes (Choudhuri, Auffret & Priest 1993a; Choudhuri, Dikpati & Banerjee 1993b; Steiner et al. 1998). These investigations suggested that there is sufficient energy flux in MHD waves to account for chromospheric heating.

Let us consider in some detail consequences of MHD wave excitation in magnetic flux tubes through the buffeting action of convective motions (granulation) in the surrounding medium. Such waves are likely to play an important role in heating the magnetic chromosphere and also possibly the corona.

4.2 The Linear Model

Consider a vertical magnetic flux tube extending through the photosphere, which is assumed to be "thin" and isothermal. It is convenient to use the "reduced" displacement, $Q(z,t)$, which is related to the physical Lagrangian displacement, $\xi(z,t)$, by $Q(z,t) = \xi_\perp(z,t)\exp(-z/4H)$, where H denotes the scale height of the atmosphere.

It can be shown that Q_α ($\alpha = \kappa$ for transverse waves and $\alpha = \lambda$ for longitudinal waves) satisfies a Klein–Gordon equation (Hasan & Kalkofen 1999, henceforth HK).

$$\frac{\partial^2 Q_\alpha}{\partial z^2} - \frac{1}{c_\alpha^2}\frac{\partial^2 Q_\alpha}{\partial t^2} - k_\alpha^2 Q_\alpha = F_\alpha \,, \tag{18}$$

where $k_\alpha = \omega_\alpha / c_\alpha$, ω_α is the cutoff frequency for the wave and c_α is the wave propagation speed in the medium and F_α is a forcing function that parameterises the impact delivered to the flux tube by a granule (for further details see HK). The speeds for the transverse and longitudinal waves are, respectively,

$$c_\kappa^2 = \frac{2}{\gamma}\frac{c_s^2}{1+2\beta} \,, \tag{19}$$

$$c_\lambda^2 = \frac{c_s^2}{1+\gamma\beta/2} \,, \tag{20}$$

where c_s is the sound speed, γ (= 5/3) is the ratio of specific heats, $\beta = 8\pi p/B^2$, p is the gas pressure inside the tube and B is the magnitude of the vertical component of the magnetic field on the tube axis.

The cutoff frequencies for transverse and longitudinal waves are, respectively,

$$\omega_\kappa^2 = \frac{g}{8H}\frac{1}{1+2\beta} \,, \tag{21}$$

$$\omega_\lambda^2 = \omega_{\rm BV}^2 + \frac{c_\lambda^2}{H^2}\left(\frac{3}{4} - \frac{1}{\gamma}\right)^2 \,, \tag{22}$$

where $\omega_{\rm BV}^2 = g^2\,(\gamma - 1)/c_s^2$ is the Brunt-Väisälä frequency (which follows from [9] or [11]) for a fully ionised isothermal plasma.

Equation (18) can be readily solved using Green's functions (for details see HK). The generic behaviour for the impulsive excitation of transverse and longitudinal waves by granular motions in the magnetic network is the same: the buffeting action due to a single impact excites a pulse that propagates along the flux tube with the kink or longitudinal wave speed. For strong magnetic fields, most of the energy goes into transverse waves, and only a much smaller fraction into longitudinal waves, a result also found by Ulmschneider & Musielak (1998). After the passage of the pulse, the atmosphere gradually relaxes to a state in which it oscillates at the cutoff period of the mode. These results show that the first pulse carries most of the energy and after this pulse has passed the atmosphere oscillates in phase without energy transport. The period observed in the magnetic network is interpreted as the cutoff period of transverse waves, which leads naturally to an oscillation at this period (typically in the 7-minute range) as proposed by Kalkofen (1997).

For weaker magnetic fields the energy fluxes in the two modes are comparable. From the absence of a strong peak at low frequencies in the power spectrum of the cell interior (CI) we conclude that both transverse and longitudinal waves must make a negligible contribution to K_{2V} bright point oscillations. The absence of the magnetic modes then implies that the waves in the CI are probably acoustic waves, and the observed 3 minute period is therefore the acoustic cutoff period – and not the cutoff period of longitudinal waves. This implies that the magnetic field structure in the CI is likely to be different from that of flux tubes in the magnetic network.

4.3 Chromospheric Heating

The above discussion has considered the buffeting of flux tubes as a single impact. In reality, we expect the excitation of waves in a tube to occur not as a single impact but continually due to the highly turbulent and stochastic motion of granules. It is interesting to examine the consequences of this interaction for chromospheric heating. Such an investigation was carried out by Hasan, Kalkofen & van Ballegooijen (2000, hereafter HKB), who modelled the excitation of waves in the magnetic network due to the observed motions of G-band bright points, which were taken as a proxy for footpoint motions of flux tubes. Using high resolution observations of G band bright points in the magnetic network, the energy flux in transverse waves was calculated in a large number of magnetic elements.

Figure 7 shows the vertical energy flux in transverse waves versus time at a height $z = 750\,\mathrm{km}$ for a typical magnetic element in the network. We find that the injection of energy into the chromosphere takes place in brief and intermittent bursts, lasting typically 30 s, separated by longer periods (longer than the time scale for radiative losses in the chromosphere) with lower energy flux. Similar pulse-like time dependence, shown in Fig. 7, has also been found by Ulmschneider (1998) and Ulmschneider & Musielak (1998). The peak energy flux into the chromosphere is as high as 10^9 erg cm^{-2} s^{-1} in a single flux tube, although the time-averaged flux is $\sim 10^8$ erg cm^{-2} s^{-1}. However, from an observational point of view, such a scenario for heating the magnetic network, would yield a high variability with time in Ca II emission, which appears incompatible with observations. A possible remedy to this difficulty would be to postulate the existence of other high-frequency motions (periods 5–50 s) which cannot be detected as proper motions of G-band bright points (HKB). Adding such high-frequency motions to the simulations of HKB results in much better agreement with the persistent emission observed from the magnetic network. For a filling factor of 10% at $z = 750\,\mathrm{km}$, the predicted flux $\sim 10^7$ erg cm^{-2} s^{-1}, which is sufficient to

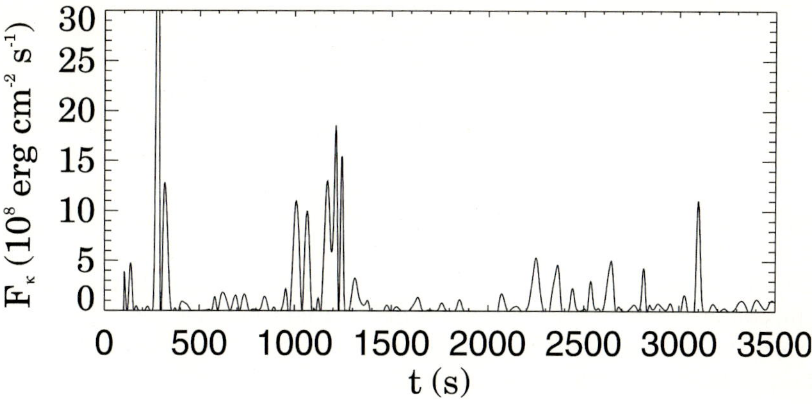

Fig. 7. Time variation of the vertical energy flux in transverse waves in a single flux tube at $z = 750\,\mathrm{km}$ due to footpoint motions taken from observations excited in an isothermal flux tube with $T = 6650\,\mathrm{K}$, $\beta = 0.3$

balance the observed radiative loss of the chromospheric network (see Model F′ of Avrett 1985). Therefore, for transverse waves to provide a viable mechanism for *sustained* chromospheric heating, the main contribution to the heating must come from high-frequency motions, with typical periods 5–50 s. HKB speculated that the high-frequency motions could be due to turbulence in intergranular lanes, but some aspects of this model require further investigation.

4.4 Nonlinear Results

The above investigations were based on a linear approximation in which the longitudinal and transverse waves are de-coupled. However, the velocity amplitude $v(z)$ for the two modes increases with height z (for an isothermal atmosphere $v \propto \exp(z/4H)$, where H is the pressure scale height), so the motions are likely to become supersonic higher up in the atmosphere. At such heights, nonlinear effects become important, leading to coupling between the transverse and longitudinal modes. Some progress on this question has been made using the nonlinear equations for a thin flux tube (Ulmschneider, Zähringer & Musielak 1991; Huang, Musielak & Ulmschneider 1995). This work has been extended to include a treatment of kink and longitudinal shocks (Zhugzhda, Bromm & Ulmschneider 1995). The above investigations have concentrated primarily on wave propagation in the photosphere and in the lower chromosphere, but have not treated the propagation of coupled transverse-longitudinal waves in the higher layers. This is clearly needed to assess whether these waves can contribute effectively to heating the magnetic chromosphere and corona. Another aspect which also needs to be examined in detail is the onset of nonlinear effects along with their implications in a vertical flux tube. These produce significant mode coupling leading to a transfer of energy between the modes, which is likely to have important consequences for the dynamics and energy transport in the solar network.

Recently, Hasan et al. (2003) carried out adiabatic calculations of nonlinear kink waves in a thin, isothermal flux tube. For the initial state, they assume that the flux tube is "thin" and initially in hydrostatic equilibrium and isothermal with a temperature $T = 6650\,\mathrm{K}$ (corresponding to a scale height $H = 155\,\mathrm{km}$) which is the same as that in the external medium. They consider a tube with a radius of 40 km and a field strength of $B = 1700\,\mathrm{G}$ at $z = 0$, corresponding to a plasma β of 0.18 (which remains constant with height). The radius of the tube increases with z as $\exp(z/4H)$.

The basic equations for adiabatic longitudinal-transverse MHD waves in a thin flux tube consist of a set of coupled differential equations (see Ulmschneider, Zähringer & Musielak 1991 for details) which are solved numerically using the method of characteristics. In the present work we adopt this method, modified to include shocks, based on the treatment of Zhugzhda, Bromm & Ulmschneider (1995). The computational domain in the vertical direction has an equidistant grid of size 5 km. The Courant condition is used to select the time step to advance the equations in time.

At the lower boundary, taken at $z = 0$, we assume that the flux tube has a transverse motion which consists of a single impulse with a velocity of the form:

$$v_x(0,t) = v_0 \exp(-[(t - t_0)/\tau]^2) , \tag{23}$$

where v_0 is the specified velocity amplitude, t_0 denotes the time when the motions have maximum amplitude and τ is the time constant of the impulse. The longitudinal component of the velocity at the base is assumed to be zero. In the present calculations we take $t_0 = 50\,\mathrm{s}$ and $\tau = 20\,\mathrm{s}$.

At the upper boundary of the computational domain (at $z = 1500\,\mathrm{km}$) we use transmitting boundary conditions, following Ulmschneider et al. (1977), and assume that the velocity amplitude remains constant along the outward-propagating characteristics. The characteristic equations are used to self-consistently determine physical quantities at the boundary.

The initial equilibrium model is perturbed with a transverse motion at $z = 0$ in the form of an impulse with a velocity given by (23). This impulse generates a transverse wave that propagates upwards with the kink wave speed c_κ, which is about 7.9 km s^{-1} for the equilibrium model. The resulting motion in the tube as a function of height and time follows from the time-dependent MHD equations for a thin flux tube.

Figures 8a and 8b shows the variation of the transverse v_x (solid lines) and longitudinal v_z (dashed lines) components of the velocity as a function of height z at various epochs of time t for $v_0 = 2.0$ km s^{-1} and $v_0 = 4.0$ km s^{-1} respectively. The numbers beside the curves denote the time t (in s). We find that low in the atmosphere, where the transverse velocity amplitude is small (compared to the kink wave speed c_κ), the longitudinal component of the velocity is negligible. As the initial pulse propagates upwards, the transverse velocity amplitude increases. Due to nonlinear effects, beginning when the Mach number $M = v_x/c_\kappa$ is as low as 0.3, longitudinal motions are generated. The efficiency of the nonlinear coupling increases with the amplitude of the transverse motions.

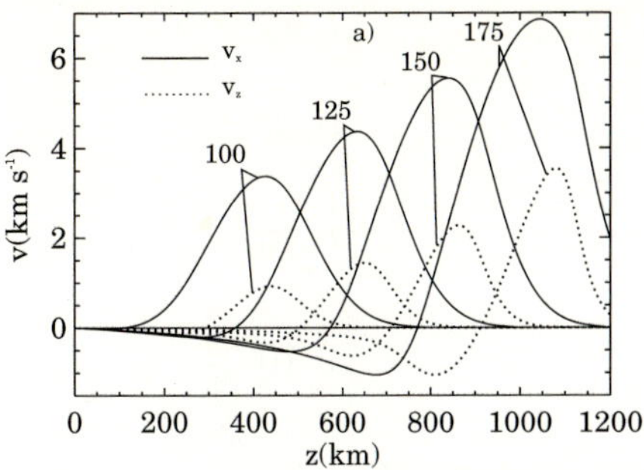

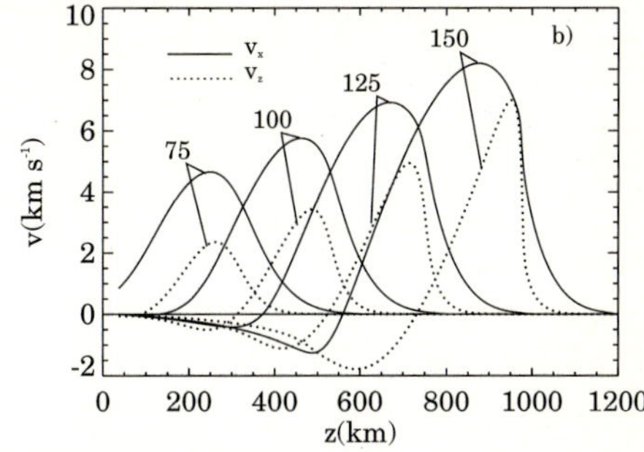

Fig. 8. Nonlinear coupling of transverse and longitudinal waves in a flux tube: The variation of the transverse v_x (solid lines) and the longitudinal v_z (dashed lines) components of the velocity as functions of height z at various epochs for (**a**) $v_0 = 2.0$ km s^{-1} and (**b**) $v_0 = 4.0$ km s^{-1} (after Hasan et al. 2003). The numbers beside the curves denote the time (in s)

When $v_x \approx c_\kappa$, the amplitudes in the transverse and longitudinal components become comparable. The longitudinal motions, being compressive, steepen with height and eventually form shocks. The steepening is clearly visible in Figure 8(b), especially at $t = 150$ s in the longitudinal component. These results are reminiscent of those found by Hollweg, Jackson & Galloway (1982), who studied the nonlinear coupling of torsional Alfvén waves and longitudinal waves in the solar atmosphere. Their results, however, did not show any wakes, which arise due to the presence of a cutoff frequency, which is absent for torsional Alfvén waves.

Let us now examine the temporal behaviour of the velocity. Figure 9 shows the variation of the transverse v_x (solid lines) and longitudinal v_z (dashed lines) components of the velocity as functions of time t at $z = 1000$ km for $v_0 = 0.5$ km s^{-1}. The vertical scale on the right corresponds to v_z. The first maxima in the velocities denote the arrival of the transverse and longitudinal components of the impulse, which travel at approximately the same speed (since $c_\kappa \approx c_T$). After the passage of the primary pulses, which eventually propagate out through the top boundary, the transverse and longitudinal components oscillate with different periods. At this stage, since the velocity amplitudes are small, the two modes essentially decouple. We find that in the asymptotic time limit, the periods of the two modes closely match their cutoff periods, which are about 490 s and 230 s for kink and longitudinal waves, respectively. Since the cutoff periods are well separated, this could provide an observational test of the model, which predicts that the signature of impulsive footpoint motions would be two distinct peaks in the wave power spectrum of network oscillations in the middle to upper chromosphere. We expect that the dominant peak with a period in the 6–7 min range would correspond to the low frequency transverse oscillations, whereas the secondary peak in the 3 min range could be identified with longitudinal oscillations. There is a hint that these features may be present in the observations of LRK93. We should, however, note that the theoretical results presented by

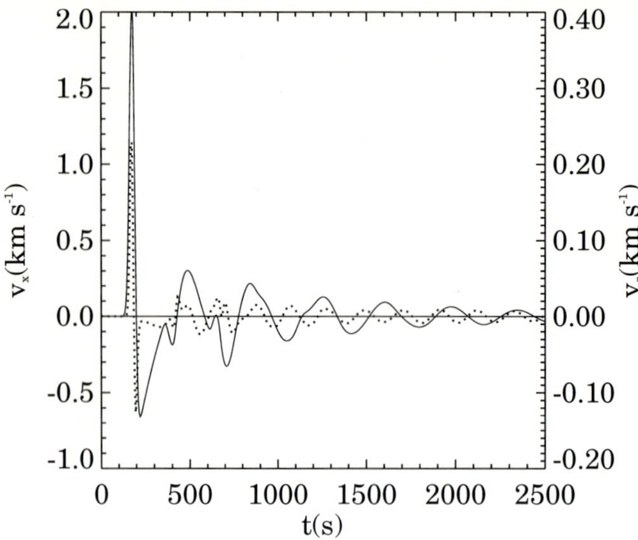

Fig. 9. Transverse velocity v_x (*solid curves*) and longitudinal velocity v_z (*dashed curves*) as functions of time t at a fixed height $z = 1000\,\mathrm{km}$ for $v_0 = 0.5$ km s^{-1}, $t_0 = 50\,\mathrm{s}$ and $\tau = 20\,\mathrm{s}$ (after Hasan et al. 2003)

us are based on the assumption of the flux tube footpoints being shaken impulsively. In reality, the footpoint motion consists of several impacts (e.g., Muller et al. 1994) that probably occur stochastically, so that the power spectrum of oscillations is unlikely to show a clear separation of peaks that would occur for a single impulse.

In a nonlinear time-dependent simulation of the evolution of an initially transverse wave pulse, Zhugzhda, Bromm & Ulmschneider (1995) found that once the kink and longitudinal shocks formed, they occurred at the same height and subsequently propagated with the same speed. This is very similar to the findings of Hollweg, Jackson & Galloway (1982), who studied the evolution of an initially torsional wave pulse. Here also torsional (switch-on) and longitudinal shocks formed at the same position and these shocks subsequently propagated with a common speed. This indicates strong mode-coupling by which transverse and torsional wave energies are converted into longitudinal energy and dissipated via the longitudinal shock (Ulmschneider private communication).

Let us summarise the main conclusions to emerge from the nonlinear calculations. When the transverse velocities are significantly less than the kink wave speed (the linear regime), there is essentially no excitation of longitudinal waves. However, at heights where $v_x \approx c_K$, longitudinal wave generation becomes efficient, leading to the modes having comparable amplitudes; a large amplitude transverse pulse, generates a longitudinal pulse, which eventually generate wakes that have low amplitudes and represent *de-coupled longitudinal and kink waves, oscillating at their respective cutoff periods.* We have examined the coupling between the two modes, and find that v_z increases quadratically with

v_x at low Mach number M (with respect to c_κ) and linearly with v_x for $M \to 1$. Transverse waves lose energy due to mode coupling. It turns out that the ratio of the wave energy in longitudinal motions to the total wave energy increases rapidly at first with the forcing transverse velocity v_0, before eventually saturating at a value of about 0.45, which is close to equipartition of energy between the two modes (Hasan et al. 2003). For a forcing amplitude of $v_0 = 1.5$ km s^{-1}, when there is almost equipartition of energy, the transverse energy flux entering the transition region is approximately 10^7 erg cm^{-2} s^{-1}. This estimate is clearly an upper bound since we need to consider two effects: firstly, footpoint motions with this velocity occur on average with a probability of around 0.1, and secondly, there is an attenuation of the flux as it propagates through the transition region, which could lead to a further reduction by a factor of about 10. We neglect the effect of area change as the flux tubes no longer flare out, but have a constant cross-section at these heights. Hence, we estimate that the net energy flux entering the corona is about 10^5 erg cm^{-2} s^{-1}, which is adequate for coronal heating. Large amplitude longitudinal waves generated in the upper photosphere, steepen and form shocks in the chromosphere. They are likely to be important for chromospheric heating.

5 Summary

The various processes contributing to solar activity, which we have discussed in the previous sections, are directly related to the effects associated with the magnetic field. This field is most likely generated through a dynamo action, which according to current understanding occurs in a shear layer just below the base of the convection zone, where strong toroidal fields in the range 10^4–10^5 G can be stably stored. Magnetic fields generated by the dynamo can be released through a hydromagnetic instability due to which flux is transported to the photosphere and appears in the activity belts with the appropriate inclination.

Magnetic fields emerge as large coherent flux tubes which fragment into the enhanced network. The emerging flux concentrations have field strengths that are unlikely to exceed the dynamic equipartition value, which is much smaller than the observed value in the network. It is generally believed that this field is further strengthened by a combination of *flux expulsion* and *convective collapse*; the latter is associated with an instability which operates preferentially in the sub-photosphere and produces intense flux tubes with fields in the kilogauss range. Support for this mechanism comes through a prediction of a relationship between the size (radius) and field strength, which shows qualitative agreement with observations. Detailed calculations have been carried out to understand thermodynamic structure of flux tubes. These have demonstrated the importance of 2-D radiative transfer effects in interpreting the observed properties of network bright points and *faculae*.

Flux tubes in the magnetic network are likely to play an important role in heating the chromosphere. Their footpoints, located in the sub-photosphere, are constantly buffeted by granules, due to which MHD oscillations are excited in the tubes. Detailed calculations confirm that the low frequency waves observed in the network can be identified mainly with kink or transverse MHD waves in a thin flux tube with a period that corresponds to the cutoff period for kink waves. In the photospheric layers, most of the energy is in kink waves – however, in the chromosphere longitudinal waves are generated due to nonlinear effects – their characteristic signature is an oscillation at the longitudinal cutoff period. Longitudinal waves, being compressible, form shocks and thereby contribute to chromospheric heating. The incompressible kink waves, on the other hand, can propagate through the chromosphere and contribute to coronal heating.

Acknowledgements

I am grateful to Drs. W. Kalkofen, P. Ulmschneider and A. van Ballegooijen for helpful discussions and to Dr. S. P. Rajaguru for help with the figures.

References

Avrett, E. H. 1985, in Chromospheric Diagnostics and Modelling, Ed. B. Lites (Sunspot, New Mexico), 67
Bogdan, T. J. 2000, Sol. Phys., 192, 373
Caligari, P., Moreno-Insertis, F., & Schüssler, M. 1995, ApJ, 441, 886
Carlsson, M., & Stein, R. F. 1995, ApJ, 440, L29
Carlsson, M., & Stein, R. F. 1997, ApJ, 481, 500
Carlsson, M., Judge, P. G., & Wilhelm, K. 1997, ApJ, 486, L63
Chapman, G. A., Cookson, A. M., & Dobias, J. J. 1997, ApJ, 482, 541
Choudhuri, A. R., & Gilman, P. A. 1987, ApJ, 316, 788
Choudhuri, A. R., Auffret, H., & Priest, E. R. 1993a, Sol. Phys., 143, 49
Choudhuri, A. R., Dikpati, M., & Banerjee, D. 1993b, ApJ, 413, 811
Collins, W. 1989, ApJ, 337, 548
Collins, W. 1992, ApJ, 384, 319
Cowling, T. G. 1933, MNRAS, 94, 39
Curdt, W., & Heinzel, P. 1998, ApJ, 503, L95
Defouw, R. J. 1976, ApJ, 209, 266
Deinzer, W., Hensler, G., Schüssler, M., & Weisshaar, E. 1984, A&A, 139, 435
Dere, K. P., & Mason, H. E. 1993, Sol. Phys., 144, 217
D'Silva, S., & Choudhuri, A. R. 1993, A&A, 272, 621
Emonet, T., & Moreno-Insertis, F. 1998, ApJ, 492, 804
Fabiani Bendicho, P., Kneer, F., & Trujillo Bueno, J. 1992 A&A, 264, 229
Fan, Y., Fisher, G. H., & DeLuca, E. E. 1993, ApJ, 405, 390
Fan, Y., Fisher, G. H., & McClymont, A. N. 1994, ApJ, 436, 907
Fan, Y., Zweibel, E. G., & Lantz, S. R. 1998, ApJ, 493, 480
Fawzy, D., Rammacher, W., Ulmschneider, P., Musielak, Z. E., & Stepien, K. 2002, A&A, 386, 971
Ferrari, A., Massaglia, S., Kalkofen, W., Rosner, R., & Bodo, G. 1985, ApJ, 298, 181
Ferriz-Mas, A., & Schüssler, M. 1993, Geophys. Astrophys. Fluid Dyn., 72, 209
Ferriz-Mas, A., & Schüssler, M. 1995, Geophys. Astrophys. Fluid Dyn., 81, 233
Fisher, G. H., Fan, Y., & Howard, R. F. 1995, ApJ, 438, 463

Fisher, G. H., Fan, Y., Longcope, D. W., Linton, M. G., & Pevtsov, A. A. 2000, Sol. Phys., 192, 119

Fontenla, J. M., Avrett, E. H., & Loeser, R. 1993, ApJ, 406, 319

Galloway, D. J., Proctor, M. R. E., & Weiss, N. O. 1978, J. Fluid Mech., 87, 243

Gilman, P. A. 1970, ApJ, 162, 1019

Grossmann-Doerth, U., Knölker, M., Schüssler, M., & Weisshaar, E. 1989, in Solar and Stellar Granulation, Eds. R. J. Rutten, & G. Severino, NATO ASI Series C Vol. 263 (Kluwer, Dordrecht) 481

Grossmann-Doerth, U., Schüssler, M., & Steiner, O. 1998, A&A, 337, 928

Hansteen, V. H., Betta, R., & Carlsson, M. 2000, A&A, 360, 742

Harvey, K. L. 1993, Ph. D. Thesis, Utrecht University, Utrecht

Hasan, S. S. 1983, in Solar and Stellar Magnetic Fields: Origins and Coronal Effects, IAU Symp. No. 102, Ed. J. O. Stenflo (Reidel, Dordrecht), 73

Hasan, S. S. 1984, ApJ, 285, 851

Hasan, S. S. 1985, A&A, 143, 39

Hasan, S. S. 1986, MNRAS, 219, 357

Hasan, S. S. 1988, ApJ, 332, 499

Hasan, S. S. 1991, in Mechanisms of Chromospheric and Coronal Heating, Eds. P. Ulmschneider, E. R. Priest, & R. Rosner (Springer, Berlin), 408

Hasan, S. S., & Kalkofen, W. 1994, ApJ, 436, 355

Hasan, S. S., & Kalkofen, W. 1999, ApJ, 519, 899 (HK)

Hasan, S. S., & van Ballegooijen, A. 1998, in Cool Stars, Stellar Systems and the Sun, Tenth Cambridge Workshop, Eds. R. A. Donahue, & J. Bookbinder, A. S. P. Conf. Series, Vol. 154, 630

Hasan, S. S., Kalkofen, W., & Steiner, O. 1999, in Solar Polarization, Proc. 2nd Solar Polarization Workshop, Eds. K. N. Nagendra, & J. O. Stenflo (Kluwer, Dordrecht), 409

Hasan, S. S., Kalkofen W., & van Ballegooijen, A. A. 2000, ApJ, 535, L67 (HKB)

Hasan, S. S., Kalkofen, W.,van Ballegooijen, A. A & Ulmschneider, P. 2003, ApJ 585, 1138

Heinzel, P., Curdt, W. 1999, in Third Advances in Solar Physics Euroconference: Magnetic Fields and Oscillations, Eds. B. Schmieder, A. Hofmann, & J. Staude, ASP Conf. Ser., Vol. 184, 201

Hollweg, J. V., Jackson, S., & Galloway, D. 1982, Sol. Phys., 75, 35

Huang, P., Musielak, Z. E., & Ulmschneider, P. 1995, A&A, 297, 579

Judge, P., Carlsson, M., & Wilhelm, K. 1997, ApJ, 490, L195

Judge, P. G., Tarbell, T. D., & Wilhelm, K. 2001, ApJ, 554, 424

Kalkofen, W. 1997, ApJ, 486, L145

Kalkofen, W., Rosner, R., Ferrari, A., & Massaglia, S. 1986, ApJ, 304, 519

Knölker, M., & Schüssler, M. 1988, A&A, 202, 275

Knölker, M., Schüssler, M., & Weisshaar, E. 1988, A&A, 194, 257

Larmor, J. 1919, J. Brit. Ass. Rep., p. 159

Lighthill, M. J. 1952, Proc. Roy. Soc. London, A211, 564

Lin, H. 1995, ApJ, 446, 421

Lin, H., & Rimmele, T. 1999, ApJ, 514, 448

Lites, B. W., Rutten, R. J., & Kalkofen, W. 1993, ApJ, 414, 345 (LRK93)

McIntosh, S. W., Bogdan, T. J., Cally, P. S., Carlsson, M., Hansteen, V. H., Judge, P. G., Lites, B. W., Peter, H., Rosenthal, C. S., & Tarbell, T. D. 2001, ApJ, 548, L237

Moreno-Insertis, F. 1986, A&A, 166, 291

Muller, R., & Roudier, Th. 1992, Sol. Phys., 141, 27
Muller, R., Roudier, Th., Vigneau, J., & Auffret, H. 1994, A&A, 283, 232
Musielak, Z. E. & Rosner, R. 1987, ApJ, 315, 371
Musielak, Z. E., Rosner, R., & Ulmschneider, P. 1989, ApJ, 337, 470
Musielak, Z. E., Rosner, R., Gail, H. P., & Ulmschneider, P. 1995, ApJ, 448, 865
Nordlund, Å. 1983, in Solar and Stellar Magnetic Fields: Origins and Coronal Effects, IAU Symp. No. 102, Ed. J. O. Stenflo (Reidel, Dordrecht) 79
Nordlund, Å., & Stein, R. F. 1989, in Solar and Stellar Granulation, Eds. R. J. Rutten, & G. Severino, NATO ASI Series C Vol. 263 (Kluwer, Dordrecht), 453
Nordlund, Å, & Stein, R. F.: 1990, in Solar Photosphere: Structure, Convection and Magnetic Fields, IAU Symp. No. 138, Ed. J. O. Stenflo, (Kluwer, Dordrecht), 191
Osterbrock, D. E. 1961, ApJ, 134, 347
Parker, E. N. 1955, ApJ, 122, 293
Parker, E. N. 1963, ApJ, 138, 552
Parker, E. N. 1978, ApJ, 221, 368
Parker, E. N. 1979, Cosmical Magnetic Fields: Their origin and their activity (Clarendon Press, Oxford)
Peter, H. 2000, A&A, 360, 761
Peter, H. 2001, A&A, 374, 1108
Pizzo, V. J., MacGregor, K. B., & Kunasz, P. B. 1993a, ApJ, 404, 788
Pizzo, V. J., MacGregor, K. B., & Kunasz, P. B. 1993b, ApJ, 413, 764
Priest, E. R. 1982, Solar Magnetohydrodynamics, (D. Reidel, Dordrecht)
Rajaguru, S. P., & Hasan, S. S. 2000, ApJ, 544, 522
Roberts, B., & Webb, A. R. 1978, Sol. Phys., 56, 5
Roberts, B., & Ulmschneider, P. 1998, in Lecture Notes in Physics (Springer Verlag, Heidelberg), Vol. 489, 75
Rutten, R. J., de Pontieu, B., & Lites, B. W. 1999, in High Resolution Solar Physics: Theory, Observations, and Techniques, Eds. T. R. Rimmele, K. S. Balasubramaniam, & R. R. Radick, ASP Conf. Ser., Vol. 183, 383
Ryutov, D. D., & Ryutova, M. P. 1976, Sov. Phys. J. E. T. P., 43, 491
Sánchez Almeida, J., & Lites, B. W. 2000, ApJ, 532, 1215
Schmitt, D. 1993, in The Cosmic Dynamo, IAU Symp. No. 157, Eds. F. Krause, K.-H. Rädler, & G. Rüdiger Eds., (Kluwer, Dordrecht), 1
Schrijver, C. J., & Harvey, K. L. 1994, Sol. Phys., 150, 1
Schüssler, M., Caligari, P., Ferriz-Mas, A., & Moreno-Insertis, F. 1994, A&A, 281, L69
Solanki, S. K. 1993, Space Sci. Rev., 63, 1
Solanki, S. K. 1999, in Solar and Stellar Activity: Similarities and Differences, Eds. C. J. Butler, & J. G. Doyle, A. S. P. Conf. Series, Vol. 158, 109
Solanki, S. K., Zufferey, D., Lin, H., Rüedi, I., & Kuhn, J. R. 1996, A&A, 310, L33
Spiegel, E. A., & Weiss, N. O. 1980, Nature, 287, 616
Spiegel, E. A., & Zahn, J. P. 1992, A&A, 265, 106
Spruit, H. C. 1976, Sol. Phys., 50, 269
Spruit, H. C. 1981, A&A, 98, 155
Spruit, H. C. 1982, Sol. Phys., 75, 3
Spruit, H. C., & van Ballegooijen, A. A. 1982, A&A, 106, 58
Spruit, H. C., & Zweibel, E. G. 1979, Sol. Phys., 62, 15
Steenbeck, M., Krause, F., & Rädler, K. H. 1966, Z. Naturforsch 21a, 369
Stein, R. F., & Nordlund, Å. 2000, Sol. Phys., 192, 91
Steiner, O., & Stenflo, J. O. 1989, in Solar Photosphere: Structure, Convection and Magnetic Field, IAU Symp. No. 138, Ed. J. O. Stenflo, (Kluwer, Dordrecht), 181

Steiner, O., Knölker, M., & Schüssler, M. 1994, in Solar Surface Magnetism, Eds. R. J. Rutten, C. J. Schrijver, NATO ASI Series C Vol. 433 (Kluwer, Dordrecht), 441

Steiner, O., Grossmann-Doerth, U., Schüssler, M., & Knölker, M. 1996, Sol. Phys. 164, 223

Steiner, O., Grossmann-Doerth, U., Knölker, M & Schüssler, M. 1998, ApJ, 495, 468

Stenflo, J. O. 1994, Solar Magnetic Fields: Polarized radiation diagnostics (Kluwer, Dordrecht)

Stix, M. 1989, The Sun (Springer Verlag, Berlin)

Takeuchi, A. 1993, PASJ, 45, 811

Takeuchi, A. 1995, PASJ, 47, 331

Takeuchi, A. 1999, ApJ, 522, 518

Thomas, J. H., & Weiss, N. O. 1992, Sunspots: Theory and Observations, Nato ASI Series C Vol. 375 (Kluwer, Dordrecht), 3

Ulmschneider, P. 1998, in Space Solar Physics, theoretical and observational issues in the context of the SOHO mission, Eds. J. C. Vial, K. Bocchialini, & P. Boumier (Springer Verlag, Berlin), 77

Ulmschneider, P., & Musielak, Z. E. 1998, A&A, 338, 311

Ulmschneider, P., Kalkofen, W., Nowak, T. & Bohn, H. U. 1977, A&A, 54, 61

Ulmschneider, P., Zähringer, K., & Musielak, Z. E. 1991, A&A, 241, 625

Unno, W., & Ando, H. 1979, Geophys. Astrophys. Fluid Dyn., 12, 107

van Ballegooijen, A. A. 1982, A&A, 113, 99

van Ballegooijen, A. A. 1983, A&A, 118, 275

Venkatakrishnan, P. 1983, J. Astrophys. Astron., 4, 135

Venkatakrishnan, P. 1985, J. Astrophys. Astron., 6, 21

Vernazza, J. E., Avrett, E. H., & Loeser, R. 1981, ApJS, 45, 635

von Uexküll, M., & Kneer, F. 1995, A&A, 294, 252

Webb, A. R., & Roberts, B. 1978, Sol. Phys., 59, 249

Weiss, N. O. 1966, Proc. Roy. Soc., A293, 310

Weiss, N. O. 1981, J. Fluid Mech., 108, 247

Wikstøl, Ø., Hansteen, V. H., Carlsson, M., & Judge, P. G. 2000, ApJ, 531, 1150

Zhugzhda, Y. D., Bromm, V., & Ulmschneider, P. 1995, A&A, 300, 302

Zwaan, C., & Harvey, K. L. 1994, in Solar Magnetic Fields, Eds. M. Schüssler, & W. Schmidt (Cambridge University Press), 27

Solar Magnetic Fields

P. Venkatakrishnan

Udaipur Solar Observatory, Physical Research Laboratory
P. B. No. 198, Udaipur 313001, India
email: pvk@prl.ernet.in

Abstract. The proximity of the Sun allows us to make detailed measurements on the properties of solar magnetic fields. The long term systematic changes in the solar magnetic field pattern indicate a global origin. A global dynamo can be sustained by the interaction of solar convection with solar rotation. The rudiments of such a dynamo mechanism are discussed. Some recent issues arising out of new theoretical and observational developments are mentioned. The importance of magnetic topology for various solar phenomena is highlighted. Finally, a few methods of measuring solar magnetic fields are described.

1 Introduction

Galileo turned a telescope to the Sun, and found that the Sun's face was not pure white, but had several dark spots on it. Careful recording of the sunspots' positions, day after day, for many years by many scientists not only revealed the solar rotation but also a curious waxing and waning of the number of sunspots with a rhythm of about eleven years. The names of Schwabe, Carrington, Maunder, Wolf, and Spörer, are associated with the exciting story of the discovery of the sunspot cycle (e.g., Schwabe 1844). What made the number of these spots increase and decrease in time? Why do spots appear darker than their surroundings? These were some of the burning questions that arose at that time. Although some progress has been made towards answering these questions, we are still far from a complete understanding.

The first step towards a physical understanding of sunspots became possible when George Ellery Hale, of Mount Wilson Observatory, detected magnetic fields in sunspots in 1908 (Hale 1908). Hale made this discovery by noting the splitting of spectral lines observed from sunspots and attributing it to the 1896 discovered Zeeman effect, named after its discoverer P. Zeeman (Nobel prize 1902). Hale managed to detect the circular polarisation of the light in spectral lines observed from sunspots that confirmed the magnetic origin for the splitting. When Hale continued his observations of sunspot fields, he found that the sunspots typically occur in bipolar groups where the preceding spots (in the direction of the solar rotation) in a given solar hemisphere always had the same polarity, and the following spots had the opposite polarity. In addition, he and his collaborators (Hale & Nicholson 1938) found that the magnetic polarities of the bipolar sunspot groups in the northern and southern hemispheres of the Sun were precisely reversed. This polarity law was seen to reverse its sign in the next consecutive 11 year sunspot cycle (see Fig. 12 of Ambastha, this volume). This pointed to the fact that the sunspot fields were part of a global magnetic field and the mechanism producing the magnetic field had to depend on global and fundamental properties of the Sun.

The theory of conducting fluids became fully developed in the middle of the 20th century, and culminated in the discovery of MHD waves by Hannes Alfvén (Alfvén 1942), for which he received the Nobel Prize in 1970. This new branch of physics began to be vigorously applied to the problem of magnetic field production in the Sun. It was realised that convection, that normally transports heat in the convection zone very efficiently, can no longer occur with the same vigour in sunspots, because the magnetic field dampens the movements of the plasma. When the energy transport is impeded, less light will reach the surface of the sunspots, making them appear dark. There was also some progress in understanding the processes responsible for the appearance and disappearance of magnetic flux with a periodicity of 11 years. But it was also seen that the magnetic fields vary on shorter time scales in a quite unpredictable fashion, producing spectacular phenomena like solar flares and coronal mass ejections. The energy for these eruptions seems to be stored in the highly twisted structure of the magnetic field in the solar corona. Presently, we can only extrapolate the coronal magnetic field from the vector magnetic field measured at the photosphere (Fig. 1). Thus, measurement of all 3 components of the solar magnetic field has become a priority for persons interested in understanding the physical causes for solar flares.

In what follows, we shall look initially at the magnetohydrodynamic approximation in stellar plasmas and use these for examining a simple model for the generation of solar magnetic fields. We shall also discuss a few interesting points about magnetic field topology in the context of solar flares and coronal heating. Finally, we shall look at some methods of measuring the solar magnetic fields.

2 Magnetohydrodynamic Approximation in Stellar Plasmas

In stellar atmospheres one usually has neutral, partially or fully ionised gases and no polariseable or magnetiseable substances. All charges and currents are explicitly taken into account, thus the electric displacement vector $\boldsymbol{D}$ is equal to the electric field strength $\boldsymbol{E}$, and the magnetic induction $\boldsymbol{B}$ is equal to the magnetic field strength $\boldsymbol{H}$ (Jackson 1969). In the stellar environment one therefore can make the following four simplifying assumptions, called *magnetohydrodynamic (MHD) approximation*:

I One has gas velocities v which are small compared with the speed of light c_L

$$\frac{v}{c_L} \equiv \widetilde{\gamma} << 1 \,. \tag{1}$$

II If L and τ are characteristic lengths and times, respectively, we assume

$$\frac{L/\tau}{c_L} = \frac{v_{\text{ph}}}{c_L} \equiv \widetilde{\beta} << 1 \,. \tag{2}$$

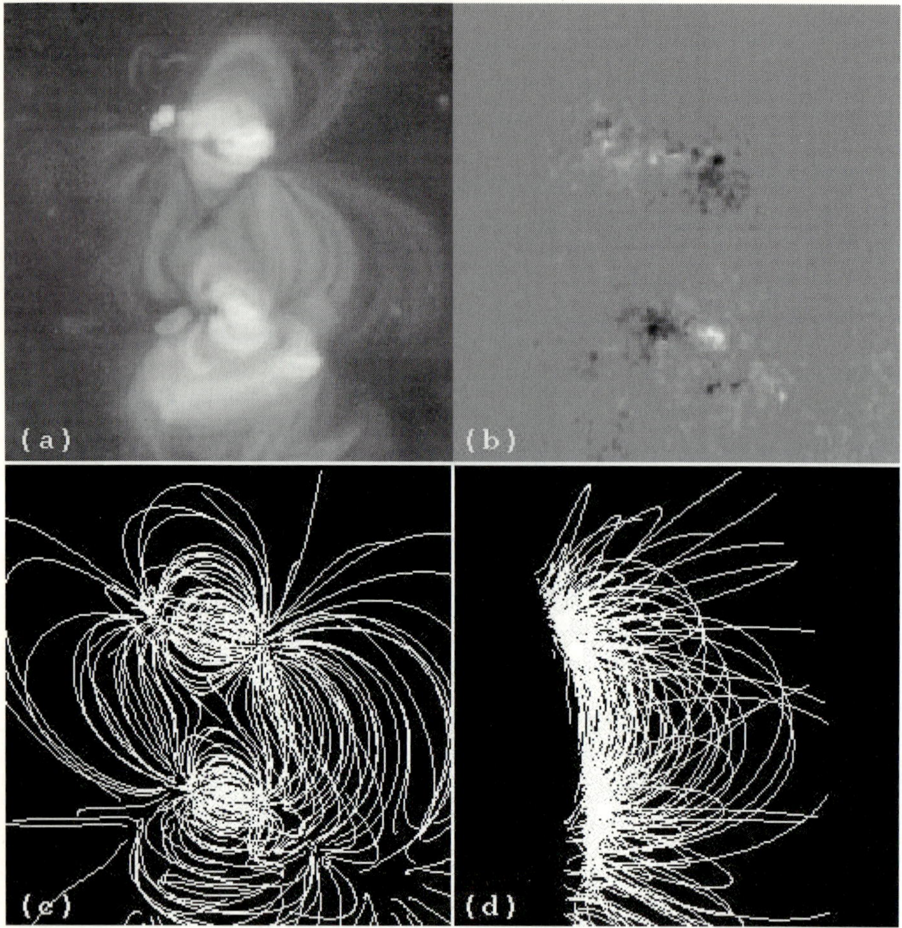

Fig. 1. Example of the extrapolation of magnetic fields. The coronal magnetic fields can be calculated under certain assumptions using the photospheric magnetic fields as a boundary conditions. **(a)** Soft X-ray image of coronal loops obtained by the Soft X-ray Telescope aboard the Yohkoh satellite on January 4, 1994. **(b)** Line-of-sight component of the underlying photospheric magnetic field, with bright and dark intensities representing N and S magnetic polarities. **(c)** Extrapolated 3-D magnetic field seen as a projection on to the solar disk. **(d)** Side view of the same 3-D field

Here $v_{\rm ph}$ is used for L/τ and represents the speed of propagation of inhomogeneities of size L in a characteristic time τ.

III One has only very weak electrical field strengths E compared to the magnetic field strength B

$$\frac{E}{B} \equiv \widetilde{\alpha} << 1 \,. \tag{3}$$

This allows us to neglect the displacement current $\frac{1}{c_L}\frac{\partial E}{\partial t}$ in Ampère's law of the Maxwell's equations since

$$\nabla \times \boldsymbol{B} \approx \frac{B}{L} >> \frac{1}{c_L}\frac{\partial E}{\partial t} \approx \frac{\tilde{\alpha} B}{c_L \tau}\frac{L}{L} = \tilde{\alpha}\tilde{\beta}\frac{B}{L} . \tag{4}$$

IV In astrophysical plasmas one always has Ohm's law, and with the above conditions as well as Maxwell's equations one can derive the generalised Ohm's law

$$\boldsymbol{J} = \sigma\left(\boldsymbol{E} + \frac{\boldsymbol{v}}{c_L} \times \boldsymbol{B}\right) . \tag{5}$$

where $\boldsymbol{E}$ and $\boldsymbol{B}$ are the electric and magnetic field strength in the rest frame and σ is the electrical conductivity. From the induction law using the generalised Ohm's law we have

$$\frac{\partial \boldsymbol{B}}{\partial t} = -c_L\left(\nabla \times \boldsymbol{E}\right) = \nabla \times \left(\boldsymbol{v} \times \boldsymbol{B}\right) - c_L \nabla \times \frac{\boldsymbol{J}}{\sigma} . \tag{6}$$

Here using Ampère's law one finds

$$c_L \nabla \times \frac{\boldsymbol{J}}{\sigma} = \frac{c_L^2}{4\pi\sigma}\nabla \times \left(\nabla \times \boldsymbol{B}\right) = \frac{c_L^2}{4\pi\sigma}\left(\nabla \underbrace{\nabla \cdot \boldsymbol{B}}_{=0} - \nabla^2 \boldsymbol{B}\right) . \tag{7}$$

This leads to the *conservation law of the magnetic field*

$$\frac{\partial \boldsymbol{B}}{\partial t} - \nabla \times \left(\boldsymbol{v} \times \boldsymbol{B}\right) = \frac{c_L^2}{4\pi\sigma}\nabla^2 \boldsymbol{B} = \lambda \nabla^2 \boldsymbol{B} , \tag{8}$$

where $\lambda = c_L^2/4\pi\sigma$ is the magnetic diffusivity.

3 Generation of Magnetic Fields

The Sun completely reverses its magnetic field in eleven years. This demonstrates that over the last 4.6 billion years the Sun has lost the fossil magnetic field which was present from the collapse of the interstellar cloud at its formation. The magnetic field presently on the Sun as well as the sunspot cycle must therefore be generated in situ by processes which depend on the structure and physics of the solar interior. For almost a century, a large effort has gone into developing a so called dynamo theory to describe how stellar and galactic magnetic fields are produced. Although such a theory is still not in place, there has been great progress both theoretically and observationally in the understanding of this problem. It is recognised that the dynamo process to generate magnetic fields needs two main ingredients: convection and rotation.

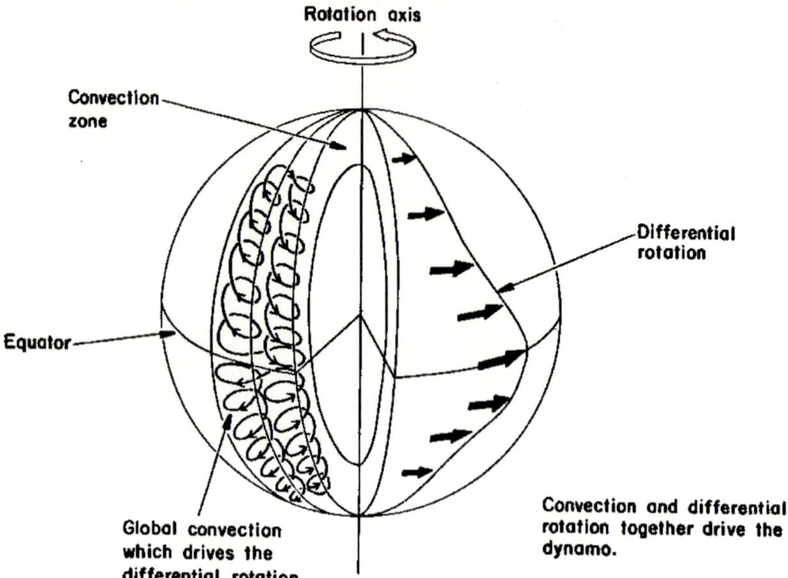

Fig. 2. Numerical simulation of solar convection and the generation of the differential rotation, after Gilman (1983)

3.1 Turbulent Dynamo and Mean Field Magnetohydrodynamics

As observations show that the solar cycle represents a temporal sequence of poloidal and toroidal field systems, the question is how these differently oriented systems arise and what role they play in their mutual generation. Here the generation of the toroidal field system seems to be reasonably well understood as a consequence of the action of the convection zone. Global numerical convection simulations by Gilman (1983) show that the solar convection zone generates the phenomenon called differential rotation, namely that the solar rotation at the equator is much faster than the rotation at higher latitudes (see Fig. 2). For recent predictions of the differential rotation in G and K stars of different rotation rates see Kitchatinov & Rüdiger (1999). This differential rotation is observationally well established as seen e.g., in Fig. 3 (Schröter 1985), (see also Antia, this volume).

How differential rotation generates toroidal magnetic field systems is shown in Fig. 4. Assume that in Fig. 4a one has a purely poloidal field configuration. Due to the action of the differential rotation these magnetic field lines are subsequently wound up around the equator (see Figs. 4b to 4d). This picture is very attractive because these wound up subsurface fields will suffer buoyancy and erupt to the surface when the field strength B increases to sufficiently high values. This naturally explains the Hale–Nicholson (Hale & Nicholson 1938) polarity law of sunspot pairs, as well as the slight tilt of the bipolar regions against the latitude circles (Joy's law, Hale et al. 1919). The problem is how from the

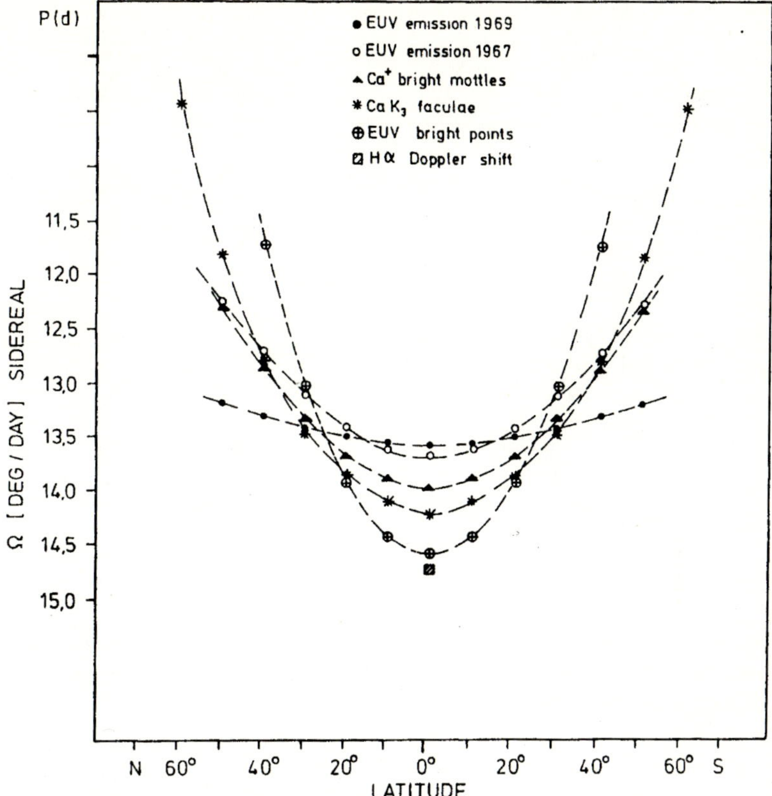

Fig. 3. Observed solar differential rotation, after Schröter (1985)

toroidal field of Fig. 4d an oppositely directed poloidal field of Fig. 4a can be generated to complete the cycle.

A step towards the solution of this problem was made when it was realised that the helical flows associated with the motion of turbulent gas bubbles in the convection zone, which rise due to buoyancy, provided a mechanism to generate poloidal fields (Parker 1955). A rising gas bubble expands as it moves into regions of lower gas pressure. This generates horizontal motions which in a rotating star are dominated by Coriolis force and lead to helical flows. Expanding flows in the northern hemisphere of a rotating star (similar to the high pressure areas in the Earth's northern hemisphere) suffer a deviation to the right, which because of the frozen-in property of the highly conducting solar gas take the magnetic field along (Fig. 5). That is, expanding flows suffer a clockwise circulation when seen from above. Therefore as shown in Fig. 5 starting from the toroidal configuration the magnetic field gains a poloidal component which is oppositely directed to the poloidal component of Fig. 4a. This idea was worked out in greater detail in

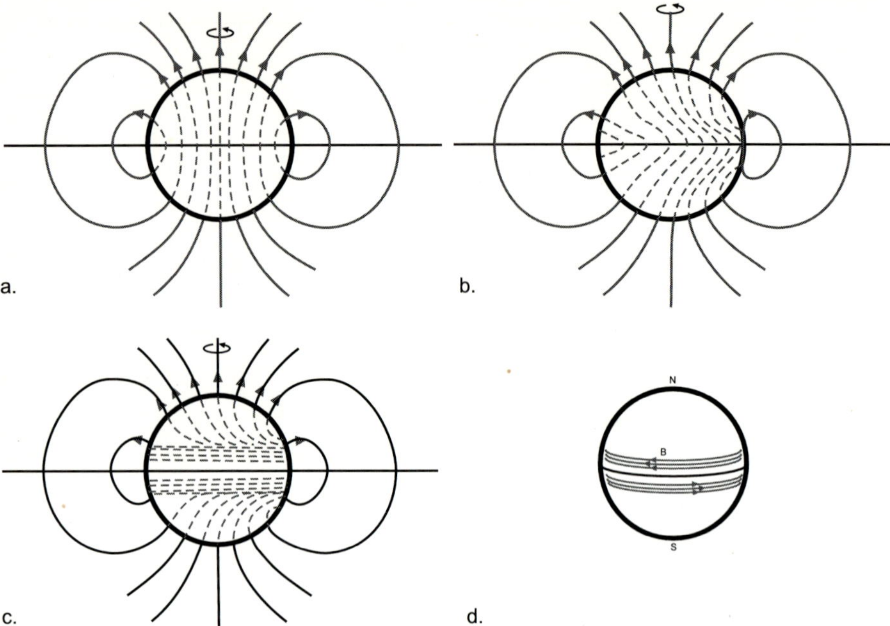

Fig. 4. Generation of toroidal magnetic fields from poloidal fields by the action of the differential rotation **(a)** Purely poloidal field. **(b)** and **(c)** Progressive winding of the field lines. **(d)** Purely toroidal fields

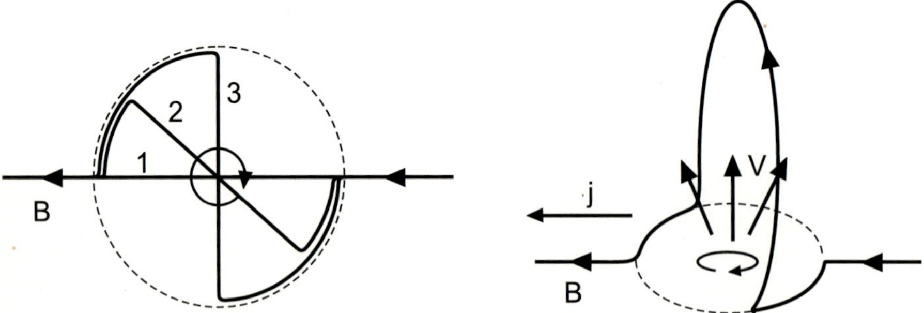

Fig. 5. Generation of poloidal fields from toroidal fields by the helical flows of rising gas bubbles in the convection zone of a rotating star. Coriolis forces acting on the expanding flows (indicated by $\boldsymbol{v}$) rotate the field ($1 \rightarrow 3$). Left: The interaction of the gas bubble with the magnetic field line is seen from the top, right: from the side

the mean field dynamo theory of Steenbeck, Krause & Rädler (1966) which in the following is shortly outlined (see also Choudhuri 1999).

With the magnetic diffusivity λ, (8) can be written

$$\frac{\partial \boldsymbol{B}}{\partial t} = \boldsymbol{\nabla} \times (\boldsymbol{v} \times \boldsymbol{B} - \lambda \boldsymbol{\nabla} \times \boldsymbol{B}) \ . \tag{9}$$

Consider a horizontally directed magnetic flux tube with field strength $\boldsymbol{B_0}$ in the convection zone which due to rising convective bubbles gets perturbed such that one has a total field strength $\boldsymbol{B} = \boldsymbol{B_0} + \boldsymbol{B'}$. Here we assume $B' < B_0$. The solar gas has a systematic (differential, meridional) motion with velocity $\boldsymbol{v_0}$ over which a turbulent fluid motion $\boldsymbol{v'}$ is superposed resulting in $\boldsymbol{v} = \boldsymbol{v_0} + \boldsymbol{v'}$, where $\boldsymbol{v'}$ is isotropic and the time average $\overline{\boldsymbol{v'}} = 0$. Introducing $\boldsymbol{B}$ and $\boldsymbol{v}$ in (9) one can identify two equations which contain the zeroth and first order terms

$$\frac{\partial \boldsymbol{B_0}}{\partial t} = \boldsymbol{\nabla} \times (\boldsymbol{v_0} \times \boldsymbol{B_0} + \boldsymbol{F} - \lambda \boldsymbol{\nabla} \times \boldsymbol{B_0}) \ , \tag{10}$$

$$\frac{\partial \boldsymbol{B'}}{\partial t} = \boldsymbol{\nabla} \times (\boldsymbol{v_0} \times \boldsymbol{B'} + \boldsymbol{v'} \times \boldsymbol{B_0} + \boldsymbol{G} - \lambda \boldsymbol{\nabla} \times \boldsymbol{B'}) \ , \tag{11}$$

where

$$\boldsymbol{F} = \overline{\boldsymbol{v'} \times \boldsymbol{B'}} \ , \tag{12}$$

$$\boldsymbol{G} = \boldsymbol{v'} \times \boldsymbol{B'} - \overline{\boldsymbol{v'} \times \boldsymbol{B'}} \ . \tag{13}$$

Here the overbar indicates time averaging. Equation (10) describes the mean behaviour and (11) the fluctuations. That this splitting is valid can be seen by adding (10) and (11). Note that if we multiply the B' and B_0 in (10) to (13) with a factor f then the equations remain unchanged. Therefore B' and F are proportional to B_0 and one can write

$$\boldsymbol{F} = \alpha \boldsymbol{B_0} - \beta \boldsymbol{\nabla} \times \boldsymbol{B_0} = \alpha \boldsymbol{B_0} - \beta \sum_{jk} \epsilon_{ijk} \frac{\partial B_{0k}}{\partial x_j} \ . \tag{14}$$

Here we use the well known ϵ operator. Let i, j, k be some permutation of 1, 2, 3, then $\epsilon_{ijk} = 1$ if i, j, k are cyclic, $\epsilon_{ijk} = -1$ if i, j, k are anticyclic and $\epsilon_{ijk} = 0$ if any two of the indices are equal. The quantities α and β are to be determined functions of $\boldsymbol{v'}$.

Note that all terms on the right hand side of (11) except one are proportional to $\boldsymbol{B'}$. Keeping $\boldsymbol{B'}$ small enough therefore leads us to

$$\frac{\partial \boldsymbol{B'}}{\partial t} \approx \boldsymbol{\nabla} \times (\boldsymbol{v'} \times \boldsymbol{B_0}) \ . \tag{15}$$

This is valid for the short time interval τ during which a convective bubble flows past the field $\boldsymbol{B}_0$ and perturbs it. Integrating one therefore finds that

$$\boldsymbol{B'} = \tau \boldsymbol{\nabla} \times (\boldsymbol{v'} \times \boldsymbol{B_0}) \ . \tag{16}$$

Expanding the double cross product, using $\nabla \cdot \boldsymbol{B}_0 = 0$ and $\nabla \cdot \boldsymbol{v}' = 0$ (because of incompressibility), one finds

$$\boldsymbol{B}' = \tau\,(\boldsymbol{B_0} \cdot \nabla)\,\boldsymbol{v}' - \tau\,(\boldsymbol{v}' \cdot \nabla)\,\boldsymbol{B_0}\,. \tag{17}$$

From this one obtains

$$\boldsymbol{F} = \overline{\boldsymbol{v}' \times \boldsymbol{B}'} = \sum_{jk} \epsilon_{ijk} \overline{v_j' B_k'} = \tau \sum_{jkl} \epsilon_{ijk} \overline{v_j' B_{0l} \frac{\partial v_k'}{\partial x_l}} - \tau \sum_{jkl} \epsilon_{ijk} \overline{v_j' v_l' \frac{\partial B_{0k}}{\partial x_l}}\,, \tag{18}$$

which can be written

$$F_i = \sum_j \alpha_{ij} B_{0j} - \sum_{jk} \beta_{ijk} \frac{\partial B_{0k}}{\partial x_j}\,, \tag{19}$$

where

$$\alpha_{ij} = \tau \sum_{kl} \epsilon_{ilk} \overline{v_l' \frac{\partial v_k'}{\partial x_j}}\,, \qquad \beta_{ijk} = \tau \sum_l \epsilon_{ilk} \overline{v_l' v_j'}\,. \tag{20}$$

Comparison with (14) where the isotropy of $\boldsymbol{v}'$ is used gives

$$\alpha = \frac{1}{3} \tau \overline{\boldsymbol{v}' \cdot (\nabla \times \boldsymbol{v}')}\,, \qquad \beta = \frac{1}{3} \tau \overline{\boldsymbol{v}' \cdot \boldsymbol{v}'}\,. \tag{21}$$

Here the factor 1/3 arises from isotropy because $\overline{v_x'^2 + v_y'^2 + v_z'^2} = 3\overline{v_x'^2}$.

The component $\boldsymbol{v}' \cdot \nabla \times \boldsymbol{v}'$ has a non-vanishing contribution in helical flows because here both $\nabla \times \boldsymbol{v}'$ and $\boldsymbol{v}'$ have components in vertical direction. It is seen that such helical flow fields arise precisely when two conditions are fulfilled for a star: the presence of convection and rotation. Indeed, it is observationally well established for late-type stars (with surface convection zones) that the magnetic field coverage is larger the more rapidly the star rotates. In the 1970's and 80's various simulations were undertaken to reproduce the sunspot cycle by assuming a flow velocity

$$\boldsymbol{v} = \Omega(r, \vartheta) r \cos\vartheta \boldsymbol{e}_\phi + \boldsymbol{v_P}\,, \tag{22}$$

assuming various dependencies of $\alpha(r, \vartheta)$ and $\Omega(r, \vartheta)$ as functions of radius r and latitude ϑ. Here Ω is the solar angular velocity and ϕ the longitude angle. $\boldsymbol{v_P}$ is a meridional gas flow which usually was neglected. Figure 6 shows an example of such a simulation by Stix (1976) over 11 years, where the left side of the figures shows the toroidal field components (dashed and solid lines indicate opposite polarities) and the right side of the figures show the poloidal fields. By adjusting the magnitude and variation of $\alpha(r, \vartheta)$ and $\Omega(r, \vartheta)$ good agreement with the observed butterfly diagram and sunspot cycle could be achieved. Here α and Ω are not independent of each other because already Parker (1955) found that these quantities must satisfy the relation

$$\alpha \frac{\mathrm{d}\Omega}{\mathrm{d}r} \leq 0\,. \tag{23}$$

While the dynamo models of Stix (1976) and others (see Choudhuri 1999) appeared to solve the problem of the magnetic field generation, the unsatisfactory feature in these models remained how to predict $\alpha(r, \vartheta)$ and $\Omega(r, \vartheta)$ from theoretical simulations or from observations.

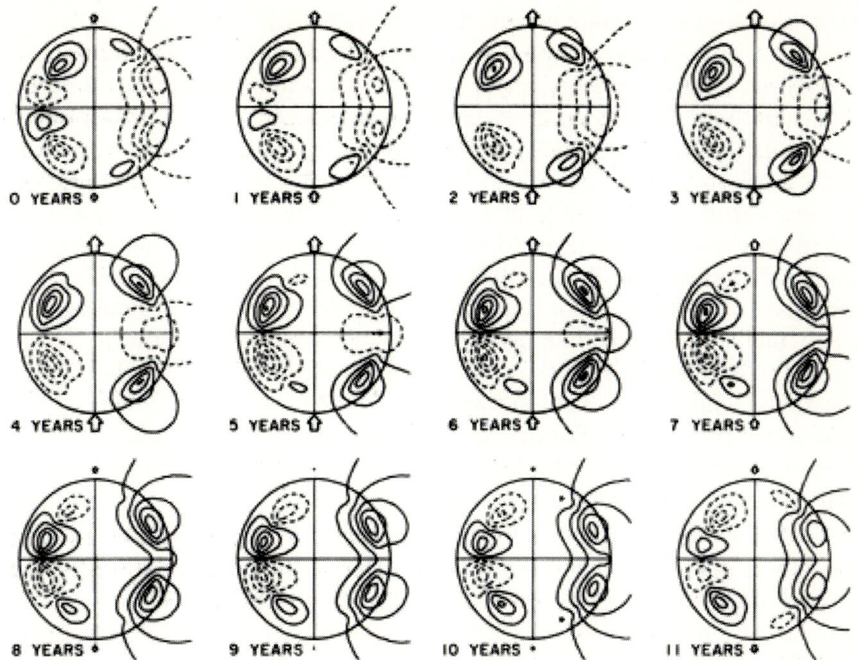

Fig. 6. Simulation of the solar magnetic field cycle using the mean field $\alpha\Omega$ dynamo theory, after Stix (1976). The right hand sides of the figures show the poloidal, and the left hand sides the toroidal fields. *Solid* and *dashed* lines indicate opposite polarities

3.2 Dynamo in the Overshoot Layer

The mean field dynamo theory appeared to be a very appealing explanation of the formation of the solar magnetic fields and the sunspot cycle. The differential rotation from the poloidal fields generated the toroidal field system and the helical turbulence in the convection zone in turn led to the poloidal field system from the toroidal fields. This fine picture, however, in the 1980's was troubled by two developments. The first was that detailed simulations of the buoyant rise of magnetic flux tubes in the convection zone became available and the second that helioseismology provided the much sought after observational constraint on the radius and latitude dependent angular velocity distribution $\Omega(r,\vartheta)$ of the Sun (see Antia, this volume).

Already Parker (1975, 1979) found that once a magnetic flux tube becomes buoyant in the convection zone, the convective instability and buoyancy reinforce each other with the consequence that the tube floats up rather quickly. A way out of this problem was found by Spiegel & Weiss (1980), van Ballegooijen (1982) and others. They suggested that the toroidal magnetic flux tube could avoid rapid buoyant eruption if it were placed in the convectively stable overshoot layer (with a thickness of about 10^4 km) below the convection zone, where the sinking convective bubbles penetrate a fraction of a scale height into the stable layer. This idea obtained strong support from the observations of the angular velocity distribution by helioseismology which discovered a strong gradient $d\Omega(r,\vartheta)/dr$ precisely at this overshoot layer called the tachocline (Antia, this volume).

However, this new picture which placed the dynamo process for the poloidal field generation into the overshoot layer produced other problems. One was the fact that $d\Omega(r,\vartheta)/dr$ could no longer be adjusted to satisfy the observed sunspot cycle and moreover was very latitude dependent which in some instances violated (23) (see Antia this volume). In addition, studying the rise of magnetic flux tubes from the overshoot layer to the surface by taking the Coriolis forces into account, Choudhuri & Gilman (1987) and Choudhuri (1989) found that for field strengths around 10^4 G these forces were so strong that the tubes did not erupt radially but rather moved parallel to the solar rotation axis and surfaced at high latitudes in contradiction to the observed range of 0 to $\pm 50°$ latitude of sunspot appearance. Only if the field strengths were increased to the range of 10^5 G did the flux tubes erupt radially in agreement with the observations. Further flux tube eruption studies by D'Silva & Choudhuri (1993) showed that only when assuming such strong fields could Joy's law be reproduced. Unfortunately, such strong fields immediately pose the problem that now the helical flows in the overshooting downward-moving turbulent gas bubbles no longer were strong enough to appreciably twist the toroidal field in order to produce the poloidal field. This meant that the whole idea of a helical turbulence driven dynamo appeared to be in difficulties.

3.3 Babcock–Leighton Picture and Hybrid Models

If convection driven helical turbulence is not the mechanism which generates poloidal fields from toroidal ones, what other processes are available? Here the investigation focused on Joy's law where the preceding spot of a bipolar group lies systematically at lower latitudes compared to the following spot. During the eruption process from toroidal fields, a significant poloidal field component has to be generated. Babcock (1961) and Leighton (1969) pointed out that the evolution of bipolar spot groups could cause the reversal of the magnetic polarity at the poles for the next sunspot cycle. They showed that this evolution is a consequence of the differential rotation and the granular flows which disperse the sunspot fields. They found that the magnetic field region of the preceding spots remains relatively concentrated while the magnetic field region of the following spots of opposite polarity disperses widely and to higher latitudes. Migration of the collected magnetic regions of the following spots to the pole, possibly aided by a poleward meridional flow, were supposed to be instrumental in the reversal of the poloidal fields. However, compared with the mathematical precision of the mean field dynamo theory, this scenario remained in a heuristic and semi-qualitative state of development despite of the numerical work by DeVore, Sheeley & Boris (1984) and Sheeley, Wang & Harvey (1989).

Choudhuri, Schüssler and Dikpati (1995, 1997) suggested a hybrid model by invoking the Babcock and Leighton ideas for the poloidal field generation Here the problem of the violation of (23) still lingered. However, by assuming a meridional circulation with a magnitude of a few m s^{-1} moving at the surface towards the pole, and therefore at the bottom of the convection zone moving towards the equator, Choudhuri, Schüssler and Dikpati's model was able to override this inequality which for a long time had almost been taken as a dogma. For more work on recent hybrid models see Dikpati & Charbonneau (1999), Charbonneau & Dikpati (2000), Nandy & Choudhuri (2001), Dikpati & Gilman (2001). It is hoped that continuing work in this direction and the application of massive MHD simulations should lead to a satisfactory dynamo theory.

3.4 Inputs from Helioseismology

We must recall that the discussion so far concerns only the kinematic dynamos, namely dynamos where the large scale velocity field is prescribed arbitrarily. In a realistic case, the field will react on the flow and a dynamical dynamo results. A consistent dynamo must therefore be compatible with the dynamical forces as well. A fully consistent computer model was developed by Gilman (1986), by including the effect of the Lorentz force and solving the equation of motion along with the induction equation. The results showed a differential rotation that was consistent with what was expected from the Taylor–Proudman theorem, namely a rotation which is constant on cylinders. This produced a butterfly diagram going from the equator to the pole which is opposite to the observed shift of sunspot emergence locations from mid-latitudes at the beginning of the cycle, towards the equator at the end of the cycle. This lack of success in the numerical modelling can be attributed to our incomplete knowledge of MHD turbulence. The recent successes of helioseismology (see Antia, Chitre, this volume) have enabled us to get an observational handle on the behaviour of solar rotation as a function of depth (Schou et al. 1998; Charbonneau et al. 1999). This observed behaviour is far removed from the constant rotation on cylinders that is expected from the Taylor–Proudman theorem. The challenge now is to be able to model turbulence properly so as to simulate the observed interior dynamics. Models of MHD turbulence can perhaps be improved by an intensive study of the interaction of solar surface magnetic fields and velocity fields on as small a scale as possible. For this, we require a large-aperture, low-polarisation telescope located at a site with a large number of clear days during the year, a large number of clear hours during the day, and a consistent seeing throughout the day.

Further, attempts to detect the interior field (Antia, Chitre & Thompson 2000) show an upper limit of 300 kG at the base of the convection zone – a value that is not sufficient for the buoyancy to act and bring the field up to the surface. On the other hand, there is a shallow layer ($\sim 0.9R_{\odot}$) where the estimated field (20 kG) is sufficient for buoyancy (Antia, Chitre & Thompson 2000). The implications of these startling results for a dynamo theory are yet to be widely discussed in the literature. However, one might add that recent simulations of magnetoconvection show that the flows could indeed anchor the field even within the convection zone (Nordlund, Dorch & Stein 2000) which means that a shallow origin for the solar magnetic fields is not ruled out.

A shallow dynamo, operating near the solar surface, tends to produce toroidal fields at high latitudes (Dikpati & Charbonneau 1999; Küker, Rüdiger & Schultz

2001; Durney 1997). But one can get around this problem by invoking a meridional circulation that pulls these high latitude toroidal fields deep below the convection zone and brings them up later at the lower latitudes (Nandy 2002; Nandy & Choudhuri 2002). One of the virtues of this scenario is that it agrees with the helioseismic inference, that strong enough magnetic fields which are naturally buoyant cannot be produced at the base of the convection zone. If a meridional flow exists, then the non-buoyant fields would be carried upwards along with the flow until they become buoyant and rise on their own at greater heights.

4 Force Free Equilibria, Topology, Reconnection and Flares

The "frozen-in field" phenomenon which is a consequence of the large conductivity and the large spatial scales in the solar plasma, leads to curious consequences. One of such by products of the "frozen-in field" nature of the solar magnetic fields is its special magnetic topology. In the case of laboratory plasmas, we are used to the concept of a magnetic field that permeates all space. The only modification occurs due to the permeability of the medium. Diamagnetic materials are interesting since they tend to avoid magnetic fields, and the flux expulsion from superconductors is a curiosity. However, the solar plasma expels flux all the time, e.g., in the case of magneto-convection, the magnetic field is expelled from the upwelling portions of the convective cells and is concentrated at the downflowing boundaries of the cells (Weiss 1981). In all such solar phenomena, the topology of magnetic fields becomes important.

Choudhuri (1999, Chap. 15) defines *magnetic topology* as follows: "If two magnetic configurations $\boldsymbol{B}_1(x)$ and $\boldsymbol{B}_2(x)$ are such that one of them can be deformed into the other by continuous displacements *without cutting or pasting field lines anywhere*, then the two configurations are said to have the same magnetic topology." For example, a uniform magnetic field has the same topology as that produced by the effect of a convective cell interacting with a uniform field. A sunspot pair of opposite polarity where all the field-lines go from one polarity to another has the same magnetic topology as a single, isolated sunspot where all the field lines go from the sunspot back to the surrounding photosphere.

One important example of change in magnetic topology occurs in the case of magnetic reconnection. In a high conductivity plasma, cutting and pasting of lines can take place within regions of large magnetic field gradients (current sheets), but the magnetic fields may be taken to be "frozen" in the plasma outside the current sheets and magnetic topologies are preserved everywhere except in the current sheet. A commonly occurring situation where current sheets form is when oppositely directed fields are brought into close proximity by the action of external flows. The field line connections get changed as a result of the magnetic reconnection (Fig. 7).

For example, two bipoles arranged in a sequential way ($-+$, $-+$) could merge into a single extended bipole ($-+$). The consequent decrease in the magnetic

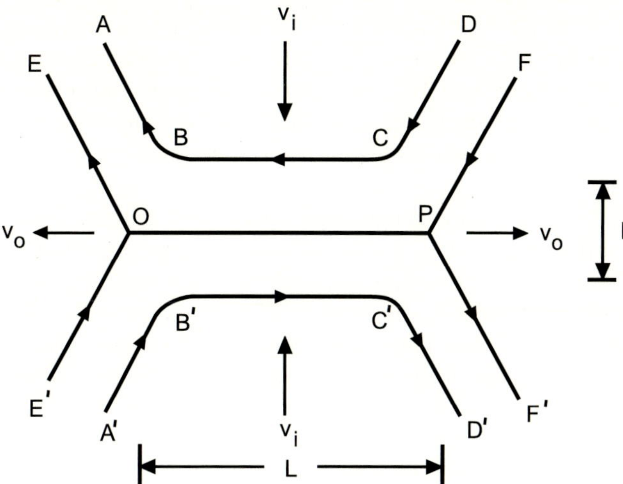

Fig. 7. A typical case of magnetic reconnection is shown. The field lines ABCD and $A^{'}B^{'}C^{'}D^{'}$ are pushed towards each other by external flows with velocity v_i, that occupy a region of size L. As they approach the line of zero field the frozen in condition no longer applies. Fluid is free to move across the field leading to a realignment or reconnection to the field lines $E^{'}OE$ and $F^{'}PF$. To condense the fluid the material is expelled with velocity v_0 over a region of size l

tension (which is inversely proportional to the radius of curvature of the magnetic loop), could sometimes lead to large force imbalances resulting in the eruption of the magnetic field configuration and the associated plasma. The motion of field lines perpendicular to the field direction can produce centrifugal acceleration of the plasma along the field direction (Venkatakrishnan 1984). The plasma acceleration could, in turn, result in the formation of a shock. A shock can often accelerate particles to high energies. These high energy particles produce non-thermal emission of high energy photons. The emission of high energy photons and energetic particles constitute a solar flare. Thus, magnetic reconnection is now recognised as an integral part of a solar flare as well as other eruptive phenomena like coronal mass ejections. These exotic solar events are described in Ambastha's contribution in this volume.

A chief motivation for the measurement of all 3 components of the photospheric magnetic field arises out of the need to guess the magnetic topology of active regions, especially in the solar corona where direct measurement of the field is at the limit of current capabilities. An important concept for the coronal magnetic fields is that of a "force-free" equilibrium. In the case of general MHD equilibrium, the Lorentz force $\boldsymbol{J}\times\boldsymbol{B}/c_L$ is balanced by the other ponderomotive forces. The Lorentz force can be subdivided into two components, the magnetic tension force $((\boldsymbol{B}\cdot\boldsymbol{\nabla})\boldsymbol{B}/4\pi)$ and the force due to the gradient of the magnetic pressure $(-\boldsymbol{\nabla}B^2/8\pi)$. In the case of active region coronal magnetic fields, the magnetic pressure force and the magnetic tension are individually larger than the other forces such as the gas pressure gradient or gravity by orders of magnitude. Even if one of the components of the Lorentz force (magnetic pressure or tension) exceeds the other by a few percent, the net magnetic force will be too large to be contained or balanced by the other available forces. Thus, the magnetic tension must almost exactly balance the magnetic pressure force in order to contain the magnetic fields within the corona.

The gradient of magnetic pressure generally acts in the outward direction from the Sun, since the fields generally decrease with height. The magnetic tension, on the other hand, acts in the downward direction since the radius of curvature of a typical active region loop points downwards. If the magnetic pressure gradient slightly exceeds the tension, then the field will expand outwards, seeking a configuration that has a lower magnetic pressure gradient to balance the tension. If the tension force is larger, then the field will submerge. The active region fields are observed to exist for weeks without significant evolution. In comparison, the dynamical relaxation time is given by L/V_A, where L is the typical length scale of active regions (30 000 km) and V_A is the Alfvén speed (300 km s^{-1}). This works out to 100 s, which is much smaller than the observed lifetime of the active region fields. This proves that the magnetic pressure and tension forces balance very precisely in the corona. Such fields, where the net Lorentz force vanishes, are called force-free fields.

A characteristic of force-free fields is that the current is directly proportional to the field, or

$$\boldsymbol{\nabla}\times\boldsymbol{B} = \frac{4\pi}{c_L}\boldsymbol{J} = \alpha\boldsymbol{B}\,. \tag{24}$$

If the factor α is constant in space, then the field is called a linear force-free field. There is a theorem called Taylor's theorem which states that the lowest energy state – to which a magnetic field relaxes via reconnections (on small scales), under a given large scale topology – is the linear force-free field. The constant α then is a measure of this lowest energy. The constant α is in turn determined by the magnetic topology, and therefore is an index of the topology of the magnetic field. It can also be shown that α is preserved along the field line.

Consider a magnetic field line that has both foot-points anchored in the photosphere. Plasma motions will twist the field line and force the field to have a given set of vector field components at each foot-point. If the field is force-free in the corona, then the force free constant α has to be same along the field line. But α is also a function of the 3 components of the vector magnetic field. Now, let us start with one set of magnetic field components at one foot-point (and thus with the corresponding value of α) and go along the field line to the other foot-point. The requirement of constant α along the field line will constrain the magnetic field components at the other foot-point to be a function of the components at the first foot-point. However, we cannot expect the twisting motions of the plasma (which determine the vector fields) at one end of the field line to be a function of the motions at the other end of the field line. We thus seem to have a paradoxical situation on our hands. According to Parker (1994), the only way of resolving this paradox is to relax the "continuity of field" condition. He suggests that magnetic discontinuities develop spontaneously in the solar corona, in response to the forcing by sub-photospheric plasma motions applied to the foot-points of the coronal loops. Parker goes a step further and suggests that this "breakdown of compatibility" and the resulting development of magnetic discontinuities accompanied by reconnection, provides the energy for heating the solar corona.

This idea seems very sound, but its observational verification must come from patient observations of the foot-point movements. Recent attempts in doing so (Schrijver et al. 1998) claim to provide verification for this concept, by establishing direct association of flux cancellation events with the events of enhancement in coronal brightness. The statistics of the flux cancellation events then provides a straightforward empirical estimate of the rate of energy deposited in the corona by such events. In my opinion, however, this is not a proper verification. The flux cancellation events could well have the required energy, but Parker's mechanism requires a change in topology to precede the creation of field discontinuities. Now, a change in topology can be monitored by looking at the evolution of helicity since this is related to the evolution of topology. A direct measure of helicity is obtained by dividing the vertical component of the electric current density J_z by the vertical component of the magnetic field B_z. Both quantities can be derived from a vector magnetogram. But the measurement of helicity on the required small scales would require advanced technology. The rapid steps taking place in the realisation of adaptive optics systems and the consequent improvement in spatial resolution of vector magnetograms lends hope to the possible verification of Parker's mechanism in the near future (e.g., Sankarasubramanian & Rimmele 2002).

5 Measurement of the Solar Magnetic Field

Let us now look at some aspects concerning the practical methods to measure the solar magnetic fields.

5.1 The Zeeman Effect

Zeeman (1897) discovered that the spectral lines from radiating atoms kept in a magnetic field showed a splitting. The difference in frequency $\Delta\nu$ of the Zeeman components is proportional to the frequency of Larmor precession of the atomic magnetic moment around the direction of the external field. The discovery of the Zeeman effect gave George Ellery Hale the much awaited tool to check whether sunspots were the sites of large magnetic field. Accordingly, Hale examined the spectra from sunspots in the Zeeman sensitive Fe lines of the Fraunhofer spectrum. To his great joy, as already said above, he found in 1908 that the spectral lines were split in sunspot regions, much in the same way as predicted by the Zeeman effect.

Considering the fact that the Zeeman effect was first seen in emission lines, and the fact that the solar lines are absorption lines (Fig. 8) which are considerably broadened by thermal motions of the hot solar plasma, it is indeed

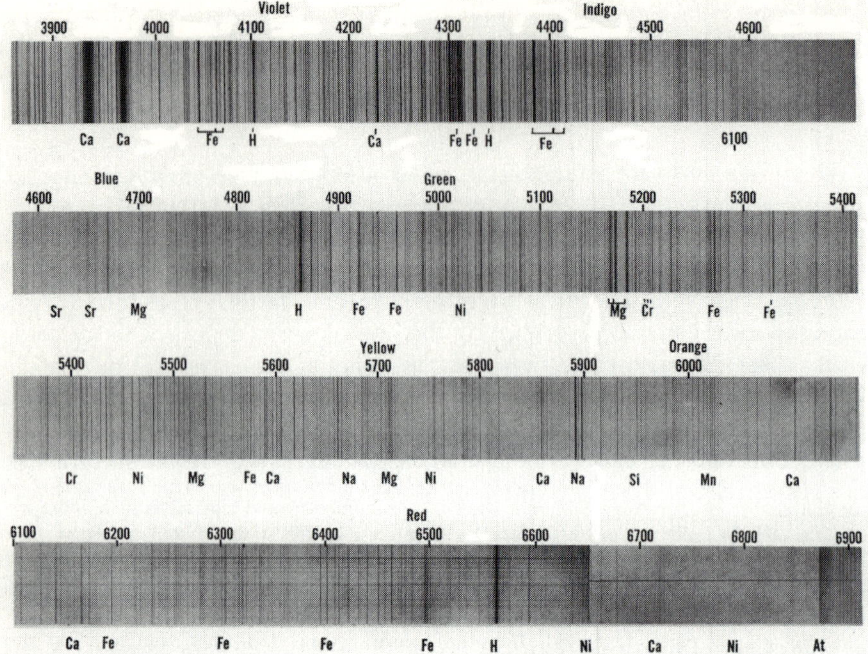

Fig. 8. Solar absorption spectrum observed with a spectrograph. Labels indicate the elements causing the spectral lines, the colours of the different parts of the spectrum are indicated, wavelengths are in Å (Mount Wilson and Palomar Observatories)

remarkable that Hale attempted to detect, and actually succeeded in detecting, the magnetic splitting of spectral lines from sunspots (See Fig. 30 of Bhatnagar, this volume).

For a strong field parallel to the line of sight, one observes a split line with two oppositely circular polarised components. While when the line of sight is perpendicular to the field, one has linear polarisation in the two split components parallel to the field and a third unshifted linearly polarised component perpendicular to the field. For weaker fields, a combination of Zeeman splitting and thermal broadening produces the following variation of polarisation along the line profile: The polarisation is zero in the continuum while it changes to right/left circular polarisation in the wings of the line with zero polarisation again at the line centre. This is the signature for the component of the magnetic field that is along the line of sight. For a purely transverse field, the Zeeman effect in an absorption line, such as observed in the solar photosphere, produces linear polarisation in the wings of the spectral line which is parallel to the azimuth of the transverse field while the line centre exhibits linear polarisation perpendicular to the magnetic field azimuth. For a field with arbitrary inclination, the line wings and line centre will show mutually orthogonal states of elliptical polarisation.

It was about 50 years later that Babcock (1953) devised an instrument that exploited the polarisation signature of the Zeeman effect and could detect much weaker fields. Still later, Severny (1964) was able to determine all 3 components of the magnetic field by measuring the linear as well as circular polarisation in Zeeman affected spectral lines. The linear polarisation signal was an order of magnitude smaller than the circular polarisation, thus making the measurement of all 3 components of the solar magnetic field even in active regions a very demanding task. However, since the topology of active region magnetic fields is very important for their eventual eruption causing flares and coronal mass ejections, the thrust of modern instrumentation has been towards greater sensitivity and reliability of the polarimetry. On the other hand, no amount of instrumental sophistry can help if the process of converting the polarisation signals into magnetic field values is not rigorous. Great progress has been attained in this aspect as well.

5.2 The Hanle Effect

Chromospheric lines are severely affected by scattering during radiative transfer and thus the effect of scattering on polarisation must be taken into account. In the case of scattering, a photon, travelling in one direction is absorbed by the atom or ion and then reradiated in another direction. The total scattered intensity along any direction is obtained as a sum of several scattering events which are not correlated. Thus, there is no net polarisation produced by scattering in general, except when there is an asymmetry introduced by geometrical effects as in the case of polarisation of skylight by Rayleigh scattering. However, in the presence of a magnetic field, the energy level of a degenerate state will be split into sub-levels. The original energy level will also have an uncertainty or spread produced by the finite life-time of the excited state according to the uncertainty principle. If the energy difference between the sub-levels is smaller than this energy uncertainty, then there will be a coherence among these sublevels. This phenomenon is called quantum interference and it produces a net polarisation in the scattering process. A magnetic field influences the polarisation so produced in two ways. First, it produces a depolarisation with increase in the magnetic field since the sublevels become more separated and the quantum interference is reduced. Secondly, the magnetic field also introduces a rotation in the plane of polarisation. These two effects are together called the Hanle effect. The Hanle effect is most effective when the Larmor precession frequency matches the natural width of the line. In other words, the Hanle effect is most effective when the time taken for a Larmor precession is of the order of the lifetime of the excited state. A rigorous treatment of radiative transfer of polarised light involving both Zeeman and Hanle effects is available (Stenflo 1994). Several new applications of the Hanle effect are also described in that book. We will now look at some basic aspects of polarimetry before going into more details on the instrumentation.

5.3 The Stokes Parameters

Visible light is part of the general electromagnetic spectrum with oscillations in the regime of 10^{15} Hz. A typical source of visible light would consist of a large number of atoms, each of which emits a wave packet (or photon) at random times. The envelope of all these waves will be a very complicated function of time. We also know that the electromagnetic waves are transverse waves with the plane of vibration orthogonal to the direction of propagation of the wave. We are free to choose any system of coordinates in this plane and assume that we can take the electric and magnetic fields to be directed along the x- and y-directions of a coordinate system. We can expect the vibrations in each of these principal directions to be not correlated with each other. Let E_x and E_y be the electric field components associated with the light waves along the x- and y-directions respectively. Each of these components has an amplitude and phase associated with the vibrations. Mathematically, one can represent such quantities in the form of complex numbers. The modulus of the representative complex number is a measure of the amplitude of the vibration. The argument of the complex number equals the phase of the vibration. Let us now represent the correlation between E_x and E_y as the ensemble average of the product of E_x with E_y. Thus the correlation $\Gamma_{xy} = \langle E_x E_y^* \rangle$ would be zero for an uncorrelated sample. This type of light is known as unpolarised light. In special situations a break of symmetry produced by a magnetic field or a geometrical asymmetry in the reflection or scattering of light introduces a non-zero correlation or a non-zero value for Γ_{xy}. This type of light is called partially polarised light. When Γ_{xy} is equal to $(\Gamma_{xx} \cdot \Gamma_{yy})^{1/2}$, then the light is said to be completely polarised. Stokes evolved a set of parameters to describe a general partially polarised beam of light as well as to define a practical means of obtaining Γ_{xy}. These parameters are known as the Stokes parameters and are given below.

$$I = \Gamma_{xx} + \Gamma_{yy} = I_x + I_y \,, \tag{25}$$

$$Q = \Gamma_{xx} - \Gamma_{yy} = I_x - I_y \,, \tag{26}$$

$$U = 2\,\mathrm{Re}(\Gamma_{xy}) \,, \tag{27}$$

$$V = 2\,\mathrm{Im}(\Gamma_{xy}) \,, \tag{28}$$

where, I_x and I_y are the intensities transmitted through a polaroid whose axis is parallel to the x- and y-axes respectively. As seen from (26), the Q parameter is obtained by measuring the difference in the intensities I_x and I_y transmitted through a polaroid whose axis is placed parallel to the X and Y axes respectively. The U parameter is obtained by taking the difference of the intensities transmitted through a polaroid with axis at respectively 45° and −45° to the x-axis.

The parameter V is the difference between the amount of right and left circular polarisation present in the beam. Now, circularly polarised light can be resolved into two linear vibrations with a phase difference of 90°. When the x-vibration is advanced with respect to the y-vibration, then the resultant electric vector moves clockwise to the observer receiving the light. This type of polarisation is called right circular polarisation. The polarisation that produces an oppositely rotating vector is called left circular polarisation. The introduction of a quarter wave retarder in the path of right circularly polarised light will make the component vibrating along the slow axis of the retarder to come in phase with the component vibrating perpendicular to the slow axis. The resultant electric vector will make an angle of 45° to the slow axis of the retarder. If the slow axis of the retarder is oriented at an angle of −45° to the x-axis, then a right circularly polarised light, after passing through the quarter wave retarder, will emerge as a linearly polarised light with vibration along the x-axis. Similarly, a left circularly polarised light will emerge as a vibration along the y-axis. Thus the difference in the intensities of the emergent linear polarisation along x- and y-axes will yield the V parameter. The reader is urged to read the excellent book on polarised light by Shurcliff (1962) for further details.

5.4 Subsystems for Polarimetry

Right from the time the light emerges from the Sun to the time it is detected by the detector, there are several subsystems through which the light is transmitted. Associated with each subsystem is a source of error, and the different schemes of polarimetry deal with the different methods of minimising these errors. The various subsystems are the atmosphere, the telescope, the polarimeter, the dispersing element and the detector respectively.

The Atmosphere

There are two ways by which the atmosphere can introduce polarimetric errors. The blurring produced by the Earth's atmosphere can dilute the actual amount of polarisation, if the source of polarised light is confined to a small region of the solar surface. This situation arises when measuring the field in quiet regions outside of active regions. Even within active regions, it is now recognised that the field is highly fragmented and complex, especially in those regions that are prolific in flare production. The dilution is not such a problem for circular polarisation since an "integrated" circular polarisation has still a physical meaning in terms of the total magnetic flux available within the observing window. The average of linear polarisation has no such physical interpretation and could sometimes lead to absurd results if the diluted polarisation is used without proper care. There is no serious remedy for this problem for ground based telescopes. Use of 2 or more spectral lines of different Zeeman sensitivities, but from the same multiplet can provide an estimate of the dilution of the polarisation caused by atmospheric blurring. The development of adaptive optics also promises to provide some relief.

A second source of atmospherically induced distortion occurs when the different polarimetric states of the beam are measured sequentially. For example, one can measure I_x first followed by I_y later, with the intention of obtaining Q from the difference of these two measurements. If the atmosphere has introduced some image distortion in between the two measurements, then the subtracted value will bear no resemblance whatsoever to the true value of Q. The remedy for this problem lies in either measuring both the orthogonal polarisation states simultaneously using two detectors, or by a fast modulation in the polarimeter allowing the fast sequential sampling of the two states and separate integration of these samples (to improve signal to noise) so that near simultaneity in the two measurements is approached.

The first method has the problem of unequal response of the detectors, which in turn is solved using the so-called Semel's technique (Semel 1967). Here, the pair of measurements is repeated with an interchange of the detectors. It can be seen that simple data processing can get rid of the problem of unequal response. The basic idea of the Semel's technique is the following. Let us make a simultaneous measurement of a pair of orthogonal polarisations, say $I+Q$ and $I-Q$, using two detectors D_1 and D_2. Let the relative response of D_2 be η times that of D_1. Let us denote the measured quantities with primes. Then,

$$(I+Q)' = I+Q \,, \tag{29}$$

and

$$(I-Q)' = \eta(I-Q) \,, \tag{30}$$

yielding the relation

$$(I+Q)/(I-Q) = \eta(I+Q)'/(I-Q)' \,. \tag{31}$$

Now, if one repeats the measurements, after interchanging the detector D_1 for the $(I-Q)$ measurement and the detector D_2 for the $(I+Q)$ measurement, then we obtain the new results,

$$(I+Q)'' = \eta(I+Q) \,, \tag{32}$$

and

$$(I-Q)'' = (I-Q) \,, \tag{33}$$

where the double primes denote the second set of observed results. This yields the new relation

$$(I+Q)/(I-Q) = (I+Q)''/(\eta(I-Q))'' \,. \tag{34}$$

The product of these two relationships yields,

$$(I+Q)^2/(I-Q)^2 = (I+Q)'(I+Q)''/((I-Q)'(I-Q)'') \,. \tag{35}$$

In this way, the factor η can be eliminated. Care must be taken that the differential response η is not far removed from unity, since the signal to noise ratio would become significantly different between the two pairs of measurements leading to uncompensated noise.

The second method is adopted in modern polarimeters like the ZIMPOL (Povel, Aebersold & Stenflo 1990), which employs a special CCD that uses synchronous charge shifting clocks to demodulate the polarisation signal that is initially encoded by modulation in the polarimeter. The best way to avoid the atmosphere is, of course, to observe from space and this is precisely what the proposed SOLAR-B mission aims to do. SOLAR-B is a Japanese mission with payloads from various other countries and is planned to be launched in 2005.

The Telescope

Solar telescopes must have a long focal length to avoid the thermal heating at the focal plane. Long steerable telescopes are expensive and have several mechanical limitations. A method of avoiding steerable telescopes is to use a flat mirror rotating about the Earth's polar axis to direct the sunlight into a stationary telescope. A single mirror that is tilted about the declination axis and is rotated with the speed of Earth's rotation, is called a heliostat (see Bhatnagar this volume). A flat mirror, rotating with half the speed of Earth's rotation with its normal always lying along the celestial equator, and using another mirror suitably placed to compensate for the different declinations of the celestial object is called a coelostat (see Bhatnagar this volume). In both cases, the oblique reflections at the flat mirrors produce instrumental polarisation. For large coelostats, one cannot calibrate for this polarisation easily since large polaroids and quarter-wave plates cannot be made with sufficient uniformity over large areas. A scheme of ellipsometry was adopted for the Kodaikanal tower telescope that was very successful in determining the instrumental polarisation quite accurately (Sankarasubramanian et al. 2000). The chief problem for off-line elimination of instrumental polarisation is that the fluctuations of light intensity caused by Poisson statistics produces a corresponding random noise in the spurious polarisation that cannot be eliminated in single measurements.

Steerable telescopes with axial symmetry produce very little polarisation (chiefly by the processes that coat the mirrors with a reflecting layer). Even this problem can be minimised using suitable design of the optics. The French–Italian instrument THEMIS is a good example of a low polarisation telescope. SOLAR-B is also being designed as a low polarisation telescope. The main problem associated with such telescopes is that they need to have a long focal length to avoid the heat loading at the image plane. Apart from this, the part of the incident solar radiation that is absorbed by the primary mirror must be efficiently removed to avoid heating it. Assuming a 10 percent absorption, this means an input of 100 W m^{-2}. For large apertures, this means a large amount of heat that needs to be removed. An innovative method of solving this problem was adopted for the MSFC (Fig. 9) vector magnetograph (Hagyard et al. 1983). A window that transmits only about 20 nm of light centred at 525 nm is attached to the entrance of a standard Cassegranian reflector. This window transmits less than 40 W m^{-2} onto the primary mirror and about 10% of this energy (4 W m^{-2}) is absorbed by the primary mirror. Such a level of heat absorption does not require any special cooling and the system has been operating quite consistently for almost 2 decades.

For less restrictive access to a wide band of wavelengths, one needs to pay special attention to the cooling of the primary mirror. If the telescope is a Cassegranian, then the secondary also needs to be cooled, and a heat trap is required at the final image plane. If the telescope is a Gregorian, then a heat trap is required at the prime focal plane.

Fig. 9. The tower housing the vector magnetograph of the Marshall Space Flight Center, Huntsville, USA, is shown along side the dome housing the new experimental vector magnetograph

The Polarimeter

As mentioned earlier, the best way to fully characterise a partially elliptically polarised light beam is to measure the four Stokes parameters. The measurement of Stokes Q and U is achieved by measuring the amount of light transmitted by two polaroids in orthogonal directions. In a typical polarimeter, these measurements are carried out, as mentioned above, either by sequentially measuring the transmitted light through a polaroid whose transmission axis is switched between the x- and y-directions, or by simultaneously measuring the two orthogonal polarisation signals after spatially separating the two components using a polarising beam splitter. The sources of error in these measurements are the inaccuracies in positioning the polaroid axis and the leakage of the orthogonal polarisation. The latter can be addressed by using polarising prisms with high extinction of the orthogonal polarisation. The former requires great care in alignment as well as good calibrations.

To measure the V parameter, we require a quarter-wave retarder. Quarter-wave retarders are available in the form of quartz or mica plates. In such plates, the optical axis is parallel to the surface of the plate that receives the light. When this axis is placed at 45° to the x- or y-axis, then plane polarised light along the x- or y-direction will be split into two components along and perpendicular to the optic axis of the retarder. The vibration perpendicular to the axis will travel with lower speed relative to the vibration parallel to the optical axis. When the two vibrations emerge from the plate, there will be a net phase retardation between them that depends on the speed difference (difference between refractive indices along the two principal directions) and the thickness of the plate. If the thickness is such that the path delay is $n + \frac{1}{4}$ wavelengths, then we obtain circular polarisation. Subsequent measurements of Q and U of this output beam will produce zero polarisation, since circularly polarised light will pass the same intensity through a polaroid placed with axis at arbitrary angle. A circularly polarised light beam, on the other hand, will pass through that wave plate and emerge as a linearly polarised beam, which can be detected in a Q measurement on the emerged beam.

Other schemes of introducing wave retardation include using KD*P crystals (see Bhatnagar this volume) which can be modulated using a high voltage, as well as liquid crystal modulators which require lower voltages for modulation. A more sophisticated modulation scheme involving piezoelastic modulators was developed for the Zurich IMaging POLarimeter (ZIMPOL) of the Institute of Astronomy of ETH, Zurich that has the advantage of very high speed modulation – a feature that is very useful to combat the atmospheric effects mentioned earlier. Slower modulators can use analysing prisms that produce both orthogonal polarisation states which can be measured simultaneously with two different detectors. The differential response of the detector can be compensated by using Semel's technique as mentioned earlier.

Dispersing Element

While the spectrograph can provide a good dispersion, the slit of the spectrograph permits only one linear section of the Sun to be measured at a time. To measure a region, the image of the Sun has to be stepped across the slit of the spectrograph. This is not a great disadvantage if the aim is to measure the active region fields that evolve on time scales of days. A spectrograph has to be kept very steady, else the spectra obtained sequentially would not be of the same location on the Sun. Off-line corrections are meaningful only along the slit direction. Motions across the slit direction can never be corrected off-line. In this case, a good solar guider is essential. The guider must be programmed such that motion corrections are done prior to each exposure and not during the exposure to avoid jerks. A great advantage of a spectrograph is that the full line profile is available. The presence of telluric lines within the frame (e.g., the 630.1 nm oxygen line near the solar Fe I line at 630.2 nm) provides additional calibration useful for velocity measurements. However spectrographs have very low overall transmission, requiring large exposure times. Two-dimensional image restoration techniques like speckle imaging cannot be used in the spectroscopic mode. In Speckle imaging, one obtains a series of short exposure pictures called specklegrams, from which the image is reconstructed using off-line processing. A spectrograph's slit, cannot capture all the "speckles" spread out into a 2-dimensional region corresponding to a feature. Thus, the slit prevents the use of the speckle imaging technique for complete reconstruction. For spectroscopy, we need an adaptive optics system for on-line compensation of the atmospherically induced image motion and blurring, and the output of such an adaptive optics system can be fed to the spectrograph.

Table 1. Spectral Lines in the solar atmosphere

Wavelength (Å)	Species	g-factor	Height (km)
4861	Hβ	1.0	800
5173	Mgb2	1.75	850
5890	Na	2.5	1000
8542	Ca II	1.1	low chromosphere
6302	Fe I	2.5	photosphere
10830	He I	1.47	1500–2000
15648	Fe I	3.0	~ 37

An alternative method is to use a narrow band tunable filter as the dispersing element. Table 1 gives a list of Fraunhofer lines originating at different heights of the solar atmosphere. As may be noticed, it covers a wide spectral range starting from near UV to near IR. Therefore, the filter must have the tuning capability in the entire spectral range. It is found that the size distribution of the solar active regions peaks at about 6 arc minutes. Consequently, by choosing the field of view of the filter to be 6 arc minutes, it is possible to use it for observing about 80% of the active regions. The thermal widths of the photospheric and chromospheric lines of interest are about 0.025 Å. The magnetic sensitivity is improved when narrower band spectral isolators are used. As may be seen from Table2, the band width of the existing filters ranges from 0.072 Å to 0.25 Å. A filter band width of 0.01 Å is necessary for optimum use in obtaining the Stokes profiles. With this band width, about 12 points in a Stokes-V profile can be obtained, which is adequate for inverting the profile to determine the magnetic field strength and direction.

Canonically, birefringent filters (see Bhatnagar, this volume) were used for imaging the solar features in desired Fraunhofer lines. The instrumental profile of these filters becomes zero at the wings, thereby minimising the continuum leak. Secondly, they also have a wide acceptance angle resulting in a wide field of view. However, the complexity, high cost and low transmission of these filters have forced the solar physicists to look for other alternatives.

One of the attractive alternatives is the use of Fabry–Perot etalons (see Bhatnagar this volume). During the past decade, a number of solar instruments based

Table 2. Summary of the Filters used in various vector magnetographs

Instrument	Bandpass (Å)	Filter-type	Observed line (Å)
MSFC (TVM)	0.125	Biref	5122.2 (Fe I)
MSFC (EXVM)	0.170	FP	5122.2 (Fe I)
BBSO	0.250	Biref	6103.0 (Ca I)
Hawaii	0.072	FP	6302.5 (Fe I)
HSOS (China)	0.125	Biref	5324.2 (Fe I)

on these etalons were commissioned for both imaging the solar features and obtaining the magnetograms as shown in Table 2. Although, during the last two decades, Fabry–Perot (FP) etalons have been extensively used in various branches of astronomy (Desai 1984; Vaughan 1989) their use in solar observation is relatively new, especially with the development of FP etalons of larger finesse (> 25) and free spectral range (FSR) of 0.4–1.0 nm (Cavallini et al. 1987; Bonaccini et al. 1989; Bonaccini & Stauffer 1990; Bendlin, Volkmer & Kneer 1992). While high finesse etalons are made by using piezo-electrically servo-controlled mechanisms, relatively low finesse and stable etalons can be made by using electro-optic materials such as Lithium Niobate ($LiNbO_3$) crystal wafers (Rust 1985). These two type of etalons have their advantages and disadvantages. While the former has high stable finesse owing to the online servo control, the latter has larger field of view and is less sensitive to ambient conditions. The major limitation of $LiNbO_3$ etalons is that it is not possible to tune to any desired line within the FSR. Hence, for universal tunability, we need to use the servo controlled piezo-electrically tuned etalons. Table 3 summarises specifications of FP etalons. This system is more compact and can be attached to the telescope. Two dimensional images at a single position of the line profile can be obtained. However, a complete spectrum can also be obtained by tuning the etalon across the line profile. Modern etalons allow quick tuning using piezo-electric devices. The peak wavelength transmitted by the etalon will be a function of position within the field of view, which needs to be calibrated. Alternately, the etalon can be placed in a telecentric beam, in which case the peak wavelength will be independent of position in the field of view.

The Detector

The CCD (Charged Coupled Device) has turned out to be the ideal detector for solar polarimetry. The most efficient use of CCDs occurs when the charge-shifting clock is synchronised with the polarisation modulation. The available area of the chip is subdivided into the active portion flanked by two masked portions that serve as storage buffers. The exposure is halted using a shutter while the charges are actually being shifted. The shutter speed generally limits the speed of the modulation. However, such detectors are difficult to manufacture, and more difficult to replace in the case of any malfunction. It would be wiser to

Table 3. Filter parameters and etalon properties

(a) Filter Parameters

Field of View	: 6 arc minutes
Bandpass	: 0.1 Å
Stability	: 0.005 Å over 8 hours

(b) Etalon Properties

Etalon diameter	: 75 mm
Usable aperture	: 60 mm
Coating	: Multilayer broad band (4500–6500 Å, 6500–8000 Å and 8000–17000 Å)
Reflectivity	: 0.95
Effective Finesse	: 50
Free Spectral Range	: 4 Å
Tuning Steps	: Piezo-electrically 0.01 Å

use commercially available CCDs and direct the two orthogonal polarisations to two separate CCDs of nearly the same efficiency. Any small difference in the efficiencies can then be very easily corrected by the Semel's technique mentioned earlier.

5.5 Conversion of Polarisation Maps into Magnetograms

The most rigorous method of converting polarisation signals into magnetic field strengths is based on the inversion of the set of Stokes profiles obtained from magnetically sensitive lines. The basis of all the inversion schemes is the following. The polarised line profiles are calculated on the basis of a few atmospheric parameters including the three components of the magnetic field and the velocity of the magnetised plasma along the line of sight. A best fit is attempted of the computed profile with the observed profile, while varying the atmospheric parameters. That set of atmospheric parameters which gives the least deviation for the computed profiles from the observed profiles is then assumed to be the output of the measurement. In recent times, it is seen that one requires more parameters to fit more detailed shapes of the line profiles. For example, a marked asymmetry between the red and blue wings of the Stokes profiles is an indication of the existence of velocity gradients in the line forming region of the atmosphere. Consequently, the inversions need to become more and more sophisticated. In fact, one of the results of such asymmetries in the Stokes profiles, has resulted in the inference of the existence of micro-scale magnetic fields in the solar atmosphere (Almeida 1999). One of the challenges of modern instrumentation is to be able to directly detect these micro-scale magnetic fields.

6 Concluding Remarks

We have seen that the solar magnetic field has a global origin and the action of convection, rotation, and differential rotation can, in principle, produce the observed evolution of this global magnetic field. The simple dynamo models that use a given velocity field to produce the magnetic field have shown those desirable properties of the velocity field that can sustain a global dynamo with the observed spatial and temporal pattern. However, the new technique of helioseismology has provided us with the empirical information on some of the important sub-surface velocity fields and this information shows difficulty in reconciling with the desired properties capable of sustaining the dynamo. This fact provides lot of scope for research and modelling efforts.

The solar magnetic field is present on a variety of spatial scales. We see larger-scale and stronger fields during active phases and smaller-scale and more randomly mixed up fields during quiet phases. Clearly, the explanation for these different modes of manifestation of the solar magnetic field is not yet available, although the solar velocity fields like differential rotation and smaller scale convective eddies must be playing a very important role. The corresponding decrease in the general coronal brightness and in the number of eruptive phenomena like flares and coronal mass ejections also indicate that the topological evolution of the field and processes like magnetic reconnection depend on the phase of solar activity.

As mentioned earlier, a consistent set of measurements of the magnetic field with large dynamic range and over a long period of time are vital to provide the basic inputs for the modelling of magneto-convection and topological evolution of the field. These measurements must be done with care to avoid the various errors in polarimetry that can arise in the different subsystems of the measurement process. The inversion of the polarisation data into the magnetic field parameters, especially in the chromosphere, also needs a lot of input from quantum scattering theory. There is a lot of work that needs to be done in a wide range of physical disciplines.

References

Alfvén, H. 1942, Nature, 150, 405

Almeida, J. S. 1999 in Solar Polarization, eds. K. N. Nagendra & J. O. Stenflo, Astrophysics and Space Science Library, vol. 243 (Kluwer Academic Press, Boston), 251

Antia, H. M., Chitre S. M., Thompson, M. J. 2000, A&A, 360, 335

Babcock, H. W. 1953, ApJ, 118, 387

Babcock, H. W. 1961, ApJ, 133, 572

Bendlin, C., Volkmer, R., & Kneer, F. 1992, A&A, 271, 817.

Bonaccini, D., & Stauffer, F. 1990, A&A, 229, 272

Bonaccini, D., Cavallini, F., Ceppatelli, G., & Righini, A. 1989, A&A, 217, 368

Cavallini, F., Ceppatelli, G., Meco, M., Paloschi, S., & Righini, A. 1987, A&A, 184, 386

Charbonneau, P., & Dikpati, M. 2000, ApJ, 543, 1027
Charbonneau, P., Christensen-Dalsgaard, J., Henning, R., Larsen, R. M., Schou, J., Thompson, M. J., & Tomczyk S. 1999, ApJ, 527, 445
Choudhuri, A. R. 1989, Sol. Phys., 123, 217
Choudhuri, A. R. 1999, The Physics of Fluids and Plasmas: An Introduction for Astrophysicists (Cambridge University Press, Cambridge)
Choudhuri, A. R., & Gilman, P. A. 1987, ApJ, 316, 788
Choudhuri, A. R., Schüssler, M., & Dikpati, M. 1995, A&A, 303, L29
Choudhuri, A. R., Schüssler, M., & Dikpati, M. 1997, A&A, 319, 362
Desai J. N. 1984, Proc. Indian Acad. Sci. (Earth Planet Sci.), 93, 189
DeVore, C. R., Sheeley, N. R., Jr., & Boris, J. P. 1984, Sol. Phys. 92, 1
Dikpati, M., & Gilman, P. A. 2001, ApJ, 559, 428
Dikpati, M. & Charbonneau, P. 1999, ApJ, 518, 508
D'Silva, S., & Choudhuri, A., R. 1993, A&A, 272, 621
Durney, B. R. 1997, ApJ, 486, 1065
Gilman, P. A. 1983, ApJS, 53, 243
Gilman, P. 1986, in Physics of the Sun, vol I, ed. P. A. Sturrock, (Reidel, Dordrecht), 95
Hagyard, M. J., Teuber, D., West, E. A., Tandberg-Hanssen, E., Henze, W., Jr., Beckers, J. M., Bruner, M., Hyder, C. L., Woodgate, B. E. 1983, Sol. Phys., 84, 13
Hale, G. E. 1908, ApJ, 28, 315
Hale, G. E., & Nicholson, S. B. 1938, Publ. Carnegie Inst. No. 498, Washington
Hale, G. E., Ellerman, F., Nicholson, S. B., & Joy, A. H. 1919, ApJ, 49, 153
Jackson, D. 1969, Classical Electrodynamics (John Wiley, New York)
Kitchatinov, L. L., & Rüdiger, G. 1999, A&A, 344, 911
Küker, M., Rüdiger, G., & Schultz, M. 2001, A&A, 374, 301
Leighton, R. B., 1969, ApJ, 156, 1
Nandy, D. 2002, PhD thesis, Indian Institute of Science, Bangalore
Nandy, D., & Choudhuri, A. R. 2001, ApJ, 551, 576
Nandy, D., & Choudhuri, A. R. 2002, Science, 296, 1671
Nordlund, Å., Dorch, S. B. F., & Stein, R. F. 2000, J. Astron. Astrophys., 21, 307
Parker, E. N. 1955, ApJ, 121, 491
Parker, E. N. 1975, ApJ, 198, 205
Parker E. N. 1979, Cosmical Magnetic Fields (Oxford University Press, Oxford)
Parker, E. N. 1994, Spontaneous Current Sheets in Magnetic Fields: with application to stellar X-rays, (Oxford University Press, New York)
Povel, H. P., Aebersold, H., & Stenflo, J. O. 1990, Applied Optics, 29, 1786.
Rust, D. 1985 in Measurement of Solar Vector Magnetic Fields, ed. M. J. Hagyard, NASA-CP 2374, 141
Sankarasubramanian, K. & Rimmele, T. 2002, ApJ, 576, 1048
Sankarasubramanian, K., Srinivasulu, G., Ananth, A. V., & Venkatakrishnan, P. 2000, J. Astron. Astrophys., 21, 241
Schou, J., Antia, H. M., Basu, S., et al. 1998, ApJ, 505, 390.
Schrijver, C. J., Title, A. M., Harvey, K. L., Sheeley, N. R., Jr., Wang, Y.-M., van den Oord, G. H. J., Shine, R. A., Tarbell, T. D., & Hurlburt, N. E. 1998, Nature, 394, 152
Schröter, E. H. 1985, Sol. Phys., 100, 141
Schwabe, S. H. 1844, Astron. Nachr., 21, 233
Semel, M. D. 1967, Ann. Astrophys., 30, 513
Severny, A. B. 1964, Izr. Kron. Astrofig. Obs., 31, 126

Sheeley, N. R., Wang, Y.-M., & Harvey, J. W., 1989, Solar Phys. 119, 323
Shurcliff, W. A. 1962, Polarized Light: Production and Use (Harvard University Press, Cambridge, MA)
Spiegel, E. A., & Weiss, N. O. 1980, Nature, 287, 616
Steenbeck, M., Krause, F., & Rädler, K.-H. 1966, Z. Naturforsch., 21a, 369
Stenflo, J. O. 1994, Solar Magnetic Fields (Kluwer Academic Publishers, Dordrecht)
Stix, M. 1976, IAU Symp. 71, 367
van Ballegooijen, A. A. 1982, A&A, 113, 99
Vaughan, J. M. 1989, The Fabry–Perot interferometer. History, theory, practice and applications (Adam Hilger, Bristol)
Venkatakrishnan, P. 1984, PhD Thesis, Bangalore University
Weiss, N. O. 1981, J. Fluid Mech., 108, 247
Zeeman, P. 1897, Phil. Mag. [5], 43, 226

The Physics of Chromospheres and Coronae

P. Ulmschneider

Institut für Theoretische Astrophysik, Univ. Heidelberg,
Tiergartenstr. 15, 69121 Heidelberg, Germany
email: ulm@ita.uni-heidelberg.de

Abstract. Towards the goal to unravel the physical reasons for the existence of chromospheres and coronae significant progress has been made. Chromospheres and coronae are layers which are dominated by mechanical heating and usually by magnetic fields. The heating of chromospheres can be explained by an ordered sequence of different processes which systematically vary as function of height in the star and with the speed of its rotation. It seems now pretty certain that acoustic waves heat the low and middle chromosphere, and MHD waves the magnetic regions up to the high chromosphere. With faster rotation, the magnetic regions become more dominant. It seems that the highest chromosphere needs additional non-wave heating mechanisms and that there possibly reconnective microflare heating comes into play. For the corona many different heating processes occur which work in the various field geometries. Here more study is needed to identify the relevance of these various processes.

1 Introduction, What Are Chromospheres and Coronae?

Already in antiquity the interested observer wondered about the *corona* (crown), the faint extended and spiky white shell around the Sun which becomes visible during a total solar eclipse, when the Moon hides the luminous solar disk (photosphere). We know today that the corona is composed of three components, the F-corona (F stands for Fraunhofer lines) which arises from dust particles which scatter the bright photospheric light spectrum with its absorption lines into the eye of the observer, the K-corona (K stands for Kontinuum) which consists of photospheric light devoid of spectral lines scattered by free electrons, and the E-corona (where E stands for emission lines among which the green and red coronal lines are the most prominent). It was B. Edlén and W. Grotrian who succeeded around 1940 to identify these coronal lines as iron lines from very high ionisation stages. The strongest green corona line at wavelength 5303 Å originates from Fe XIV, while the red one at 6375 Å from Fe X. The presence of such lines demonstrated that the corona is a gas with temperatures of many millions K. This high temperature explains why the energy of the electrons is so large that without great effort, by collisional impact, they strip away 9 to 13 electrons from the iron atoms. It also explains why the scattering of the photospheric light by the fast moving electrons results in a continuous spectrum.

While the corona extends to many solar radii the *chromosphere* is a layer of only 2 to 3 thousand km thickness which becomes visible near the start and end of a total eclipse. The chromosphere got its name from the prominent red emission of the Hα line of neutral hydrogen at 6563 Å. The chromosphere is a layer where the temperature rises from photospheric values of between 4000 and 6000 K to about 20 000 K and where neutral hydrogen is still present. In the region of a few 100 km thickness between the chromosphere and corona, called *transition layer*, hydrogen becomes ionised and the temperature increases from 20 000 to millions of K.

Observations show that essentially all late-type stars with surface convection zones have chromospheres and that most of them also possess coronae. The presence of chromospheres in stars other than the Sun is indicated e.g., by emission cores in the Ca II H and K as well as the Mg II h and k lines while coronae are detected by e.g., the presence of C IV and Mg X lines as well as the X-ray emission. The surprising fact is that the coronal temperatures (of 1 to 6 million K) are almost as high as the temperatures in the solar core, where the nuclear processes take place and where the energy for the solar luminous radiation is generated. Also surprising is the fact that the density in the short extent of the photosphere and chromosphere decreases by 8 orders of magnitude (from $\rho = 3 \times 10^{-7}$ to 3×10^{-15} g cm^{-3}) which is similar to the 9 orders of magnitude density decrease (from $\rho = 200$ to 2×10^{-7} g cm^{-3}) which occurs from the solar core to the solar surface but over a much larger distance of 700 000 km.

With an underlying photosphere, where the temperature decreases from values close to the effective temperature (of 5770 K for the Sun) to a minimum temperature of about 4000 K, the question is why are these chromospheric and coronal layers so hot and why are they so different from the photosphere? This question has to be answered by realising that a single main-sequence star like our Sun is uniquely characterised by specifying four independent parameters and that isolated stars such as our Sun, except in their initial T-Tauri phase, are not influenced by the surrounding interstellar medium. With the next stars several light years away, it is clear that not only the solar interior and the photosphere, but also the chromosphere and corona must be uniquely determined by these four parameters. We therefore face the great challenge to unravel the physical reasons how on basis of four parameters the average structure of the chromospheres and coronae are precisely and uniquely determined. This aim has so far not been fully realised but significant progress has been made towards this goal and towards the clarification of the physics of chromospheres and coronae.

That four parameters are sufficient to uniquely characterise a main-sequence star can be seen by considering the state when after its pre main-sequence evolution the star reaches the Zero Age Main Sequence (ZAMS) phase. Here it is homogeneous and can be described by its total mass $M_\star$ and metallicity (metal content or chemical composition) Z_{m}. Although stars can have peculiar element abundances, the majority of stars can be well characterised by a single metallicity ratio Z_{m} which gives the stellar abundances of the elements heavier than He (called metals) in terms of the solar metal abundances (see below). The elapsed time t_{S} since the ZAMS phase is then a crucial third parameter which characterises the present state of the star.

The low rotation rate of late-type main-sequence stars indicates that they have lost their primordial magnetic field and it is generally agreed that the present magnetic field of these stars is newly generated from the *dynamo mechanism* which works efficiently only when rotation and convection are both present. The rotation period $P_{\rm Rot}$ thus is a fourth essential parameter which characterises a star. There is the difficulty that there is no unique rotation period for a star. As discussed in the contributions of Antia and Venkatakrishnan (this volume), our Sun due to the differential rotation has a sidereal surface rotation rate which varies from 15° to 11° per day between 0 and 60° latitude. This amounts to a latitudinal variation of the rotation period from 24 to 33 days. However, the Sun with an average rotation period of $P_{\rm Rot} = 28 \pm 4$ days can still be considered to belong to the group of slow rotators (see Fig. 45).

One might have the fortunate case, that for some single stars $P_{\rm Rot}$, due to a similar rotational braking history, is a known function of $t_{\rm S}$ which would reduce the basic parameters to only three. However, because more than 70% of the stars are members of multiple systems where unknown amounts of orbital angular momentum was converted into spin angular momentum, $P_{\rm Rot}$ usually should be taken as an independent parameter. For practical purposes these four independent parameters $M_\star$, $Z_{\rm m}$, $t_{\rm S}$, $P_{\rm Rot}$ can finally be expressed by four more convenient parameters $T_{\rm eff}$, g, $Z_{\rm m}$, $P_{\rm Rot}$, where $T_{\rm eff}$ is the effective temperature and g the surface gravity.

2 Heating Mechanisms

The conservation of energy in a gas element somewhere in the chromosphere can be written as

$$\frac{\partial}{\partial t}\left(\frac{1}{2}\rho v^2 + \rho c_V T + \rho\phi\right) + \boldsymbol{\nabla}\cdot\rho\mathbf{v}\left(\frac{1}{2}v^2 + c_V T + \frac{p}{\rho} + \phi\right)$$
$$= \left.\frac{\mathrm{d}Q}{\mathrm{d}t}\right|_{\rm Mech} + 4\pi\kappa\,(J - B)\ . \qquad (1)$$

Here $\frac{1}{2}\rho v^2$ is the kinetic, $\rho c_V T$ the internal and $\rho\phi$ the potential energy density (in erg cm^{-3}), where the gravitational potential is given by $\phi = -GM_\odot/r$ with G, $M_\odot$, r being the gravitational constant, the solar mass, and the radial distance of the gas element from the centre of the Sun, respectively. ρ is the density, p the gas pressure, c_V the specific heat at constant volume, T the temperature and v the flow speed of the solar wind. The three energy flux components in (erg cm^{-2} s^{-1}) are the kinetic energy flux $\rho\mathbf{v}\frac{1}{2}v^2$, the enthalpy flux $\rho\mathbf{v}\,(c_V T + p/\rho)$ and the potential energy flux $\rho\mathbf{v}\phi$. On the right hand side, we have the heat addition by mechanical heating in which we also include magnetic heating, and radiative heating. For simplicity, radiative heating is written in terms of a gray (i.e., frequency independent) Rosseland opacity κ (e.g., given by (10)) times the mean intensity J minus the source function for which we take the frequency integrated Planck function B.

As the chromosphere and the corona surround the solar surface since billions of years and as the mass flux ρv of the solar wind is undetectably small at chromospheric and low coronal heights, one can neglect both terms of the left hand side of (1) and gets essentially a balance of mechanical heating and radiative cooling:

$$\left.\frac{\mathrm{d}Q}{\mathrm{d}t}\right|_{\mathrm{Mech}} = 4\pi\kappa\,(B - J)\ . \tag{2}$$

The mean intensity and the Planck function can be written

$$J = \frac{1}{2}\frac{\sigma}{\pi}T_{\mathrm{eff}}^4\ , \qquad B = \frac{\sigma}{\pi}T^4\ , \tag{3}$$

where σ is the Stefan–Boltzmann constant where the factor 1/2 in J comes from the fact that above the solar surface there is only outgoing radiation.

If we neglect mechanical heating, then radiative heating alone must be zero. This is the condition of *radiative equilibrium* which is the physical basis of stellar photospheres (in deeper photospheric layers, however, convective energy transport must be considered in addition to the transport by radiation). In radiative equilibrium one has $J = B$, that is, the photospheric temperature should decrease in outward direction until a boundary temperature of $T = 2^{-1/4}T_{\mathrm{eff}} \approx 0.8T_{\mathrm{eff}}$ is reached. For the Sun with $T_{\mathrm{eff}} = 5770\,\mathrm{K}$ one would have a boundary temperature of $T \approx 4850\,\mathrm{K}$.

From the observed high temperatures of 20 000 to millions K in chromospheres and coronae one must have using (3) that $B \gg J$, which implies large and persistent amounts of mechanical heating to satisfy (2). In contrast to photospheres, *chromospheres and corona are thus characterised as layers which require large amounts of mechanical heating.*

Table 1 summarises the mechanisms which are thought to provide a steady supply of mechanical energy to balance the chromospheric and coronal energy losses. The term heating mechanism comprises three physical aspects, the *generation* of a carrier of mechanical energy, the *transport* of mechanical energy into the chromosphere and corona and the *dissipation* of the energy in these layers. Table 1 shows the various proposed energy carriers which can be classified into two main categories: *hydrodynamic* and *magnetic* mechanisms. The magnetic mechanisms can be subdivided further into wave- or *AC-mechanisms* and current sheet- or *DC-mechanisms.* Also in Table 1 the mode of dissipation of these mechanical energy carriers is indicated. For more details see Narain & Ulmschneider (1990, 1996), as well as Ulmschneider (1996)

Ultimately, these mechanical carriers derive their energy from the nuclear processes in the stellar core from where it is transported in the form of radiation and convection to the stellar surface. In late-type stars the mechanical energy generation is caused by the gas motions of the surface convection zones. These gas motions are largest in the regions of smallest density near the top boundary of the convection zone. Due to this the mechanical energy, particularly the wave energy, is generated in a narrow surface layer.

Let us now discuss the various mechanisms. All of them have been found to work in terrestrial applications and thus should also work in a stellar environment. However, the problem is to identify which of these mechanisms are

Table 1. Heating mechanisms

energy carrier	*dissipation mechanism*
hydrodynamic heating mechanisms	
acoustic waves, $P<P_A$ pulsational waves, $P\geq P_A$	shock dissipation shock dissipation
magnetic heating mechanisms	
1. alternating current (AC) or wave mechanisms	
slow mode MHD waves, longitudinal MHD tube waves	shock dissipation
fast mode MHD waves	Landau damping
Alfvén waves (transverse, torsional)	mode-coupling resonance heating turbulent heating compressional viscous heating ion-cyclotron resonance heating Landau damping
magnetoacoustic surface waves	mode-coupling phase-mixing resonant absorption
2. direct current (DC) mechanisms	
current sheets	reconnection (turbulent heating, wave heating)

dominant and if so, in what regions and situations do these various mechanisms work. At the present time the heating processes in the corona and even in the high chromosphere are not yet fully clarified and for instance there is the debate whether AC or DC-mechanisms are more important in these layers. Also there is the question of what is the role of surface waves propagating along magnetic flux tubes. Only careful studies can clarify the hierarchy of the heating mechanisms in chromospheres and coronae.

3 Hydrodynamic Heating Mechanisms

There are two types of hydrodynamic mechanisms, *acoustic waves* and *pulsational waves*. The convection zone like every turbulent flow field generates *acoustic waves* which propagate in all directions. Even if the wave energy due to radiative damping is not fully conserved, acoustic waves which propagate in outward direction will suffer a large growth of the wave amplitude due to the strong density decrease. Because of nonlinear processes the wave profile gets distorted and shocks form which then dissipate the wave energy and heat the atmosphere (see Fig. 1). *Pulsational waves* which occur e.g., in Mira stars are

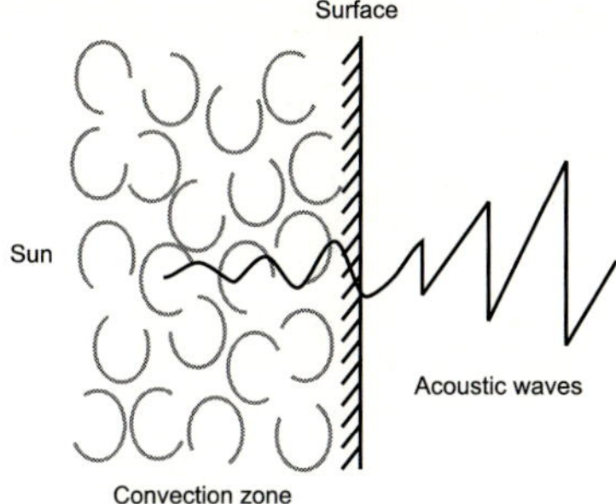

Fig. 1. Acoustic waves and the heating by shocks

another hydrodynamic heating mechanism which propagate to the outer stellar atmosphere where they also dissipate through shocks.

Acoustic waves have periods smaller than the acoustic cut-off period

$$P_A = \frac{4\pi c_S}{\gamma g} , \tag{4}$$

where c_S is the sound speed, g the acceleration due to gravity and $\gamma = 5/3$ the ratio of specific heats. Pulsational waves have periods $P \geq P_A$. Typical values for the acoustic cut-off period for the Sun ($g = 2.74 \times 10^4$ cm s^{-2}, $c_S = 7$ km s^{-1}) are $P_A \approx 190$ s and for Arcturus ($g = 50$ cm s^{-2}, $c_S = 6$ km s^{-1}) $P_A \approx 1.5 \times 10^5$ s.

For *late-type stars* of spectral type F to M, acoustic waves are generated by turbulent velocity fluctuations near the top boundary of the stellar convection zones. The development of acoustic waves into shock waves is shown in Fig. 2. The left panel of this figure shows a calculation of a monochromatic, radiatively damped acoustic wave of period $P = 45$ s and initial energy flux $F_A = 2 \times 10^8$ erg cm^{-2} s^{-1}. It is seen that the wave grows from a small amplitude up to a point where sawtooth shocks form which attain a limiting strength. This limiting strength for a given star depends only on the wave period (larger periods give stronger shocks). Also shown in Fig. 2 (right panel) is a calculation with an

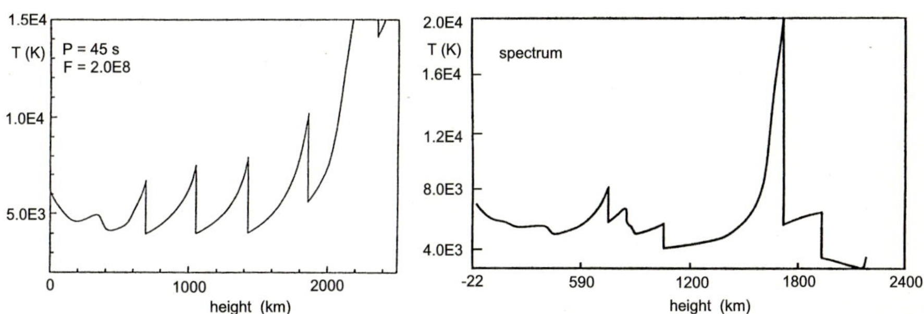

Fig. 2. Left: monochromatic, radiatively damped acoustic wave with period $P = 45$ s and initial energy flux $F_A = 2 \times 10^8$ erg cm^{-2} s^{-1}. Right: propagating acoustic wave spectrum of the same initial energy

acoustic spectrum. Here shocks of different strengths are generated and one has the tendency that strong shocks eat up the weak shocks.

In *early-type stars* of spectral type O to A, where surface convection zones are absent, it is the intense radiation field of these stars which generates acoustic disturbances and amplifies them to strong shocks. This mechanism works as follows (Fig. 3). Consider a gas blob in the outer atmosphere of an early-type star. The line-opacity κ_ν as function of frequency ν of the outer atmosphere at rest is shown dashed in Fig. 3. Also shown is the intensity I_ν of the photospheric stellar radiation field which has an absorption line at that frequency. If by chance the gas blob acquires a slight outward velocity, then relative to the dashed profile, its line-opacity κ_ν (solid) is Doppler shifted towards the violet and photons from the violet wing of the stellar absorption line get absorbed in the blob, imparting more momentum and thus accelerating it even more (Fig. 3). This results in a line-opacity κ_ν shifted further out of the region of the photospheric absorption line, where additional photons accelerate the blob, etc. This process is called *radiative instability* and results in a powerful acceleration of gas blobs which leads to the formation of strong shocks with X-ray emitting post-shock regions and intense local heating leading to the coronal emission of these early-type stars.

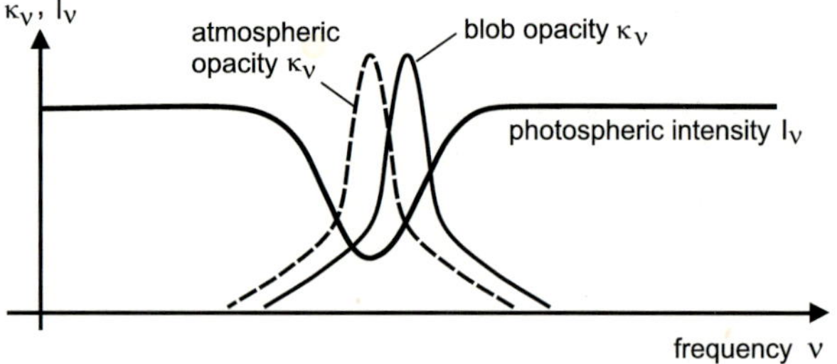

Fig. 3. Radiative instability for an accelerating gas blob in an intense radiation field

The other hydrodynamic heating mechanism is *pulsational waves*. Pulsational waves are prominent in Mira-star pulsations, but also in other late-type giants. These pulsations are generated by the *κ-mechanism*. The κ-mechanism (here κ refers to the opacity) functions similarly as the internal combustion engine in motorcars (see Fig. 4). In the internal combustion engine a reactive gas mixture is compressed in a pulsational motion and is ignited at the moment of strongest compression, resulting in a violent decompression. The timing of the ignition ensures that the pulsational motion is amplified. In the κ-mechanism the opacity of stellar envelope material increases (due to the adiabatic temperature and pressure increase) when the star contracts in a pulsational motion. The opacity increase leads to an increased absorption of radiation energy and thus to a large heat input into the contracted envelope layers. The overheated envelope layer subsequently reacts by rapid expansion, thus driving the pulsation. These

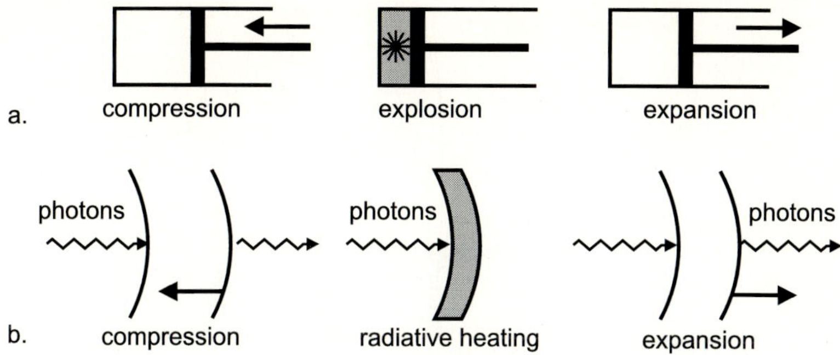

Fig. 4. (**a.**) Generation of pulsation by a gasoline engine, (**b.**) by the κ-mechanism

pulsational waves propagate to the outer stellar atmosphere where they form shocks.

This and related processes, also with different drivers (e.g., the ϵ-mechanism where the nuclear energy generation is enhanced, see e.g., Kippenhahn & Weigert 1990), work also for nonradial oscillations. Any process which kicks on the basic pulsational and vibrational modes of the outer stellar envelope belongs to the category of pulsational wave mechanisms. For the Sun the 3 min oscillation is such an example of a basic resonance which is generated by transient events produced in the convection zone (3 min shock waves). Unfortunately, a systematic study of this heating mechanism for late-type stars is missing at the present time.

4 Magnetic Heating Mechanisms

It is observationally well established that isolated strong vertical magnetic fields (flux tubes) exist outside sunspots, particularly at the boundaries of supergranulation cells (e.g., Solanki 1993; Stenflo 1994; Schrijver & Zwaan 2000), and there they give rise to the chromospheric network emission in the chromospheric spectral lines. It is found that the cross-section of the magnetic flux tubes increases with height (see Hasan, this volume). At an altitude of about 1500 km the individual flux tubes fill out the entire available space and form the *magnetic canopy.* In the corona the field strength is $B \approx 10$–100 G. At the surface of the Sun the field strength in an isolated flux tube of the chromospheric network is $B \approx 1500$ G. One has horizontal pressure balance

$$p_i + \frac{B^2}{8\pi} = p_e \ , \tag{5}$$

where $p_{\rm i}$ is the gas pressure inside the flux tube and $p_{\rm e}$ the gas pressure in the non-magnetic region outside. At the solar surface at $z = 0$ one finds $p_{\rm e} = 1.2\times 10^5$ dyn cm^{-2} in the Vernazza, Avrett & Loeser (1981) model C. If the tube were empty, that is, $p_{\rm i} = 0$, one would have $B = B_{\rm eq} = \sqrt{8\pi p_{\rm e}} = 1740$ G. This is

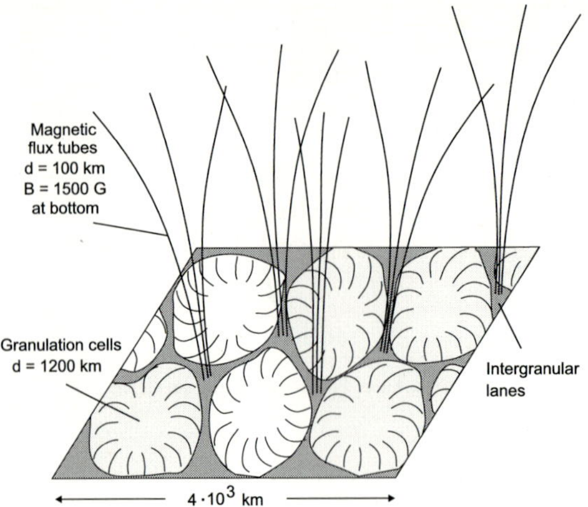

Fig. 5. Surface region of the Sun with granulation cells and magnetic flux tubes

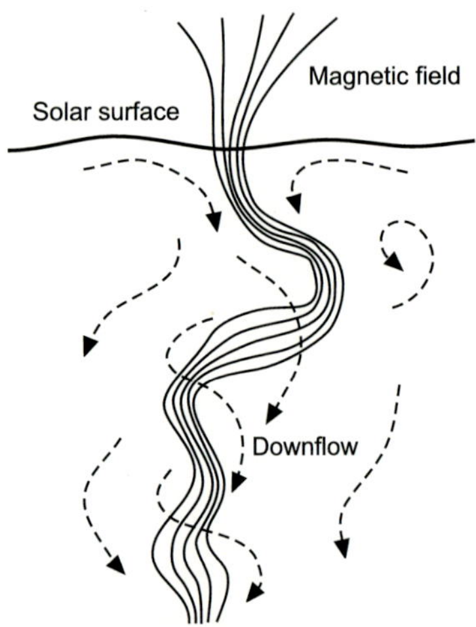

Fig. 6. Cyclonic turbulent downflows generate longitudinal, transverse and torsional MHD waves, modified after Parker (1981). The squeezing of the field lines produces longitudinal waves, the shaking transverse waves and the twisting torsional waves

called *equipartition field strength.* Actually the tube is not empty but has a gas pressure of about 1/3 to 1/6 of the outside pressure.

For a discussion of the magnetic heating mechanisms consider a surface region on the Sun (Fig. 5). It is seen that the granulation flows, which are produced by rising turbulent gas bubbles in the convection zone, concentrate the magnetic fields into magnetic flux tubes in the intergranular lanes, where the gas flows back into the Sun. This flow is not laminar, and as it converges from a large area to the downflow region and because the Sun rotates, the flow experiences Coriolis forces and generates tornados (Fig. 6). A magnetic flux tube therefore sits

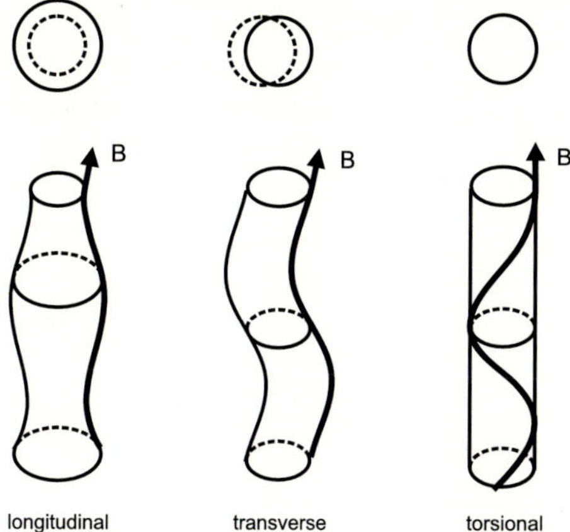

Fig. 7. The three possible wave modes in magnetic flux tubes. Cross-sections are shown at the top

in the centre of a tornado of downflowing gas. As the flows are turbulent there is a lot of squeezing, shaking and twisting of the flux tubes. These external perturbations generate three types of magnetohydrodynamic (MHD) wave modes (see Fig. 7). The squeezing produces *longitudinal MHD waves*, the shaking *transverse Alfvén waves*, and the twisting by turbulent cyclonic flows *torsional Alfvén waves*. Longitudinal tube waves cause cross-sectional variations of the tube, they are essentially acoustic tube waves. Because of this similarity they dissipate via shocks. The transverse and torsional Alfvén waves do not show a cross-sectional variation of the tube. In principle, the transverse and torsional waves can also form shocks (see Figs. 9, 10) but usually they are more difficult to dissipate (see e.g., Narain & Ulmschneider 1990, 1996).

A typical time scale is the Alfvén transit time $t_{\rm A}$:

$$t_A = \frac{l_{||}}{c_A} = l_{||}\frac{\sqrt{4\pi\rho}}{B}\ , \tag{6}$$

where $l_{||}$ is the length of a magnetic loop or flux tube and $c_{\rm A}$ is the Alfvén speed. The magnetic waves are generated by *rapid* ($t << t_{\rm A}$) velocity fluctuations outside the tube. These fluctuations are produced in the convection zone, but also by sudden events (see Sect. 4.10). Above the canopy the waves encounter a more or less homogeneous medium and other wave modes, *fast* and *slow mode* MHD waves are possible.

If the motions of the convection zone are *slow* ($t >> t_{\rm A}$), then instead of waves, stressed magnetic structures are built up which contain a large amount of energy. Here often magnetic fields of opposite polarity are brought together and form current sheets. The energy of the stressed fields is then released by *reconnection*, where the field lines break open and reconnect in such a way that the field geometry afterwards is simpler. These reconnection processes usually occur suddenly like in a *flare* where the magnetic field energy of a large spatial region is released in seconds. Smaller reconnection events are called *microflares* (see Sect. 4.10). The local release of energy generates waves in turn. However, same as for waves the ultimate source of the DC-heating mechanism is the convection zone.

Let us now discuss the magnetic heating processes in detail.

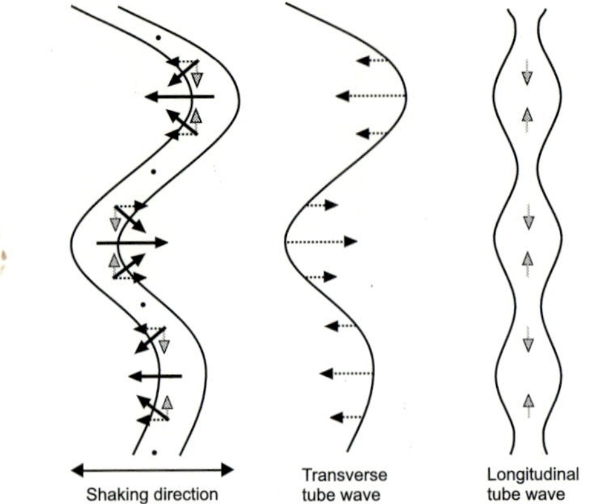

Fig. 8. Mode-coupling between transverse and longitudinal waves which results from the shaking of the flux tube

4.1 Mode-Coupling

This mechanism is not a heating process by itself, but converts wave modes, which are difficult to dissipate by non-linear coupling into other modes, where the dissipation is more readily achieved. Typical cases are the conversion of transverse or torsional Alfvén waves into acoustic-like longitudinal tube waves which dissipate their energy by shock heating. Figure 8 shows an example of such a process when a magnetic flux tube is shaken. It is seen that the magnetic tension force which is directed towards the centre of curvature can be split into longitudinal and transverse components. The longitudinal force components act to compress and expand the gas in the tube such that a longitudinal wave of twice the frequency is generated.

Mode-coupling is particularly efficient when the transverse waves are very stochastic in nature as is expected from observation (Muller et al. 1994) and from wave generation calculations (e.g., Musielak & Ulmschneider 2001). Figure 9 (after Zhugzhda, Bromm & Ulmschneider 1995) shows the generation and development of a longitudinal shock wave pulse produced by mode-coupling from a transverse wave pulse. While the propagation speeds of the transverse and longitudinal waves are very different these authors find that when the shocks appear, both the transverse shocks (also called kink shocks) and longitudinal shocks form at the same height and subsequently propagate with a common speed. This indicates strong mode-coupling.

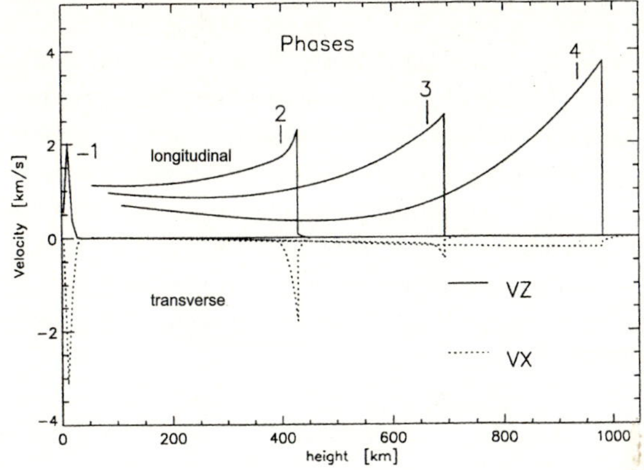

Fig. 9. Longitudinal wave pulse and shock generated by mode-coupling from a transverse wave pulse, after Zhugzhda, Bromm & Ulmschneider (1995). Different wave phases are labelled 1 to 4

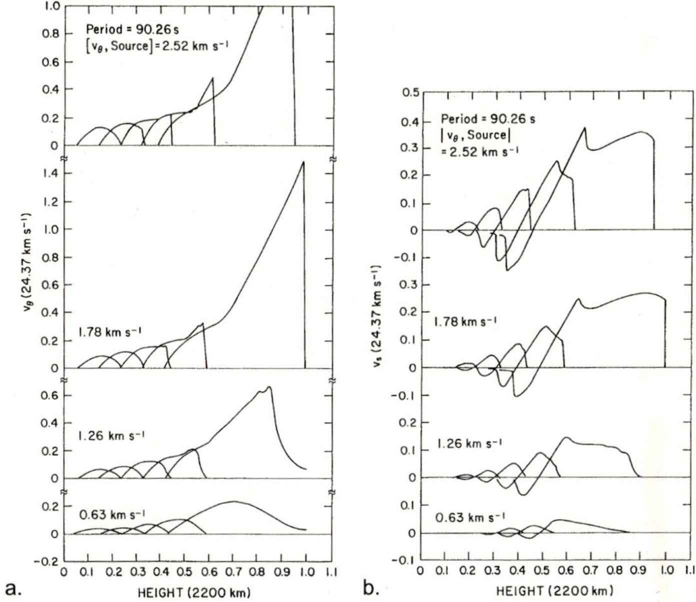

Fig. 10. Longitudinal waves and shocks generated by mode-coupling from torsional wave pulses. Different wave phases are indicated. (**a.**) propagation of torsional wave pulses of various initial amplitudes, (**b.**) corresponding longitudinal waves, after Hollweg, Jackson & Galloway (1982)

A similar process of mode-coupling occurs when torsional wave pulses propagate (Hollweg, Jackson & Galloway 1982). Figure 10a shows the propagation of various torsional wave pulses. The longitudinal waves generated by mode-coupling are displayed in Fig. 10b. It is seen that the torsional shocks (also called switch-on shocks) and longitudinal shocks form at the same height and subsequently propagate with a common speed indicating strong mode-coupling. These calculations have to be taken with some caution because it is not well known whether the thin flux tube approximation describes these situations well. Three-dimensional time-dependent work by Ziegler & Ulmschneider (1997) on swaying magnetic flux tubes in the solar atmosphere shows that there is extensive leakage of the transverse wave energy into the outside medium. Thus the true magnitude of the longitudinal wave energy generation by mode-coupling is presently not well determined.

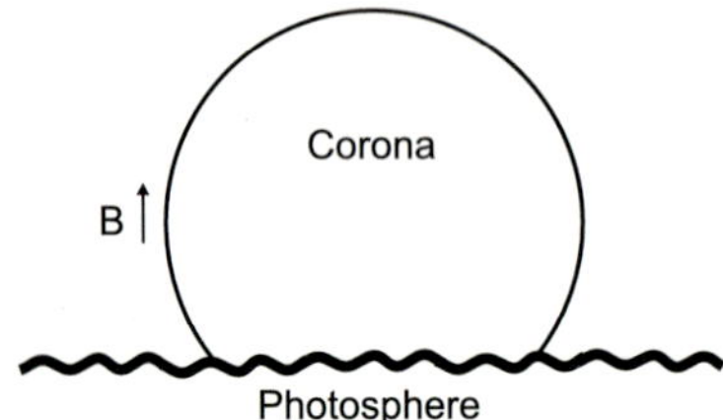

Fig. 11. Resonance heating in coronal loops

4.2 Resonance Heating

Resonance heating occurs, when upon reflection of Alfvén waves at the two foot points of the coronal loops, one has constructive interference (see Fig. 11). For a given loop length $l_{\|}$ and Alfvén speed c_{A}, resonance occurs, when the wave period is $mP = 2l_{\|}/c_{\mathrm{A}}$, m being a positive integer. Waves which fulfill the resonance condition are trapped and after many reflections are dissipated by Joule-, thermal conductive or viscous heating.

4.3 Turbulent Heating

In a turbulent flow field with high Reynolds number there are bubbles of all sizes. The energy usually is put into the largest bubbles. Because of the large inertial forces, the moving big bubbles are ripped apart into smaller bubbles, and these in turn into still smaller ones etc. This process is called turbulent cascade. A turbulent flow field can be described by three characteristic quantities, density ρ, bubble size $l_{\mathrm{k}} = 2\pi/k$, and the mean velocity u_{k} of such bubbles. k is the wavenumber. It is easily seen, that from these three quantities only one combination for a heating rate can be formed

$$\Phi_k = \rho \frac{u_k^3}{l_k} \qquad \left[\frac{\mathrm{erg}}{\mathrm{cm}^3\,\mathrm{s}}\right] . \tag{7}$$

If there are no other losses, such as radiation, all the energy which is put in at the largest bubbles must reappear in the smaller bubbles etc. Thus if $k1, k2, \ldots$

represents a series of smaller and smaller bubbles one must have $\Phi_{\rm k1} = \Phi_{\rm k2} = \cdots = $ const. This implies

$$u_k \sim l_k^{1/3} , \tag{8}$$

which is the *Kolmogorov law.* The range $l_{\rm k1}, \ldots, l_{\rm kn}$ of validity of this law is called the *inertial range.* Consider what happens if $l_{\rm k}$ becomes very small. The viscous heating rate is given by $\Phi_{\rm V} = \eta_{\rm vis}(du/dl)^2 \approx \eta_{\rm vis}u_{\rm k}^2/l_{\rm k}^2 \approx \eta_{\rm vis}l_{\rm k}^{-4/3}$, which goes to infinity for $l_{\rm k} \to 0$. Here $\eta_{\rm vis}$ is the coefficient of viscosity. Thus at some small enough scale, viscous heating sets in and the inertial range ends. It is seen that turbulent heating lives from the formation of small scales. One can visualise the process as follows. Because of the continuous splitting of bubbles into smaller sizes, with the velocities decreasing much less rapidly, one eventually has close encounters of very small bubbles with large velocity differences where viscous heating dominates.

As the fluctuations generated in the turbulent convection zone produce acoustic and MHD waves it is of interest to deduce from the inertial range an estimate of the frequency range of the generated waves. If $k_0 \approx 2\pi/H$ is the wavenumber of the scale where the energy is put into the turbulence, with H being the scale height, we have for the size $l_{\rm k}$ of the smallest bubble (where viscosity ends the cascade) $\eta_{\rm vis}u_{\rm k}^2/l_{\rm k}^2 = \rho u_{\rm k_0}^3/l_{\rm k_0}$ and $u_{\rm k}^3/l_{\rm k} = u_{\rm k_0}^3/l_{\rm k_0}$. From this we obtain

$$l_k = \left(\frac{\eta_{\rm vis} l_{\rm k_0}^{1/3}}{\rho u_{\rm k_0}}\right)^{3/4} . \tag{9}$$

With $l_{\rm k_0} = H = 150\,{\rm km}$, $u_{\rm k_0} = 1$ km s^{-1}, $\rho = 3 \times 10^{-7}$ g cm^{-3}, $\eta_{\rm vis} = 5 \times 10^{-4}$ dyn s cm^{-2}, one finds $l_{\rm k} = 2.9\,{\rm cm}$ as well as $u_{\rm k} = 290$ cm s^{-1} and derives a maximum frequency of $\nu_k = u_{\rm k}/l_{\rm k} = 100\,{\rm Hz}$ or a period of $P = 1/100\,{\rm s}$. This estimate is somewhat idealised as small bubbles become transparent to radiation. In this case the temperature excess, which drives the convection, is exchanged directly via radiation. Thus it is expected that the optical depth limits the bubble size. Assuming that the smallest bubble has an optical depth of $\tau = l_{\rm k}\kappa = 0.1$, where

$$\frac{\kappa}{\rho} = 1.38 \times 10^{-23} p^{0.738} T^5 \quad {\rm cm}^2\ {\rm g}^{-1} , \tag{10}$$

is the gray H$^-$ opacity, $T = 8320\,{\rm K}$ the temperature and $p = 1.8 \times 10^5$ dyn cm^{-2} the gas pressure, one finds $l_{\rm k} = 6.3 \times 10^4$ cm, $u_{\rm k} = 1.6 \times 10^4$ cm s^{-1}, $\nu_k = 0.25\,{\rm Hz}$ and $P = 3.9\,{\rm s}$ for the smallest bubble.

So far we have discussed turbulent heating in a non-magnetic environment. When there is a magnetic flux tube one has to take into account the tube geometry, the frozen-in condition and the MHD wave modes. Figure 12 shows how the turbulent dissipation of a torsional Alfvén wave is pictured (after Heyvaerts & Priest 1983; Hollweg 1983). Shearing motions in azimuthal direction generate closed magnetic loops (similarly to the growth of Kelvin–Helmholtz type instabilities) over the tube cross-section, which decay into smaller tubes etc. and are ultimately dissipated by reconnection.

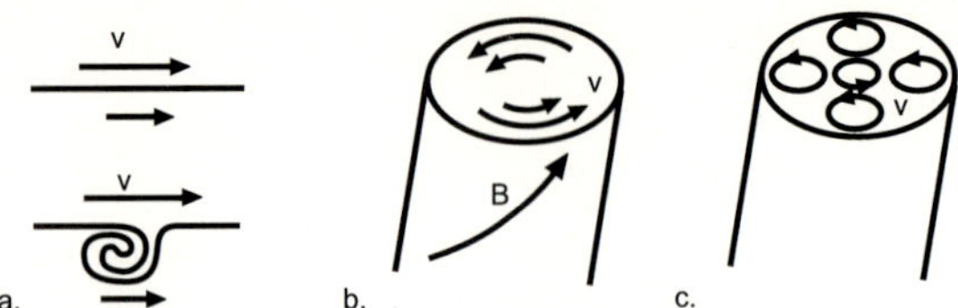

Fig. 12. Turbulent heating in magnetic flux tubes

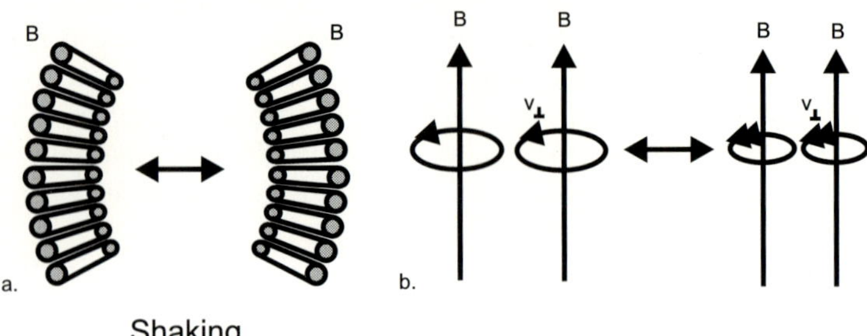

Fig. 13. Compressional viscous heating in helical fields. (**a.**) Cross-section (gray) of a magnetic flux tube with greatly exaggerated winding. The shaking of the spring leads to compressions and expansions of the tube. (**b.**) The field strength changes from the cross-section variations affect the gyro-frequency

4.4 Compressional Viscous Heating

Compressional viscous heating, proposed by Strauss (1991), is a very promising mechanism for coronal regions where the collision rate becomes small and the gyro-frequency gets much larger than the collision-frequency. In the presence of magnetic fields the particles have no restriction moving parallel to the magnetic field **B**, but cannot move freely perpendicular to it. In this direction when there are few collisions, Lorentz forces cause them to orbit around the field lines with the gyro-frequency $\Omega_{\mathrm{L}} = qB/mc$, where c is the light velocity, q the charge and m the mass of the gyrating particle.

Swaying an axial magnetic flux tube sideways with velocity $\mathbf{v}_\perp$ results in a transverse Alfvén wave which is incompressible ($\boldsymbol{\nabla}\cdot\mathbf{v}_\perp = 0$) to first order. This is different for tubes with helicity, where one has $\boldsymbol{\nabla}\cdot\mathbf{v}_\perp \approx \dot{\rho}/\rho$ (see Fig. 13). With an increase of the density, the magnetic field is compressed and the gyro-frequency increased. Gyrating around the field lines more quickly in a narrower space, the ions collide more readily with each other, and generate velocity components in other directions as well, which constitutes the heating process.

4.5 Ion-Cyclotron Resonance Heating

This type of damping also occurs at coronal heights, where due to the low density the collision rate becomes small. Due to the decreasing magnetic field strength B with height, high frequency torsional or transverse Alfvén wave propagating along the field lines will come to regions where the wave frequency becomes

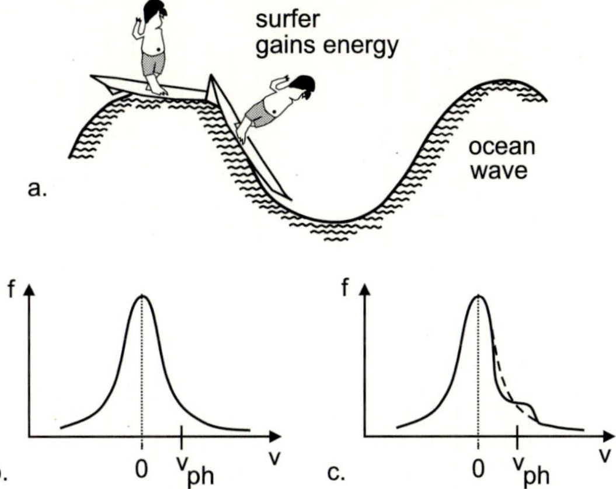

Fig. 14. Landau damping and the analogy to surfing. (**a.**) Surfing the ocean waves. (**b.**) Particle distribution function. (**c.**) Around its phase speed the wave modifies the distribution function

equal to the gyro-frequency Ω_{L} of protons and ions like e.g., O^{5+}. Here ion-cyclotron resonance heating occurs. The concerted gyrations of the ions caused by the wave action leads to intersections of the orbits. This results in collisions which convert the wave energy into random motions, constituting the heating process. Ion-cyclotron resonance heating has been proposed e.g., by Tu & Marsch (1997, 2001a, 2001b) to explain the different types of parallel and perpendicular temperatures (with respect to the magnetic field **B**) of protons, alpha particles and oxygen ions and their distribution with distance from the Sun in the solar wind.

4.6 Landau Damping

Landau damping is a third damping process which occurs at coronal heights where the collision rates are small. As Chen (1974) has explained, this mechanism is analogous to surfing on ocean waves (see Fig. 14a). When surfing, a surfboard rider launches himself in propagation direction into the steepening part of an incoming wave and gets further accelerated by this wave. In Landau damping, the propagating wave accelerates gas particles which, due to their particle distribution function, happen to have similar direction and speed as the wave. Because a distribution function normally has many more slower particles than faster ones (Fig. 14b), the wave looses energy to accelerate the slower particles (solid line in Fig. 14c). This gained energy is eventually shared with other particles in the process to reestablish the distribution function (dashed line in Fig. 14c), which constitutes the heating mechanism.

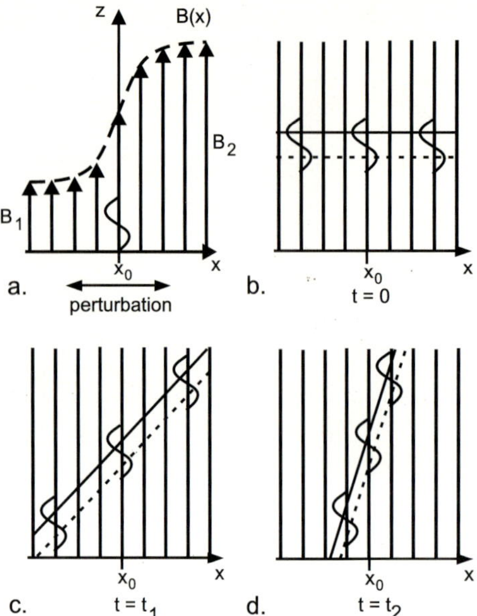

Fig. 15. Resonant absorption. In a field pointing in z-direction, where the field strength varies in x-direction. (**a.**) resonant absorption of a surface wave, (**b.**) wave fronts at time $t = 0$, (**c.**) and (**d.**) at later times t_1 and t_2

4.7 Resonant Absorption

In the process of resonant absorption one considers magnetoacoustic surface waves in a magnetic field $\mathbf{B}$ which points in z-direction, and varies from $\mathbf{B}_1$ to $\mathbf{B}_2$ in x-direction (see Fig. 15a). The surface wave, with its field perturbation $\delta B = B'_x$ in x-direction, has a phase speed $v_{\rm ph} = ((B_1^2 + B_2^2)/(4\pi(\rho_1 + \rho_2))^{1/2}$ such that at an intermediate position x_o, the phase speed becomes equal to the local Alfvén speed $c_{\rm Ao} = B(x_o)/\sqrt{4\pi\rho(x_o)}$. In Fig. 15b consider the wave fronts of the peak (solid) and trough (dotted) of a surface wave. Because to the right of x_o, the Alfvén speed is larger and to the left smaller, the wave fronts at a later time get tilted relative to the phase propagating with speed $c_{\rm Ao}$ (Fig. 15c). At a still later time (Fig. 15d) the wave fronts get tilted even further and approach each other closely at the position x_o. This leads to small scales and intense heating by reconnection at that field line.

4.8 Phase-Mixing

For phase-mixing (Fig. 16) one considers a magnetic field geometry similar to that in Fig. 15, however, the field perturbation $\delta B = B'_y$ of the wave is now in y-direction, perpendicular to the x- and z-directions. As the Alfvén speeds of two closely adjacent regions x_0 and x_1 are different, it is seen that after propagating some distance Δz, the fields $B'_y(x_0)$ and $B'_y(x_1)$ will be very different, leading to

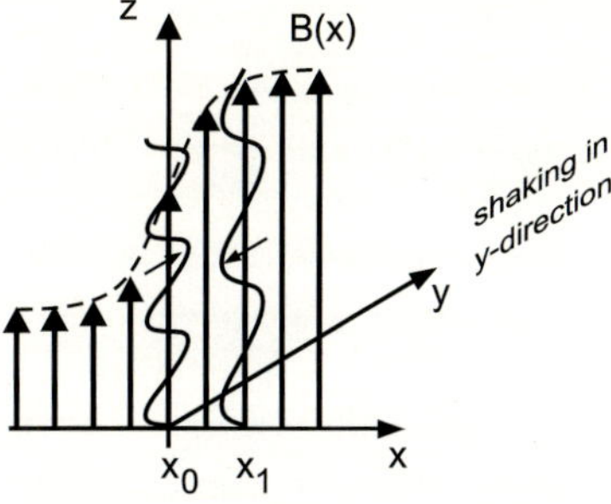

Fig. 16. Phase-mixing of a surface wave (shaking in y-direction)

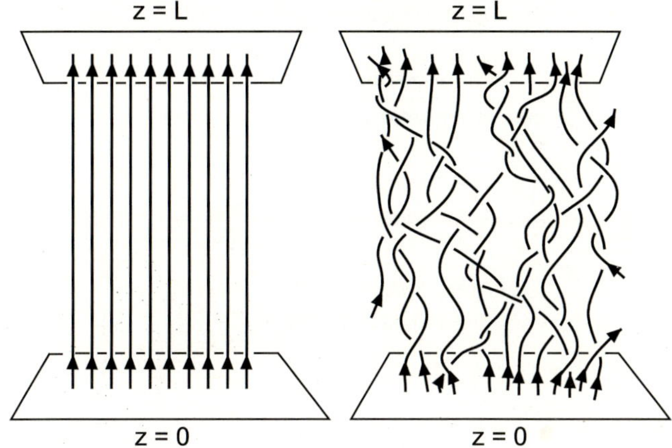

Fig. 17. Left: Magnetic fields in coronal loops, initial axial field. Right: tangled fields after considerable foot point motions, after Parker (1991)

a current sheet and strong dissipation. Here again it is the appearance of small scale structures which leads to the dissipation.

4.9 Reconnection in Current Sheets

Let us now discuss DC-mechanisms. In Fig. 17 the magnetic flux tube bundle which represents a closed coronal loop is plotted such that the two ends of the loop, which both originate in the photosphere, are displayed on the top and bottom of the figure. This way the minimum energy configuration, where the field lines are all straight and parallel, can be shown in the left panel. As the granular flows in the photosphere will displace the foot points of the individual flux tubes of the bundle, they get entwined such that after a while a complicated mesh of flux tubes results (right panel). This braided mesh has more stored energy than the minimum energy configuration and there is no chance that the flows will ever precisely reverse such that the minimum energy configuration reappears. On the contrary, the continuing motions will store more and more energy in the magnetic field until this can no longer go on.

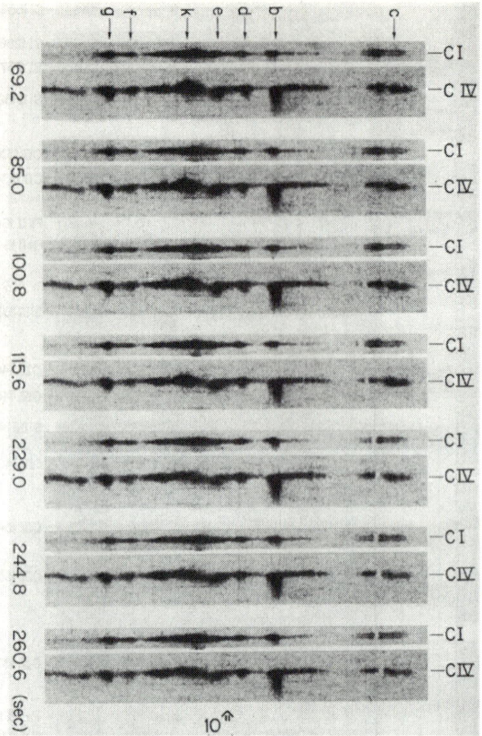

Fig. 18. Various outbreaks of C IV emission from high-velocity turbulent events in the transition layer observed on the Sun, after Brueckner (1981)

At many locations in the web of field lines, oppositely directed fields occur, giving rise to local current sheets, which by reconnection (in the form of numerous microflares) release the magnetic field energy (see also Narain & Ulmschneider 1990, 1996, and recently Mitra-Kraev & Benz 2001). The energy is dissipated both directly and via the generation of waves and turbulence. Note that similarly to the wave mechanisms, reconnection happens in small scale regions.

Such small scale reconnective events of different magnitude have been observed on the Sun by Brueckner (1981) (see Fig. 18) and Brueckner & Bartoe (1983) as sudden velocity shifts in the C IV ($T \approx 10^5$ K) transition layer line with velocities of 250 km s^{-1} and even 400 km s^{-1}. These sudden velocity shifts have been called turbulent events and high velocity jets.

The question whether microflares are a significant coronal heating mechanism and what its importance is as compared to wave heating (DC- versus AC-heating) has also been studied by Wood, Linsky & Ayres (1997) by observing C IV and Si IV transition layer lines on stars other than the Sun (see Fig. 19). They found that the total line profile can be explained as a combination of a

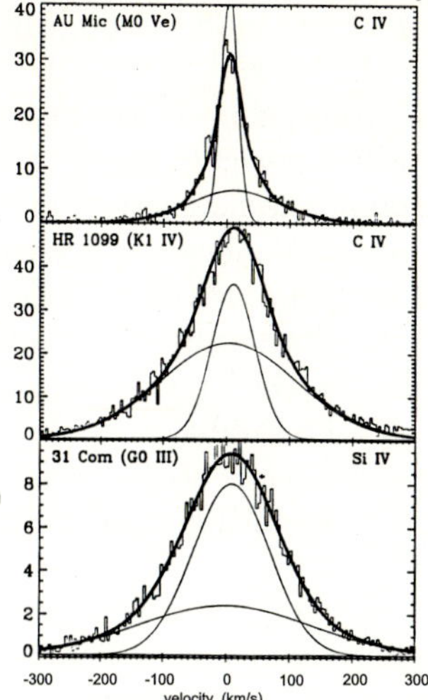

Fig. 19. Surface fluxes in transition layer lines (in 10^4 erg cm^{-2} s^{-1} Å^{-1}) of giants and main sequence stars, after Wood, Linsky & Ayres (1997)

broad component, attributed to microflares, and a narrow component attributed to wave heating.

Another example of current sheet formation and DC-heating is seen in Fig. 20, after Priest (1991). It shows arcade systems, which from slow motions get laterally compressed and develop a current sheet. Here oppositely directed fields reconnect. Similar systems of approaching magnetic elements of opposite polarity and large scale field annihilation are thought to be responsible for the heating of X-ray bright points.

5 Acoustic Energy Generation

After discussing the extensive list of proposed heating mechanisms we now ask which of these mechanisms are the important ones for chromospheres and coronae. Of most mechanisms we already know that they work in terrestrial laboratory settings and therefore should also work on the Sun, given the right situation and magnetic field geometry. Due to the large density decrease from the photosphere to the corona the heating requirements as function of height are very

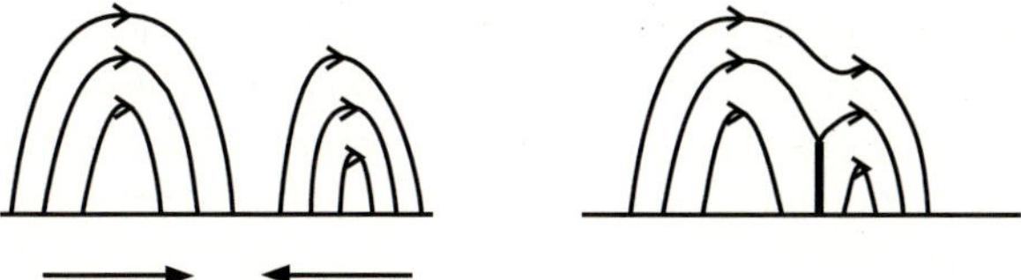

Fig. 20. Formation of current sheets in arcade systems, after Priest (1991)

different and therefore some of the mechanisms will be more important than others.

In addition there are variations due to the magnetic field geometry. Magnetic surface waves, for instance, dissipate energy which propagates through a ring shaped cross-section around the magnetic flux tube. The heating per volume of solar atmosphere is probably much less efficiently done by surface waves than by body waves such as longitudinal tube waves or transverse and torsional Alfvén waves which employ the entire tube cross-section for transportation and dissipation of the energy. Finally, the importance of a heating mechanism depends on how effectively it can be produced in the convection zone. One thus has the tedious task to investigate each of the proposed heating mechanisms in detail in order to understand how much energy it carries and where and how it dissipates this energy.

Ideally the identification problem could be solved by a full scale simulation of the convection zone including the magnetic field together with the generation of the different wave types and the formation of current sheets (like e.g., in Fig. 17). In such a simulation one would have to compute the motions and development of the physical variables on very different geometrical scales, from the supergranulation size of about 50 000 km down to about a few m where the shock dissipation and current sheet dissipation happens. While undoubtedly such simulations will be attempted in the future they are at the moment beyond our numerical capabilities. So far energy generation calculations are available only for a few mechanisms, all of them waves, and of the waves only for the three types: acoustic waves, longitudinal MHD waves and transverse Alfvén waves.

Let us concentrate on those heating mechanisms where the mechanical energy generation can be computed and start with the hydrodynamic mechanisms. To calculate acoustic fluxes of the Sun and other stars one must first compute a convection zone model. As convection zones do not depend on rotation, such a model can be computed by specifying only three parameters: $T_{\rm eff}$, g and $Z_{\rm m}$. Actually, because the convection zone calculations use the *mixing-length theory*, a fourth parameter, the mixing-length parameter α, has to be specified. This dimensionless quantity α is the ratio of the mixing-length L to the scale height H and from solar and stellar observations one typically has $\alpha \approx 2.0$ (for references about the choice of α see Theurer, Ulmschneider & Kalkofen 1997). However, α is not a basic parameter, and in the future, numerical convection zone calculations will specify its precise value and replace the mixing-length theory altogether.

With density ρ, convective velocity u, sound speed $c_{\rm S}$ and scale height $H = \Re T/\mu g$ versus height z, provided by the convection zone models, acoustic fluxes $F_{\rm A}$ can be computed using either the simple so called *Lighthill-* or *Lighthill–Proudmann formula*, or the more elaborate Lighthill–Stein theory (Musielak et al. 1994). Here $\Re$ is the universal gas constant and μ the mean molecular weight. Figure 21a shows acoustic fluxes for stars with (solar) population I abundances ($[Z_{\rm m}] = 0$ and $\alpha = 2.0$) for a wide range of late-type stars after Ulmschneider, Theurer & Musielak (1996), while Fig. 21b displays acoustic fluxes for stars with

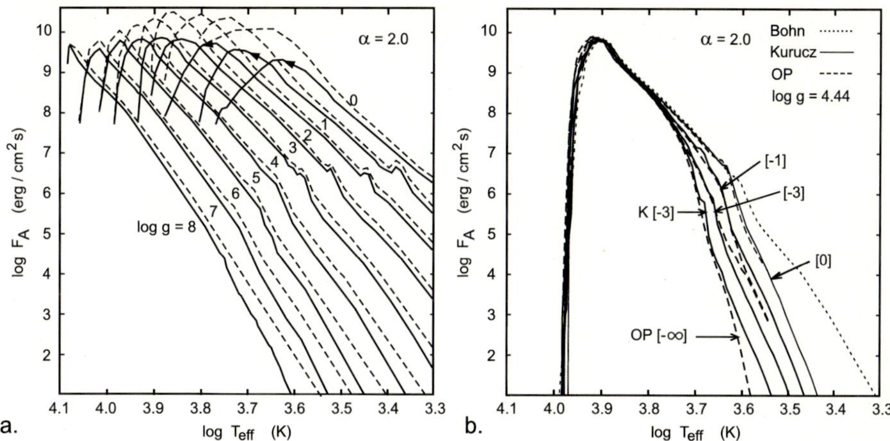

Fig. 21. Acoustic fluxes $F_{\rm A}$ for stars versus $T_{\rm eff}$, (**a.**) for given $\log g$ and $\alpha = 2.0$ (*solid*), the fluxes using the Lighthill formula are shown *dashed*, after Ulmschneider, Theurer & Musielak (1996), (**b.**) for $\log g = 4.44$ and different metal abundances $[Z_{\rm m}]$, after Ulmschneider et al. (1999). In **b.** Bohn opacities are shown *dotted*, Kurucz opacities *solid* and OP opacities *dashed*

1/1, 1/10, 1/100 and 1/1000 solar metal abundances ($Z_{\rm m} = [0], [-1], [-2], [-3]$ and $\alpha = 2.0$) after Ulmschneider et al. (1999).

Both the Lighthill formula and the Lighthill–Stein theory use the experimentally and theoretically well established energy distribution of free turbulence described by a Kolmogorov energy spectrum. Lighthill and Proudman derived a simple formula

$$F_A = \int 38 \frac{\rho u^8}{c_S^5 H} \, {\rm d}z \;, \tag{11}$$

with the famous u^8-dependence for *quadrupole sound generation.* This u^8-law has been successfully reproduced in terrestrial experiments as shown in Fig. 22. Also one finds in Fig. 21a that the total acoustic fluxes using the elaborate Lighthill–Stein theory are surprisingly close to values given by the Lighthill formula which is due to the similar Kolmogorov energy spectrum used. Actually in the Lighthill–Stein theory a slightly different extended Kolmogorov spatial spectrum with a modified Gaussian frequency factor (eKmG spectrum) has been employed (Musielak et al. 1994).

In Fig. 21a one sees that the acoustic flux $F_{\rm A}$ rises rapidly with $T_{\rm eff}$ and g. This is explained from the fact that the convective velocity u increases when a star gets hotter and/or its gravity decreases. The large variation is due to the u^8-dependence. That $F_{\rm A}$ varies with metallicity (in Fig. 21b) is a consequence of the location of the top boundary of the convection zone in a star. That boundary is the layer where rising convective bubbles reach the stellar surface and radiate their temperature excess directly into space. In cool stars with large $Z_{\rm m}$ the opacity is large and thus the top of the convection zone lies at shallow layers where convection employs large velocities to transport the stellar flux $\sigma T_{\rm eff}^4$. For

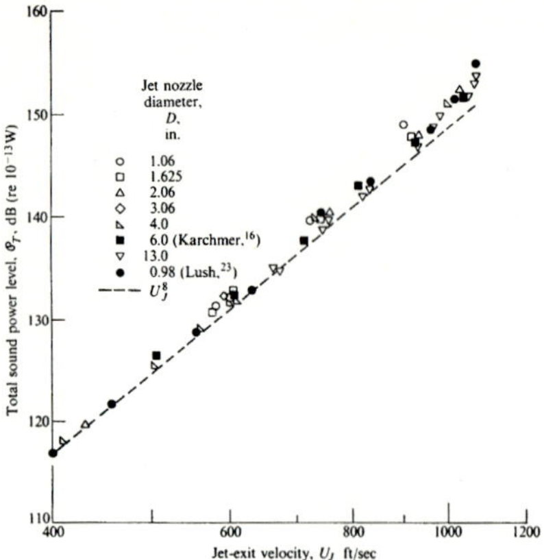

Fig. 22. Lighthill's u^8 power law in terrestrial applications, after Goldstein (1976)

lower metal content the opacity decreases and the boundary moves to layers of higher density where the convective velocity is smaller. This explains why F_A decreases for lower Z_m. For hot stars the opacity is mainly due to hydrogen and not to metals and thus F_A does not vary with Z_m.

6 Theoretical Chromospheres

The Lighthill–Stein theory not only provides the total acoustic flux F_A (erg cm^{-2} s^{-1}) but also an acoustic frequency spectrum. For monochromatic calculations the maximum of this spectrum determines the wave period $P(s)$ which can be used together with F_A to compute the propagation of acoustic waves into the chromosphere (Buchholz, Ulmschneider & Cuntz 1998). For this computation one first needs an initial radiative equilibrium atmosphere model for the star, which depends on T_eff, g and Z_m. The wave calculation and the initial atmosphere computation both employ frequency-dependent radiation by the H$^-$ continuum and single Mg II k, Ca II K and Lyα lines, taking into account departures from local thermodynamic equilibrium (NLTE). Multiplying the losses from single lines with scaling factors, the total chromospheric losses are simulated.

Wave calculations require an initial atmosphere model. This initial model is constructed by using a (time-independent) standard temperature correction procedure (Cuntz et al. 1999), from which one usually obtains an outwardly decreasing temperature profile. The time-dependent wave code introduces the acoustic wave in this model by applying a velocity perturbation at the lower boundary:

$$v = -v_0 \sin\left(\frac{2\pi}{P}t\right) , \qquad (12)$$

where $v_0 = \sqrt{2F_A/(\rho\, c_S)}$ is the wave amplitude, and the minus sign is taken to minimise switch-on effects. Switch-on effects are large transient events which occur when a numerical wave calculation is started and the initial atmosphere reacts strongly to the incoming wave. Here ρ is the density and c_S the sound speed at the bottom. One usually takes monochromatic waves instead of the full acoustic spectrum because of the greater simplicity of the computation and because the results can be more easily analysed.

Figure 23 shows a snapshot from a wave calculation for the Sun with $P =$ 45 s and $F_A = 1 \times 10^8$ erg cm^{-2} s^{-1}. The temperature profile of the initial atmosphere is labelled T_0. The snapshot is shown at a time where many waves have already gone through the atmosphere. It is seen that the wave at first has a rapid amplitude growth with height, and that near 500 km height shocks form, which quickly attain a saw tooth shape. For monochromatic waves the shocks reach a constant limiting strength (magnitude of the shock jump) which depends on the wave period. For small wave amplitudes the magnitude of the shock jump is directly proportional to the wave period. After shock formation the mean temperature (after time-averaging, (dash-dot-dash)) increases in outward direction and at about 2000 km rises rapidly. This temperature behaviour is called a *classical chromosphere.*

The reasons for the rapid rise at great height are the following. When a shock traverses a gas element, its temperature abruptly increases. Subsequently the temperature gradually decreases again due to radiative cooling. When the temperature jump has not been completely radiated away before the next shock arrives, then the mean temperature in the gas element rises. Therefore, after a long time, a mean temperature is established in the gas element such that the heat brought by the temperature jump is exactly removed by radiative cooling until the next shock arrives, a state called dynamical equilibrium. Now at great height where the mean temperatures are high, the typical emitters Ca II, Mg II and H I get ionised to Ca III, Mg III and H II, respectively. By destroy-

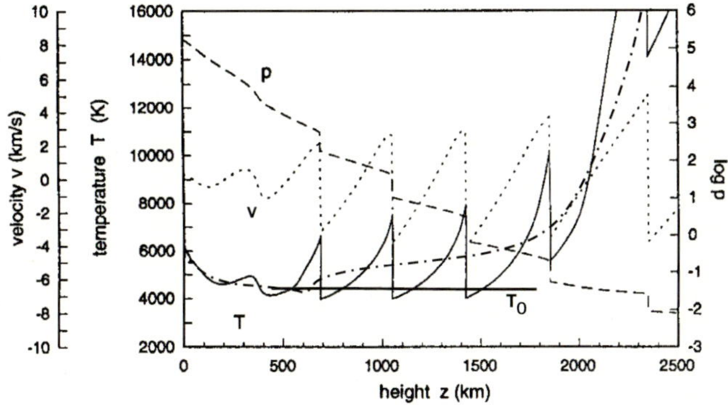

Fig. 23. Acoustic wave calculation. The velocity v, temperature T and $\log p$ in dyn cm^{-2} are shown as function of height z

ing the cooling mechanism the shock heating thus becomes unbalanced (heating catastrophe). The reaction of the atmosphere is that the temperature shoots up rapidly into the transition layer and corona.

To compare such theoretical chromospheres with observations one computes the line profiles of the Mg II h and k as well as Ca II H and K lines. Figure 24 shows the observed Ca II K line from the Utrecht solar atlas. Because the chromosphere is a very tenuous layer overlying the photosphere it is only visible as emission peaks in the line core. The inset of Fig. 24 indicates the terminology of the spectral features of the K line, the minima K_1, the emission peaks K_2 and the central absorption feature K_3.

The Ca II line profile mirrors the temperature profile of the atmosphere. This is seen in Fig. 25. The opacity, shown in Fig. 25b varies rapidly with frequency. From radiative transfer theory the monochromatic intensity at a wavelength difference $\Delta\lambda$ from line centre is equal to the source function (essentially the Planck function) at optical depth unity for that wavelength, that is, one has $\tau_\lambda = -\int \kappa_\lambda \, \mathrm{d}z = 1$. Here z is the height in the atmosphere (Fig. 25) and κ_λ the opacity. At $\Delta\lambda = \pm 10$ Å the opacity is low and one can look into very deep layers of the star where one has high photospheric temperatures, and where one obtains high intensities in the wings of the line. At $\Delta\lambda = \pm 0.4$ Å one has much higher opacity and one can look only as far as the temperature minimum (K_1 in Fig. 24). There the Planck function is small and one gets a low intensity. Finally at $\Delta\lambda = \pm 0.2$ Å one has very high opacity and looks only as far as the chromosphere where the temperature is high and one gets a high intensity (K_2 in Fig. 24). This very nice mapping of the temperature profile onto the line profile breaks down in the innermost line core, where NLTE effects decouple the source function and thus intensity from the temperature distribution, which leads to the K_3 intensity dip at line centre (Fig. 24). From this it is clear that the emission cores of the Ca II, Mg II and Lyα lines are generated by a chromosphere.

Figure 26 shows wave calculations by Buchholz & Ulmschneider (1994) for three main-sequence stars of spectral type F0V, G5V and K5V. On basis of the

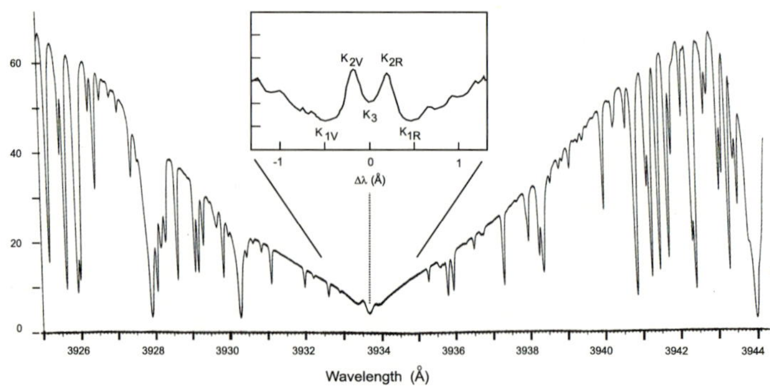

Fig. 24. Ca II K line profile from the Utrecht solar atlas, inset: terminology of the Ca II K core region

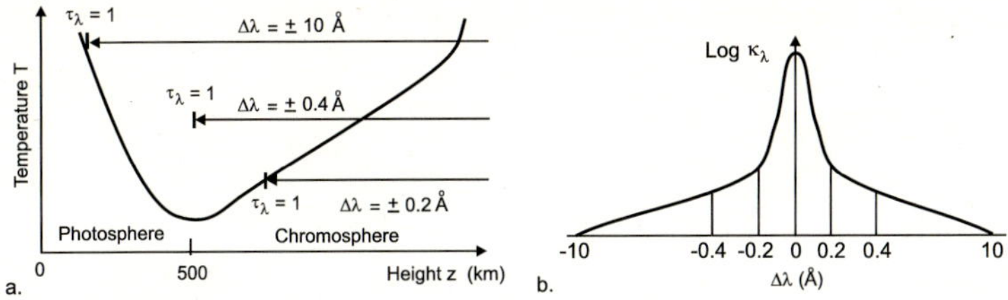

Fig. 25. Ca II line formation. (**a.**) temperature versus height, (**b.**) line opacity versus wavelength difference from line centre

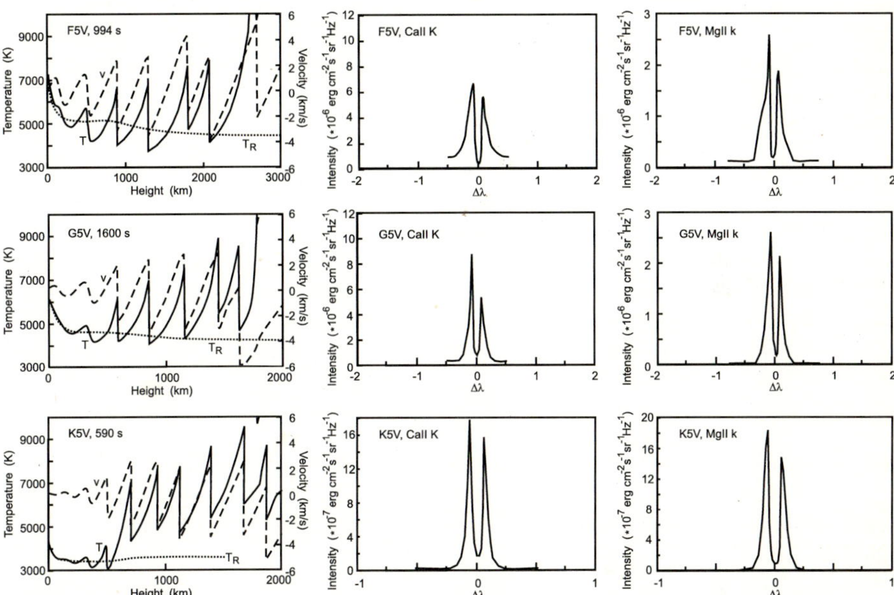

Fig. 26. Chromosphere models with acoustic shock waves (left panels), Ca II K line core profiles (middle panels) and Mg II k line core profiles (right panels) simulated for these chromosphere models for three different main sequence stars, after Buchholz & Ulmschneider (1994). $T_{\rm R}$ is the initial radiative equilibrium temperature distribution at the start of the calculation

theoretical chromosphere models the Ca II K and Mg II k lines were simulated (Fig. 26). Integrating the monochromatic fluxes in these emission cores, total chromospheric emission fluxes in these lines can be computed and compared with stellar observations as shown in Fig. 27. These observations (Rutten et al. 1991) measure the energy flux of the emission cores of the Ca II and Mg II lines and find that all late-type stars have at least a minimal core emission indicating a chromosphere. This empirically found lower envelope of stellar chromospheric emission is called *basal flux line* (heavy solid in Fig. 27). For the significance of the other lines in the figure see Rutten et al. (1991), and for more details on the basal flux line see also Fawzy et al. (2002b). Note that the basal flux

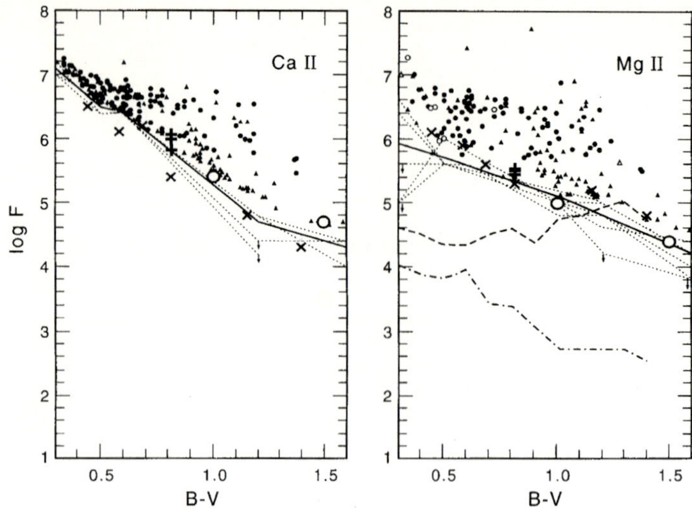

Fig. 27. Comparison of the chromospheric emission from acoustically heated theoretical chromosphere models with observations in Ca II and Mg II lines. (x) mark main-sequence stars of solar metal abundances, (o) giant stars of solar abundances, while (+) mark giants with 1/1, 1/10 and 1/100 solar metal abundances, after Cuntz, Rammacher & Ulmschneider (1994) and Buchholz, Ulmschneider & Cuntz (1998). The empirical basal flux line is marked heavy solid

line is found to be the same for main-sequence stars, giants and for stars with low metallicity. But there are also stars in Fig. 27 which show a much higher emission and therefore there also is an upper envelope called *saturation limit* at high chromospheric emission.

It is seen in Fig. 27 that the theoretical Ca II and Mg II fluxes agree quite well with the basal flux line. Note that the theoretical calculations are completely ab initio computations, based only on the three parameters $T_{\rm eff}$, g and $Z_{\rm m}$, and that there are only these three basic parameters through which acoustic energy generation in the stars can vary. Therefore, the agreement with the purely observational basal flux line not only for main-sequence stars (varying $T_{\rm eff}$), but also for giants (varying g) as well as low metallicity stars (varying $Z_{\rm m}$) suggests that *acoustic waves are the basic heating mechanism for stellar chromospheres.*

But the stars with emission fluxes higher than the basal flux need an additional (magnetic) heating mechanism. Indeed it is found that the chromospheric emission strongly depends on rotation and that the greater the rotation the higher is the chromospheric emission (rotation–emission correlation, see Fig. 45 below). As the dynamo mechanism depends on convection and rotation, the more rapid the star rotates, the greater is the magnetic flux that covers the star. The chromospheric emission higher than the basal flux line thus involves the fourth basic stellar parameter $P_{\rm Rot}$.

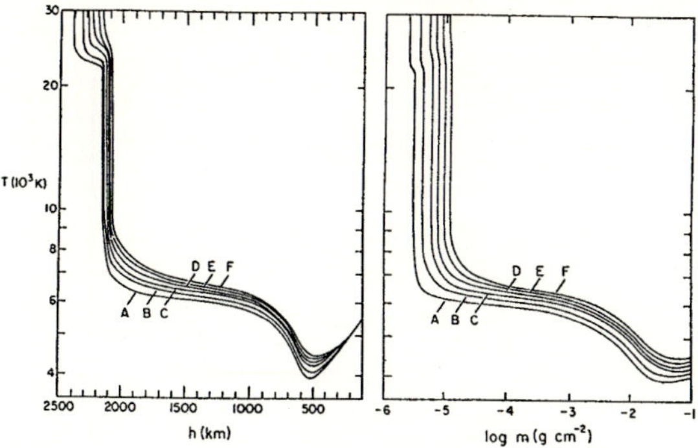

Fig. 28. Semi-empirical solar chromosphere models by Vernazza, Avrett & Loeser (1981)

7 Semi-empirical Chromosphere Models

Vernazza, Avrett & Loeser (1981) generated semi-empirical solar chromosphere models by selecting temperature distributions which optimally predict observed spectral features (mainly UV continua but also lines) and by assuming hydrostatic equilibrium to obtain the density and pressure distribution (see Fig. 28). Non-solar semi-empirical chromosphere models are based almost exclusively on the intensity profiles of the Ca II and Mg II lines. Examples of such models are shown in Fig. 29. Here in a first step with the stellar parameters T_{eff} and g, radiative and hydrostatic equilibrium photosphere models are computed (dashed in Fig. 29). Then in a second step the simulated Ca II K line profiles based on various outward temperature distributions (solid) are optimally fitted to the observed line wings.

It should be noted that all of these models, both semi-empirical and early theoretical models, in addition to an enhanced photospheric temperature distribution invariably show a classical chromosphere with a monotonic outwardly increasing temperature (Figs. 28, 29, 23). Clearly semi-empirical and theoretical models cannot fully agree, because for the semi-empirical modelling a smooth monotonic temperature and emission distribution versus height is assumed, while in the theoretical calculations the temperature varies inhomogeneously with the radiation primarily concentrated in the hot regions behind the shocks. However, by time averaging, this emission is smoothed out such that it appears monotonic. This smoothing can also be thought as the result of a large number of independent shock wave propagations averaged over the wide expanse of the stellar surface.

Such wave effects can explain for instance why pure Ca II line semi-empirical models give lower temperatures than pure Mg II line models because the ultraviolet Mg II emission in the post shock region has a stronger temperature dependence than the Ca II emission. Aside of these effects, both semi-empirical and theoretical models agree reasonably well. However, this picture of classical chromospheres in the last few years has been challenged by Carlsson & Stein

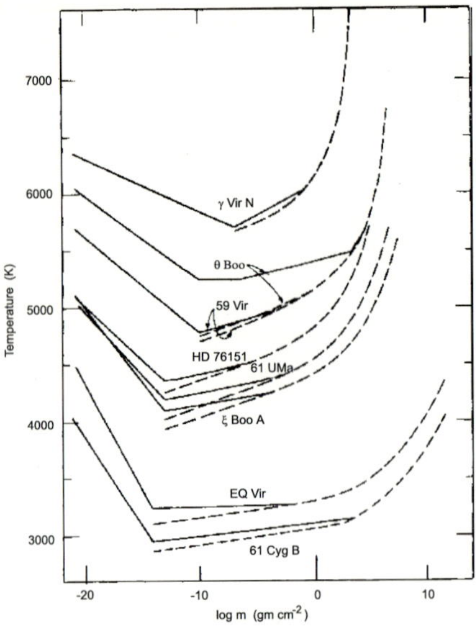

Fig. 29. Semi-empirical stellar chromosphere models based on Ca II K line fits (*solid*), after Kelch, Linsky & Worden (1979). Radiative equilibrium photosphere models are shown *dashed*

(1995) who argued that the time-dependence in the chromosphere is much more extreme than previously thought and that over most of its height the excess chromospheric emission is due to strong solitary waves which propagate over the outwardly decreasing radiative equilibrium temperature distribution.

8 Extremely Time-Dependent Chromospheres

To firmly establish the presence of propagating acoustic waves which explain the observed time-dependent profile variations and phase relations of chromospheric lines, Carlsson & Stein (1994, 1997) employed a hydrodynamic code which incorporates the time-dependent treatment of the rate equations (which determine the populations of the energy levels), the hydrogen ionisation and the radiative transfer equations for H^-, hydrogen and various chromospheric lines (Ca II, Mg II). In order to decouple themselves from the uncertainties of the generation of acoustic waves in the convection zone, the authors decided to use observed velocity fluctuations in a low-lying Fe I line (at 3966.8 Å formed at a height of about 250 km) observed by Lites, Rutten & Kalkofen (1993) as input for their wave code. The idea was to use an observed input and to simulate the perturbation caused by this wave input at greater height, and to explain the complicated line shifts and profile brightenings in the core of the Ca II H line at 3968 Å which was recorded simultaneously with the Fe line. This velocity input and its frequency spectrum is shown in Fig. 30.

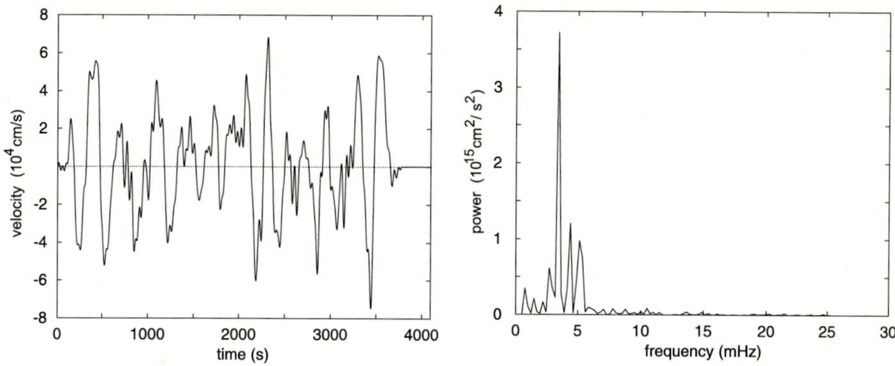

Fig. 30. Left: Velocity. Right: power spectrum in an Fe I line observed by Lites, Rutten & Kalkofen (1993, see Theurer, Ulmschneider & Kalkofen 1997) at a height of $z =$ 250 km

The result of the simulation was impressive, as the authors were able to reproduce quite well the complicated time-dependent core behaviour of the Ca II H line which is formed at a height of about 1500 km. This showed that indeed propagating acoustic waves are essential for explaining the time-dependent behaviour of the chromosphere. The success of this simulation proved even more impressive when the frequency-dependent phase differences between velocity fluctuations in spectral lines originating from different heights were compared.

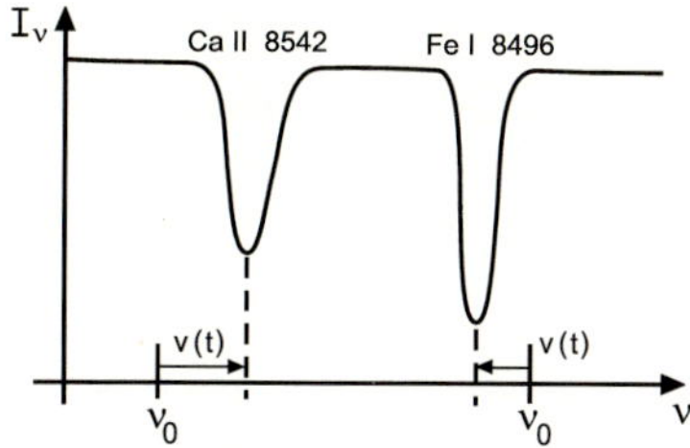

Fig. 31. Simultaneous Doppler measurements of velocity fluctuations in a low lying Fe I line and a Ca II IRT line at great height

To explain these phase differences, we discuss the Doppler shift observations of Fleck & Deubner (1989). Figure 31 shows how at some point on the slit of a spectrograph (and thus at a given spatial location on the Sun) the Doppler shifts relative to the rest frequency ν_0 of two lines are measured. The lines form at different atmospheric heights. These Doppler shifts allow to infer the time-dependent velocities $v(t)$ at the formation heights of these lines. Fleck & Deubner observed a low lying Fe I line formed at 200 km and one of the Ca II IRT (infrared triplet) lines, which on basis of semi-empirical chromosphere models originates at 1500 km. After long time-series of velocity fluctuations $v(t)$ in these two lines

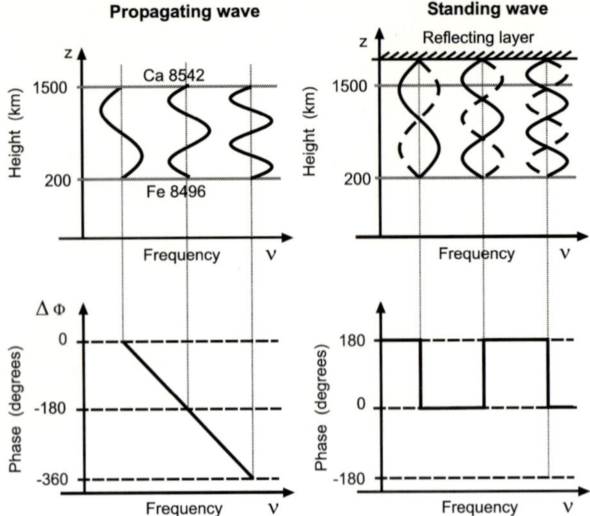

Fig. 32. Phase differences between velocity fluctuations in different spectral lines, for propagating and standing acoustic waves

are recorded, Fourier transforms

$$F(\omega) = |\tilde{v}(\omega)| \, \mathrm{e}^{\mathrm{i}\varphi(\omega)} = \int_{-\infty}^{+\infty} v(t)\mathrm{e}^{-\mathrm{i}\omega t}\mathrm{d}t \, , \tag{13}$$

can be computed for each line and a cross-correlation can be formed

$$CC(\omega) = |\tilde{v}_{\mathrm{Ca}}(\omega)\tilde{v}^*_{\mathrm{Fe}}(\omega)| \, \mathrm{e}^{\mathrm{i}\Delta\varphi(\omega)} \, , \tag{14}$$

where $\Delta\varphi(\omega) = \varphi_{\mathrm{Ca}}(\omega) - \varphi_{\mathrm{Fe}}(\omega)$ is the phase difference between acoustic velocity fluctuations detected at the two different heights. The frequency dependence of this phase difference provides information about the nature of the acoustic waves. Figure 32 shows the different phase behaviour of propagating and standing acoustic waves. On the left panel for a certain frequency, the wavelength λ of the acoustic wave just fits into the distance interval Δz between the heights where the cores of the Ca and Fe lines are formed. Here one has a phase difference of 0°. If the acoustic frequency is increased and $3/2\lambda$ fit into Δz then a phase difference of $-180°$ is found, and if 2λ fits, the phase difference is $-360°$ etc.. One sees that for propagating waves one has the relation

$$\Delta\varphi = -360° \left(\frac{\Delta z}{\lambda} - 1 \right) = -360° \left(\frac{\Delta z \, \nu}{c_{\mathrm{phase}}} - 1 \right) , \tag{15}$$

where c_{phase} is the phase speed for which the sound speed can be taken. That is, for propagating waves one has a linear dependence of the phase difference on frequency ν. The right panel shows the behaviour of standing waves. As standing waves have infinite phase speed, the phase differences can be either zero or $-180°$ and as function of frequency must undergo sudden phase jumps between these two values.

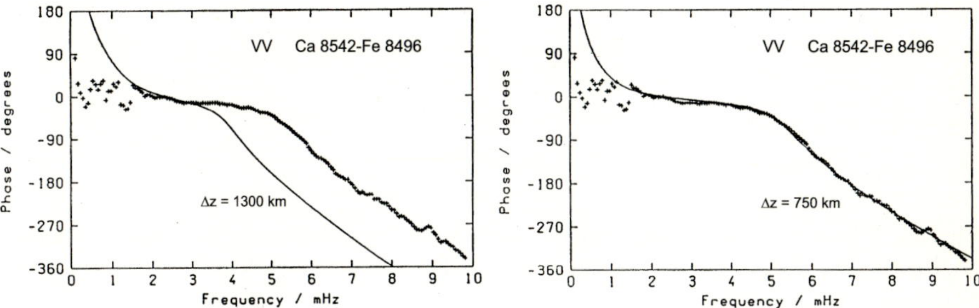

Fig. 33. Observed phase differences between velocity fluctuations in an Fe I line and a Ca II IRT line as function of frequency, after Fleck & Deubner (1989)

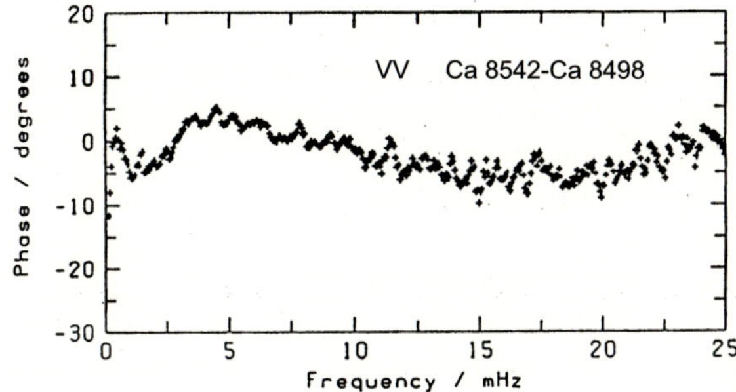

Fig. 34. Observed phase differences between velocity fluctuations in two Ca II IRT lines as function of frequency, after Fleck & Deubner (1989)

Figure 33 shows the observed phase differences between the Fe and Ca lines and theoretically computed phase differences on basis of the known sound speed and heights of formation of the two lines. The linear decrease of the phase difference with frequency indicates that acoustic waves of frequency $\nu = \omega/2\pi >$ 5 mHz or wave periods less than 200 s propagate. However, comparison between the two panels shows that theory and observation fit perfectly only if the theoretical height interval $\Delta z = 1300$ km is reduced to 750 km. The big surprise came when in Fig. 34 the Ca II 8542 Å and Ca II 8498 Å IRT lines were compared. The latter line was thought to originate at 1200 km, and thus should have a height interval of $\Delta z = 300$ km relative to the former line. For this height interval the observed phase differences were much too small. This, Fleck & Deubner attributed to a large phase speed (taking $c_{\text{phase}} = \infty$ in (15)). They concluded that at 800 km there is a 'magic height' below which acoustic waves are propagating

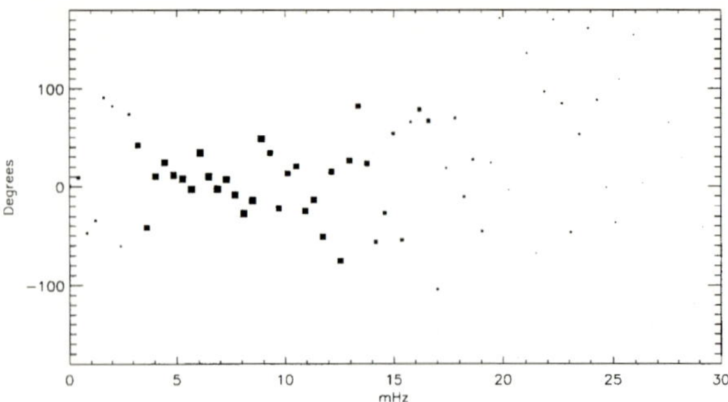

Fig. 35. Theoretical phase differences between velocity fluctuations in two Ca II IRT lines as function of frequency (larger symbols have higher coherence), after Skartlien, Carlsson & Stein (1994)

and above which they are standing. This idea causes problems for the acoustic heating, because standing waves do not heat and thus the energy losses of the middle and high chromosphere could not be balanced.

This picture of standing waves above 800 km was shattered by the Skartlien, Carlsson & Stein (1994) simulations. In their computations they also simulated the Ca II IRT lines and in agreement with the observations found essentially zero phase differences (with some scatter) between these lines as seen in Fig. 35. The reason for this behaviour lies in the nature of the acoustic wave which they discovered in their computation. As seen in Fig. 30, their acoustic input spectrum consists of low frequencies and from these a strong solitary shock formed in their calculation which propagated through the atmosphere and caused the complicated line profile variations in the Ca II H line. After this shock has propagated through the atmosphere, another solitary shock forms etc.. A snapshot of this shock is seen in Fig. 36. It propagates (from the right to the left) on top of the outwardly decreasing (radiative equilibrium) temperature profile (dashed). Behind the shock the temperature rapidly cools down to radiative equilibrium temperatures. To see why this shatters the 'magic height' picture one must realise that the line cores of the Ca II IRT lines form in the post shock region and thus emerge from a thin common height interval Δz behind the shock.

While $\Delta\varphi \approx 0$ in (15) was interpreted by Fleck & Deubner to mean $c_{\text{phase}} = \infty$, Stein & Carlsson essentially found that $\Delta\varphi = 0$ comes from $\Delta z = 0$ because both IRT lines are formed at the same height. This very impressive result, which also removed a stumbling block towards a solution of the chromospheric heating problem, gave support to the Carlsson & Stein (1994, 1997) picture and led the authors to draw far reaching conclusions. They claimed that on basis of their calculations, *classical chromospheres do not exist* and that chromospheres are extremely time-dependent phenomena, in which solitary shocks ever once in a while propagate on top of an outwardly decreasing radiative equilibrium temperature distribution (Carlsson & Stein 1994). Time-averages of their wave calculations resulted in an outwardly decreasing temperature profile (see Fig. 37).

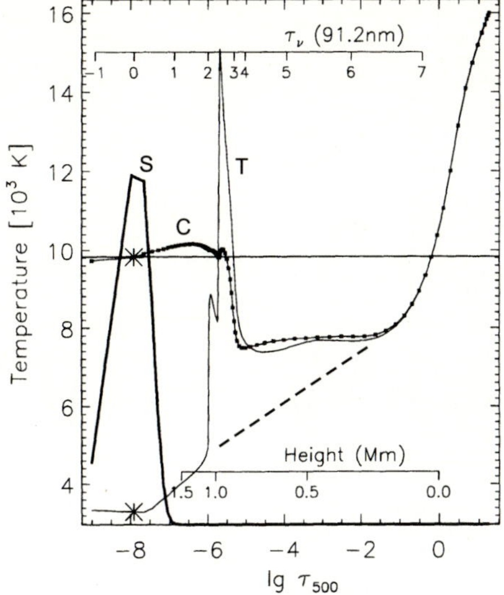

Fig. 36. Snapshot of the temperature (T) and Lyman continuum source function (S) distribution, and the intensity contribution function (C), after Carlsson & Stein (1994)

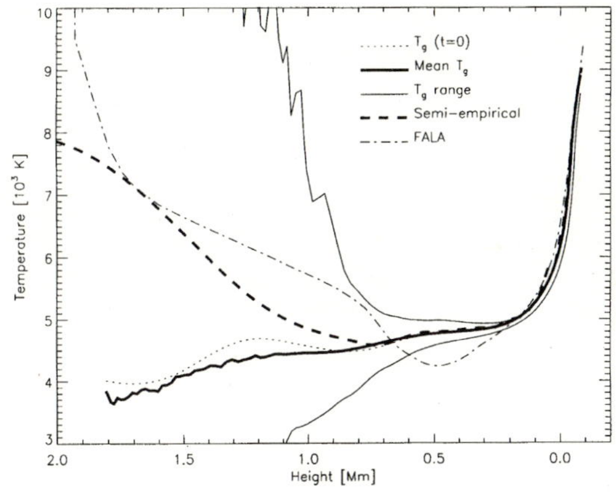

Fig. 37. Time-averaged temperature, after Carlsson & Stein (1994)

This new extremely time-dependent picture shatters the semi-empirical solar and stellar models.

The present author and others (Theurer, Ulmschneider & Kalkofen 1997; Kalkofen, Ulmschneider & Avrett 1999) think that this new view of stellar chromospheres raises a number of unanswered questions and that the old picture of classical chromospheres with an outwardly rising mean temperature although clearly in need of modification should not be discarded prematurely, particularly at great heights in the chromosphere where the kinetic temperature rises to transition layer and coronal values.

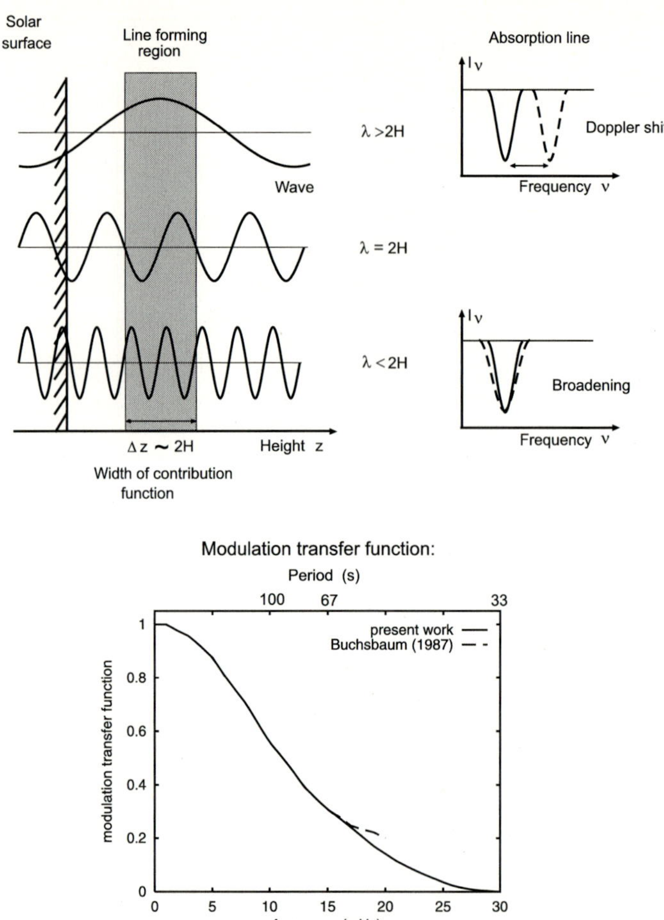

Fig. 38. Influence of the line contribution function on the observability of acoustic waves, and the modulation transfer function (see Theurer, Ulmschneider & Kalkofen 1997)

One problem with the calculation of Carlsson & Stein most likely is that they took an inadequate acoustic spectrum and therefore used only about 1/10 of the available acoustic energy. To see why taking observed velocity fluctuations leads to a severe underestimation of the available acoustic energy consider the relation between the acoustic wavelength λ and the width Δz of the line contribution function, that is, the height interval over which a spectral line forms. This *contribution function* even for weak lines is not much smaller than about two scale heights $\Delta z = 2H$. In Fig. 38 acoustic waves of different frequency are seen to propagate through the line forming region. The case of $\lambda \gg \Delta z$ results in a full Doppler signal, while for $\lambda \ll \Delta z$ only line broadening is generated (Fig. 38). The critical wavelength where Doppler shifts can no longer be observed occurs when the width of the line contribution function becomes equal to the wavelength of

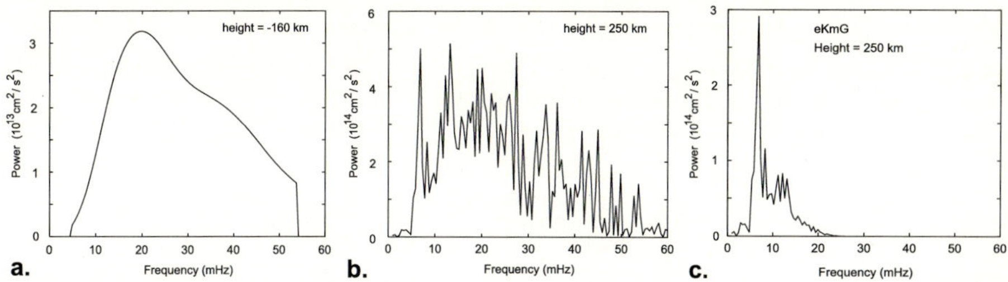

Fig. 39. (**a.**) Acoustic wave spectrum computed in the convection zone at height -160 km. (**b.**) the spectrum after propagation of the wave at height 250 km. (**c.**) spectrum of case **b.** after applying the modulation transfer function. It shows what actually can be observed from this acoustic spectrum as Doppler shifts in a spectral line at 250 km height

the acoustic wave, that is, when $\lambda = \Delta z = 2H$, and for the Fe I line this is at a wave period of about $P = 50$ s. Buchsbaum (1987) a student of Deubner has evaluated a modulation transfer function (see Fig. 38 lower panel) which tells which fraction of a physically present acoustic wave velocity fluctuation can be observed as Doppler shift fluctuation.

To determine the amount of acoustic wave flux actually present at the height of the Fe I line, Fig. 39 shows a computation by Theurer, Ulmschneider & Kalkofen (1997) where the acoustic energy spectrum generated in the convection zone at -160 km height is propagated to the height level of the Fe I line at about +250 km. The acoustic spectrum present at 250 km (Fig. 39b) is then folded with the modulation transfer function to show what a terrestrial observer will measure as Doppler shift (Fig. 39c). It is seen that only the low frequency component is actually observed (Kalkofen, Ulmschneider & Avrett 1999) which contains about 5 to 10% of the actually present acoustic energy. Most of the energy is in high frequencies and cannot be detected (compare Figs. 39b, 39c with Fig. 30). As the low frequency component of the acoustic spectrum is present both in the observations and in the theory, it is clear that the findings of Carlsson & Stein about the solitary shock and the explanation of the Ca II H line core and Ca II IRT phase behaviour will remain unchanged. However, the large amount of acoustic energy in the high frequency component invariably changes the chromospheric heating picture.

9 Realistic Chromospheres

The above discussions show that for realistic chromospheres one must include strongly time-dependent acoustic wave effects, employ a powerful hydrodynamic wave code and use the full acoustic spectrum as input. Figure 40 shows a time sequence of a full acoustic spectrum calculation by Theurer (1998) which, however, did not yet include the full time-dependent treatment of the hydrogen ionisation. It is seen that similarly to Carlsson & Stein strong solitary shocks form and propagate through the atmosphere. But there are also numerous smaller shocks which contribute to the heating of the chromosphere. In Theurer's calculation the strong shock, because of its much greater speed, cannibalises many smaller shocks which enhances the solitary shock picture.

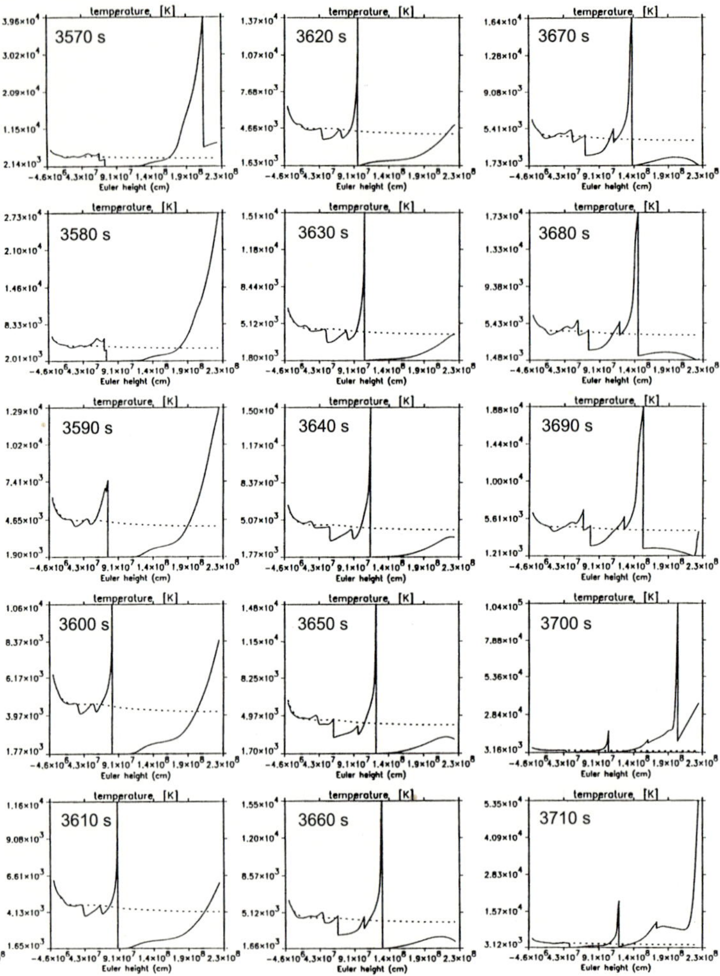

Fig. 40. Series of temperature profiles of a time-dependent acoustic wave computation using an acoustic spectrum in the solar atmosphere, after Theurer (1998)

Despite the general agreement with each other of the theoretical simulations of chromospheres by Carlsson & Stein (1994, 1997), Theurer (1998) and Rammacher & Ulmschneider (2003), these computations are not yet fully realistic. Probably a severe shortcoming of all present calculations is that they are one-dimensional (1D) computations. As pointed out by Kalkofen (2003), acoustic waves propagate spherically and the computations should therefore be carried out in a funnel-type geometry. That the acoustic energy generation is nonuniformly distributed over the solar surface has recently been shown by Wunnenberg, Kneer & Hirzberger (2002) but was already suggested long ago by Kuperus (1972). The realistic situation is that one has a large number of acoustic sources at discrete locations distributed all over the solar surface from which spherically propagating acoustic waves emanate.

One-dimensional wave computations can only very poorly model this situation. The difficulty is not to simulate the funnel-type geometry in which the wave energy spreads over a progressively larger cross-section, but to take into account the fact that for each energy loss due to a wider tube cross-section there must be an energy gain from neighbouring funnels which cannot be modelled in the 1D calculation. There is another unrealistic property of 1D acoustic wave computations. In the calculations of Theurer (1998) and Rammacher & Ulmschneider (2003) the strong solitary shock gets much of its power from cannibalising many weaker shocks. That the merging of shocks results in a single stronger shock is a special property of 1D calculations. In a three-dimensional (3D) situation, shocks from waves propagating in different funnels can overlap only at an oblique angle: they amplify only at certain points or lines and thus do not form single merged shocks.

Yet, 3D simulations with many discrete acoustic sources using adequate physics (radiation and ionisation treatments) are presently beyond our computational power. As in a realistic situation the shocks typically are less strong (due to the funnel-type area growth) and stay unmerged (due to intersection at an oblique angle) it appears that presently 1D monochromatic wave calculations are probably the best choice to simulate the general chromospheric heating. This does not contradict the fact that there occasionally are strong shocks.

Figure 41 shows a comparison of the mean chromospheric temperatures from two acoustic wave calculations using monochromatic waves and one with a full acoustic spectrum. The mean temperatures are produced by time-averaging temperature distributions like in Figs. 23 and 40. It is seen that the monochromatic wave calculations generate classical chromospheres. Yet the acoustic spectrum calculation at greater height also shows a classical chromospheric temperature rise (see also Theurer, Ulmschneider & Cuntz 1997).

To discuss the treatments of the hydrogen ionisation consider in Fig. 42 the population of energy levels of important chromospheric atoms before and after a strong shock travels through a gas element. For simplicity we take a hydrogen atom with two bound levels 1 and 2 and a continuum level 3. In the cold phase, before the shock arrives, the populations n_1, n_2, n_3 are concentrated to the ground state, while after the shock has passed the high temperature in the post-shock phase is supposed to lead to an increased population of the higher atomic levels. This increased population by transitions to the ground state should produce enhanced radiation losses from the atmosphere. This behaviour is indeed found when the populations are calculated (called NLTE treatment) using the

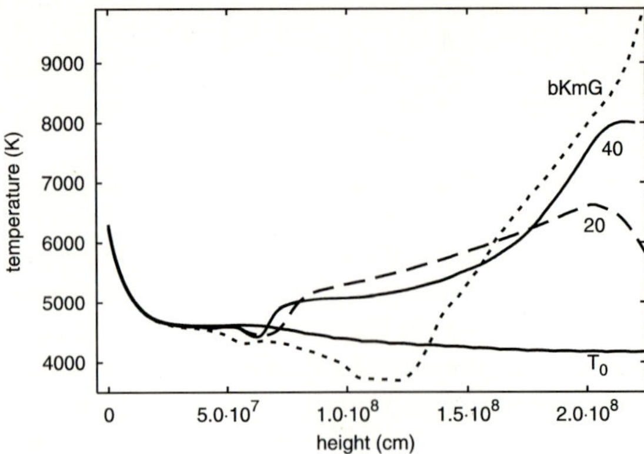

Fig. 41. Averaged solar temperature profile of acoustic wave computations, after Theurer (1998). Monochromatic calculations have periods $P = 20$ and 40 s, bKmG displays a computation with an acoustic spectrum

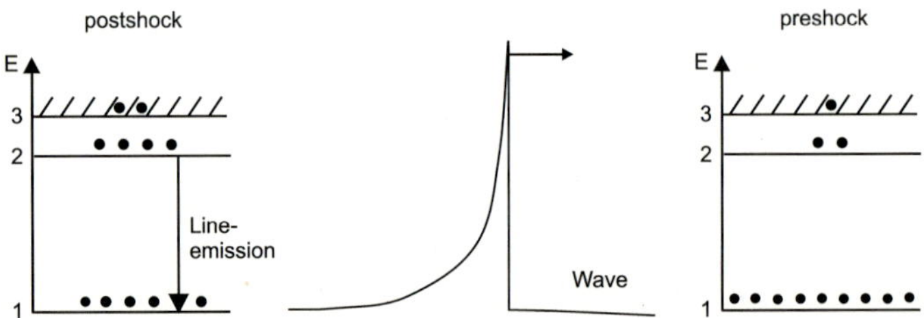

Fig. 42. Change of energy level populations before and after transit of a strong shock

individual radiative R_{ij} and collisional C_{ij} processes and solving *statistical rate equations* like e.g., the following for the ground level 1 of the three level atom:

$$n_2(R_{21} + C_{21}) + n_3(R_{31} + C_{31}) - n_1(R_{12} + C_{12} + R_{13} + C_{13}) = 0 \,. \tag{16}$$

Here the two left hand terms are the gains of electrons from the upper levels 2 and 3 and the third term is the loss of electrons from level 1. In most of our work we solve this equation consistently with the instantaneous temperature. In reality, however, the increase of the population of the high levels, particularly the continuum level 3, takes a finite time. Here it must be realised that the bound-bound transitions are very rapid while the bound-continuum transitions are slower by many orders of magnitude. That means it is not good enough to solve the statistical rate equations, one must actually solve the *time-dependent rate equation*

$$\frac{\mathrm{d}n_1}{\mathrm{d}t} = n_2(R_{21} + C_{21}) + n_3(R_{31} + C_{31}) - n_1(R_{12} + C_{12} + R_{13} + C_{13}) \,. \tag{17}$$

The time-dependence delays the buildup of the population of the high levels. This delay can be so severe that the level populations essentially decouple from the spiky temperature variation in the wave. Therefore the consequence of solving the statistical rate equations instead of the time-dependent rate equations is

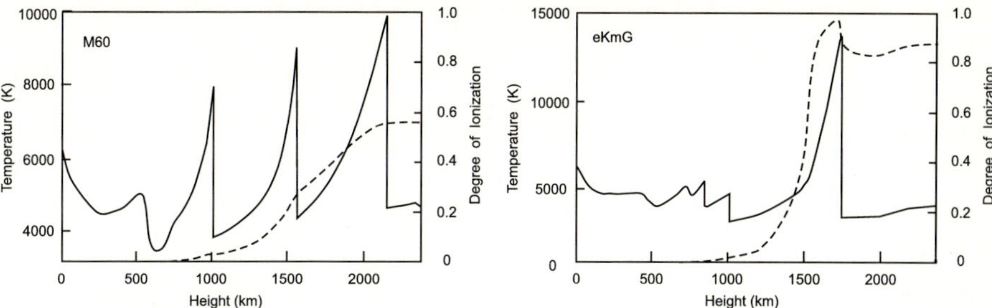

Fig. 43. Temperature (*solid*) and degree of ionisation (*dashed*) in acoustic wave calculations with time-dependent hydrogen ionisation and a flux $F_A = 1 \times 10^8$ erg cm^{-2} s^{-1}. Left: Monochromatic acoustic wave with period $P = 60$ s (M60). Right: acoustic wave with an eKmG spectrum, after Rammacher & Ulmschneider (2003)

that one greatly overestimates the radiation losses in strong shock computations (this error is less severe for weak shocks which occur in a monochromatic wave computation). In wave calculations which treat the hydrogen ionisation with a time-dependent rate equation the large relaxation time of the continuum transitions ($t_{\mathrm{Rel}} \approx 1/R_{31}$) leads to a degree of ionisation in the middle and high chromosphere which practically is decoupled from the temperature fluctuations. Figure 43 shows snapshots of acoustic wave calculations of this type. In the monochromatic wave calculation on the left panel it is seen that the degree of ionisation is practically unaffected by the temperature fluctuations at the shocks. Only if a shock is very strong as in the wave calculation with an acoustic spectrum (right panel) does the degree of ionisation in the post-shock region react to the temperature jump in the way expected from Fig. 42.

Surprisingly, time-averaging the fluctuating temperature in the gas elements (Fig. 44, left panel) gives a classical chromosphere (like in Figs. 41, 23) only for the monochromatic wave, while for the acoustic spectrum calculation essentially no temperature rise is found. This situation is very different when the ionisation temperature is considered (Fig. 44 right panel). The ionisation temperature T_{i} is the temperature which satisfies the NLTE Saha-equation

$$\frac{n_1}{n_3} = n_e \left(\frac{h^2}{2\pi m_e k T_{\mathrm{i}}} \right)^{3/2} \mathrm{e}^{\frac{E_H}{kT_{\mathrm{i}}}} , \tag{18}$$

when the time-averaged number densities n_1, n_3 and n_e are used. Here h is the Planck constant, m_e the electron mass, k the Boltzmann constant and E_{H} the ionisation energy of hydrogen. The ionisation temperature shows a classical temperature rise in the chromosphere together with the rapid rise in the transition layer. That the acoustic spectrum calculation shows a more pronounced temperature rise is explained from the fact that this wave leads to a higher degree of ionisation (compare the two cases in Fig. 43).

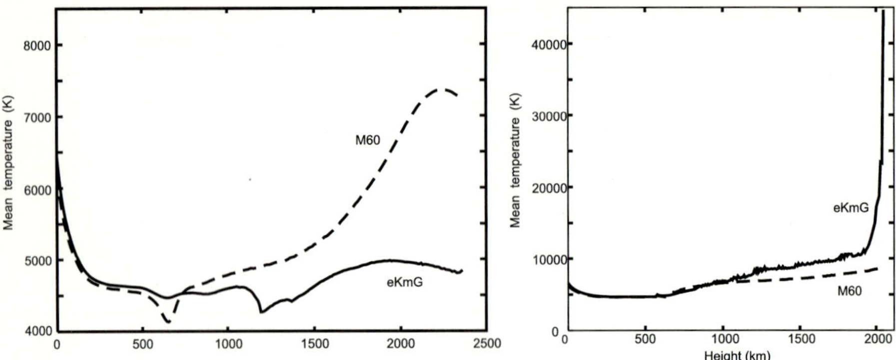

Fig. 44. Different mean temperature evaluations in the wave calculations of Fig. 43. Left: temporal averaging. Right: Ionisation temperature from the Saha equation using averaged number densities n_1, n_3, n_e, after Rammacher & Ulmschneider (2003)

10 Magnetic Chromospheres

We now consider magnetic chromospheres. As discussed above, the great problem to understand magnetic chromospheres is that many heating mechanisms exist. A basic task therefore is to identify which of these are the relevant ones. This is difficult as only for a few magnetic mechanisms is it presently known how much mechanical energy is available. While the situation is relatively good for magnetic body waves (although calculations of the energy generation of torsional Alfvén waves are still missing) a computation of the amount of energy available for e.g., micro flare heating is lacking, as is an evaluation of the energy generation of surface waves.

Two important observational effects can be used to identify the magnetic heating mechanisms: the so called *rotation–chromospheric emission correlation* and *magnetic flux–chromospheric emission correlation.* Figure 45 shows that in a given range of colour B − V (or T_{eff}) the chromospheric Ca II emission is strongly correlated with the stellar rotation period. Rapidly rotating stars have large chromospheric emission while the emission of the stars with longest rotation period decreases to the basal flux emission level. The other correlation is seen in Fig. 46 where the observed excess (relative to the basal flux) Ca II emission intensity is plotted against the measured magnetic flux. These observations indicate that higher magnetic flux results in greater chromospheric emission. Combining the two correlations one finds that the more rapidly a star rotates the more magnetic flux it has on its surface.

As discussed above it is found that on the Sun the magnetic field appears in sunspots and plage regions but also in a large number of thin isolated magnetic flux tubes. These tubes are distributed more or less uniformly over the entire solar surface and show a concentration towards the boundaries of supergranulation cells. In the Ca II K and He II 304 Å lines as well as other chromospheric and transition layer lines the supergranulation boundaries show intense emission associated with the magnetic field, called *chromospheric network.*

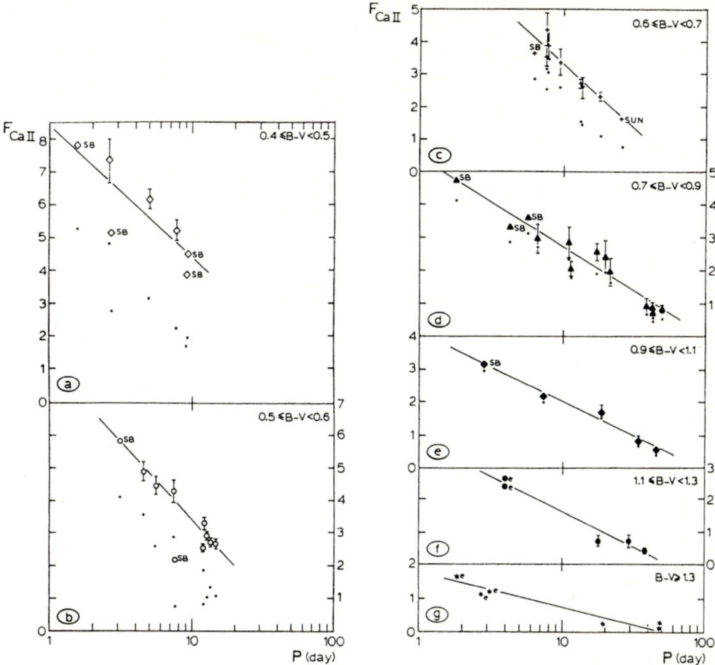

Fig. 45. Rotation–chromospheric Ca II emission correlations, after Rutten (1986). The panels a to g are for stars in different B − V ranges of width 0.1 from B − V = 0.4 to greater than 1.3

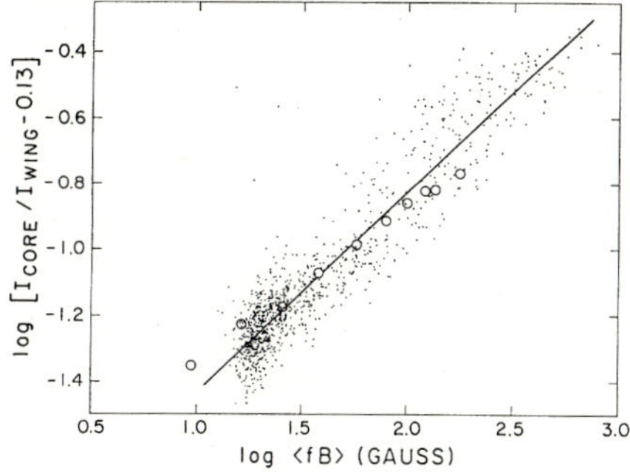

Fig. 46. Magnetic flux–chromospheric Ca II emission correlation for the Sun, after Schrijver et al. (1989)

Carrying most of the stellar magnetic flux $\Phi_M = |B|A_M$ these magnetic flux tubes have a magnetic filling factor $f = A_M/A_*$, defined as the ratio of the area covered by magnetic fields A_M to the total area A_* at the stellar surface. The Sun has a small filling factor $f \approx 0.02$ (see Fig. 47) while rapidly rotating stars can have $f > 0.4$. The diameter of these flux tubes grows, until in the middle chromosphere (for the Sun at around 1500 km height) the magnetic fields fill out

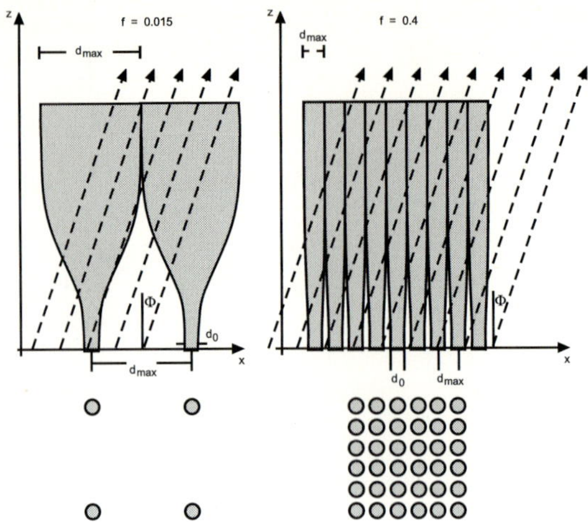

Fig. 47. Flux tube models of different filling factors f, for the solar case (left panels) and the case $f = 0.4$ (right panels). The lower panels show the cross-sections at the stellar surface

the entire available space (Fig. 47). It was also mentioned above that perturbed by external turbulence at the top of the convection zone, longitudinal, transverse and torsional MHD waves are generated in these tubes.

Note that presently there is no way to specify the magnetic flux Φ_{M} for stars from first principles on basis of the four parameters T_{eff}, g, Z_{m} and P_{Rot}. This awaits the successful development of a dynamo theory. However, it is possible to progress by constructing magnetic chromosphere models of stars by assuming different magnetic filling factors f. These models can be characterised by four parameters T_{eff}, g, Z_{m} and f. We suppose that similar to the Sun the magnetic field on other stars is also dominated by a large number of thin isolated magnetic flux tubes and that the field strength of these tubes at the stellar surface is $B = 0.85 B_{\mathrm{eq}} = 0.85\sqrt{8\pi p}$ where p is the external gas pressure. In addition we assume that the diameter of the flux tubes at the stellar surface is about a scale height. Figure 47 shows flux tube models for stars of the same T_{eff} and g but different filling factors f computed with these assumptions.

Using such tube models, magnetic wave energy fluxes and wave spectra can be calculated for longitudinal and transverse MHD waves. For longitudinal wave fluxes using an analytical approach see Musielak, Rosner & Ulmschneider (1989, 2000, 2002), Musielak et al. (1995), and with a numerical approach Ulmschneider & Musielak (1998), Ulmschneider, Musielak & Fawzy (2001). For transverse wave fluxes using an analytical approach see Musielak & Ulmschneider (2001, 2002a, 2002b), and for the numerical approach see Huang, Musielak & Ulmschneider (1995) with a correction discussed in Ulmschneider & Musielak (1998). Figure 48 displays and compares some of these fluxes. It is found that roughly transverse waves are by a factor of 30 more efficiently generated than longitudinal waves.

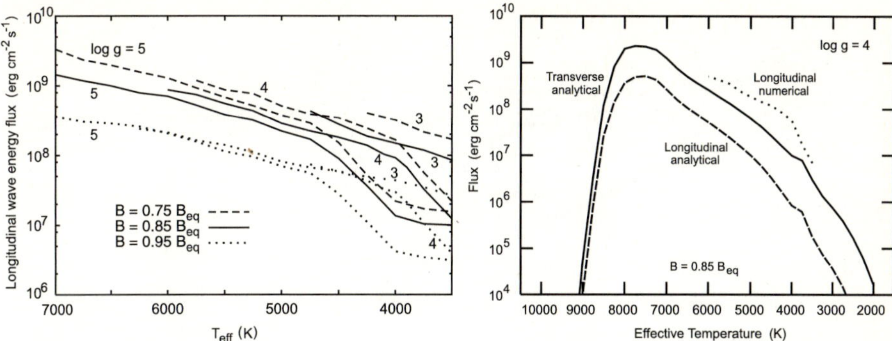

Fig. 48. Longitudinal tube wave fluxes (numerical), after Ulmschneider, Musielak & Fawzy (2001) (left), longitudinal and transverse tube wave fluxes, after Musielak & Ulmschneider (2002a) (right)

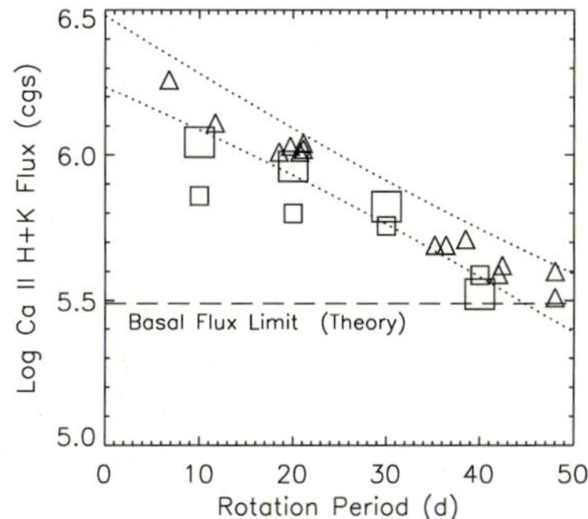

Fig. 49. Rotation–Ca II core emission flux relation of a sample of K2V stars (triangles) compared to that of simulated core emissions on basis of MHD wave heating (squares), after Cuntz et al. (1999). The *dotted* lines indicate the observed emission–rotation correlation similar as in Fig. 45, the *dashed* line gives the theoretical basal flux which the stars would have if they did not rotate

On basis of these fluxes and spectra one calculates the propagation of longitudinal MHD waves along the flux tubes together with the shock heating. Here the thin flux tube approximation was used. In addition, employing acoustic wave computations, the heating of the medium outside the tubes is computed. Same as for the acoustic wave calculations discussed above, monochromatic magnetic waves were employed with periods taken from the maximum of the longitudinal wave energy spectrum. Similarly as for our pure acoustic wave computations, using a multi-ray transfer code, the Mg II and Ca II line profiles emerging from the forest of magnetic flux tubes are then evaluated using ray-paths as shown in Fig. 47, and finally the emission core fluxes were compared with observations.

The result of such a comparison for stars of spectral type K2V are shown in Fig. 49. Here magnetic field strengths and rotation periods have been measured and the generation, propagation and heating of longitudinal waves was calculated. Large squares assume a uniform distribution of the magnetic tubes over

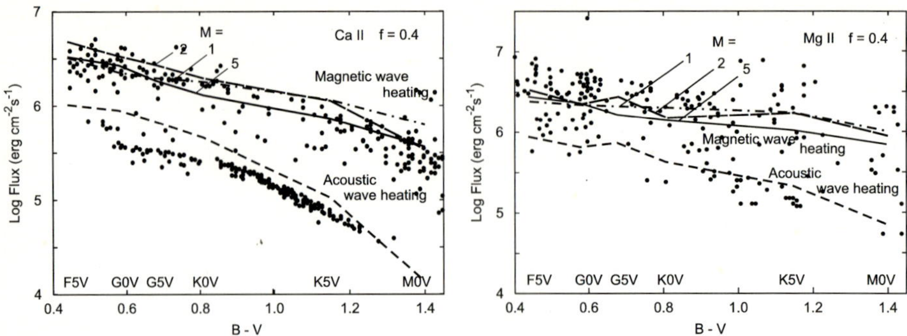

Fig. 50. Empirical core emission fluxes of Ca II (dots, left) and Mg II lines (dots, right), compared with theoretical fluxes for pure acoustic wave heating and for magnetic wave heating in flux tubes with an area filling factor of $f = 0.4$, after Fawzy et al. (2002b)

the stellar surface while small squares are for distributions with a pronounced network structure. It is seen that for the K2V stars, the observed rotation–chromospheric emission correlation is relatively well reproduced by the theoretical simulations.

Figure 50 shows a more extensive comparison of theoretical emission fluxes with observations for late-type stars by Ulmschneider et al. (2001) and Fawzy et al. (2002a, b). Shown dashed are theoretical models with pure acoustic wave heating for stars with no magnetic fields ($f = 0$). They agree fairly well with the observed emission of basal flux stars. The discrepancy for F5V to G5V stars for the Ca II emission is attributed to the large errors in subtracting the photospheric background emission in the observations and to inaccuracies of the theoretical computations (see Fawzy et al. 2002a, b).

Theoretical MHD wave heating models are indicated solid. To show the maximum that MHD wave heating can generate, a maximum magnetic filling factor of $f = 0.4$ was assumed. The reason for this is that with $f = 0.4$ the main energy carrying convective bubbles (with diameters of a scale height) still fit in the space outside the flux tubes to permit efficient MHD-wave generation (see Fig. 47). To allow for additional wave energy by mode-coupling from transverse waves the longitudinal wave flux in some cases were multiplied by factors $M = 1$ to $M = 5$ (see Fig. 50). A maximum factor $M = 5$ due to mode-coupling was assumed to be realistic, particularly in view of the fact that transverse waves are much more efficiently produced as mentioned above. As the amount of longitudinal wave energy generation due to mode-coupling cannot yet be reliably computed the calculations with different values of M show the theoretical uncertainty at present.

When comparing the theoretical emission fluxes for the magnetic cases with the observed Ca II and Mg II saturation limits it has to be noted (like in the acoustic case) that they are obtained from fully ab initio calculations which only specify the four parameters $T_{\rm eff}$, g, $Z_{\rm m}$ and f. Thus the theoretical steps to construct the convection zone and magnetic flux tube models, the wave generation process, the wave propagation and 2D radiative transfer in the spectral lines are expected to lead to considerable cumulative errors. It is therefore interesting that the magnitude of the emission and its $T_{\rm eff}$-dependence as well as the emission variation with the filling factor appears to be in relatively good agreement with the observations. Such an agreement would not be expected if the wave heating mechanism were wrong or only a minor contribution to the total chromospheric heating. We conclude therefore that the agreement confirms that longitudinal tube waves are the main heating mechanism of magnetic chromospheres.

Figure 50 also shows that while the variation of the filling factor covers the entire chromospheric emission variability (between the basal flux line and the saturation limit) for Ca II, there appears to be a gap between the theoretical simulations and the observed maximum of the Mg II emissions. Because the Ca II emission is generated at lower chromospheric heights than the Mg II emission we feel that this gap indicates that in the highest chromosphere another non-wave magnetic heating mechanism comes into play. It is possible that this missing magnetic heating mechanism consists of microflare reconnective heating proposed by Parker (see Sect. 4.10). This verdict, however, must await the study of the torsional waves and of the role of surface waves.

11 Conclusions

Chromospheres and coronae are hot outer layers of late-type stars which are dominated by mechanical heating. Similarly as for the stellar interior and the stellar atmosphere (photosphere) the average behaviour of these layers and the magnitude of their variability should be describable by only 4 basic parameters, effective temperature $T_{\rm eff}$, gravity g, metallicity $Z_{\rm m}$ and surface rotation period $P_{\rm Rot}$. The connection of the chromospheric and coronal structures with these 4 parameters has so far not been completely unravelled but significant progress has been made.

1. There are two classes of heating mechanisms (see Table 1): hydrodynamic and magnetic mechanisms. The latter are further subdivided in AC-mechanisms (waves) and DC-mechanisms (current sheets). All suggested heating mechanisms are thought to work for stars, given the right situation and magnetic field geometry, because they are known to work in terrestrial applications. However, it is important to identify the main processes for the individual stellar layers and magnetic regions.

2. The turbulent flow fields of the surface convection zones of late-type stars generate acoustic waves. Propagating to the outer layers these waves form shocks which heat the chromospheres. Acoustically heated chromospheres depend only on three basic parameters $T_{\rm eff}$, g, $Z_{\rm m}$, as convection zones do not vary with rotation.

3. From the computed acoustic energy fluxes, theoretical shock heated chromospheres can be constructed and the Ca II and Mg II line core emission simulated. These simulations reproduce the observed basal flux line, that is, the lower limit of chromospheric emission for main sequence stars, giants and low

metallicity stars. This shows that acoustic waves are very likely the main heating mechanism for the low chromospheres of late-type stars.

4. When propagating from the convection zone, the low frequency part of the acoustic wave spectrum generates strong solitary shocks, which explain the observed solar Ca II H line profile variations and phase relations between the Ca II IRT lines. The propagating wave spectrum also generates numerous weaker shocks which together with the strong shocks produce a classical chromosphere, that is, a layer with an outwardly rising mean temperature distribution, most easily seen in the degree of ionisation and the ionisation temperature.

5. In the high chromosphere the dominant cooling mechanism (H I Lyα, Ca II H+K+IRT and Mg II h + k lines) get destroyed by the ionisation of H, Ca II and Mg II. This causes the heating to become unbalanced, which generates the transition layer temperature rise to the corona.

6. Rotation (described by the fourth basic parameter P_{Rot}) together with convection produces the magnetic fields of late-type stars. Since the dynamo theory is so far not sufficiently developed to predict the stellar magnetic flux from the 4 basic parameters, observations must be used to unravel the magnetic heating mechanisms. These observations are mainly the rotation–chromospheric emission and magnetic flux–chromospheric emission correlations, the magnitude and T_{eff}-dependence of the basal and saturation limits of chromospheric emission, and magnetic field strength measurements, although the latter tend to measure sunspot and plage fields and not the large numbers of small scale flux tubes where most of the magnetic flux is.

7. For various flux tube models and filling factors f of magnetic flux covering the star, longitudinal and transverse MHD wave energy fluxes can be computed and the wave propagation along these magnetic flux tubes as well as in the non-magnetic regions performed. This generates two-component stellar chromosphere models and allows to simulate the chromospheric emission. For K2V stars, where magnetic field observations are available, the observed rotation–emission relation could be reproduced.

8. For simulations using a maximum filling factor of $f = 0.4$ and a maximum amount of longitudinal and transverse MHD wave energy it was found that for Ca II the observed saturation limit of chromospheric emission could be reproduced, but that the limit for Mg II could not be reached. This indicates that the magnetic regions of the middle and upper chromosphere are heated by MHD wave dissipation, but that at the top of the chromosphere another non-wave heating mechanism, possibly reconnective microflare heating operates. This conclusion is tentative as the role of the heating by torsional Alfvén waves and surface waves must be studied.

9. To clarify the zoo of coronal heating processes much further work remains to be done.

Acknowledgements

The author thanks the Deutsche Forschungsgemeinschaft for generous provision of funds for projects Ul57/28-1, Ul57/30-1 and Ul57/32-1.

References

Brueckner, G. E., 1981, in Solar Active Regions, Skylab Solar Workshop III, ed. F. Q. Orrall, (Colorado Assoc. Univ. Press) 113

Brueckner, G. E., & Bartoe, J.-D. F., 1983, ApJ, 272, 329

Buchholz B., & Ulmschneider P., 1994, in Cool Stars, Stellar Systems and the Sun, ed. J. P. Caillault, ASP Conf. Series 64, 363

Buchholz, B., Ulmschneider, P., & Cuntz M., 1998, ApJ, 494, 700

Buchsbaum, G., 1987, Diplom thesis, Univ. Würzburg, Germany

Carlsson, M., & Stein, R. F., 1994, in Chromospheric Dynamics, Proc. Mini-Workshop, ed. M. Carlsson, Inst. Theor. Astroph., Oslo, p. 47

Carlsson, M., & Stein, R. F., 1995, ApJ, 440, L29

Carlsson, M., & Stein, R. F., 1997, ApJ, 481, 500

Chen, F. F., 1974, Introduction to Plasma Physics (Plenum Press, New York)

Cuntz, M., Rammacher, W., & Ulmschneider, P., 1994, ApJ, 432, 690

Cuntz, M., Rammacher, W., & Ulmschneider, P., Musielak, Z. E., Saar, S. H., 1999, ApJ, 522, 1053

Fawzy, D., Rammacher, W., Ulmschneider, P., Musielak, Z. E. & Stępień, K., 2002a, ApJ, 386, 971

Fawzy, D., Ulmschneider, P., Stępień, K., Musielak, Z. E., & Rammacher, W. 2002b, ApJ, 386, 983

Fleck, B., & Deubner, F.-L., 1989, A&A, 224, 245

Goldstein, M. E., 1976, Aeroacoustics (McGraw-Hill, New York) p. 94

Heyvaerts, J., & Priest, E. R., 1983, A&A 117, 220

Hollweg, J. V., 1983, in Solar Wind, ed. V, M. Neugebauer, NASA CP-2280, 5

Hollweg, J. V., Jackson, S., & Galloway, D., 1982, Sol. Phys., 75, 35

Huang, P., Musielak, Z. E., & Ulmschneider, P., 1995, A&A, 297, 579

Kalkofen, W., 2003, in Current Theoretical Models and High Resolution Solar Observations: Preparing for ATST, eds. A. A. Pevtsov & H. Uitenbroek, ASP Conference Series Vol. 286, 385

Kalkofen, W., Ulmschneider, P., & Avrett, E. H., 1999, ApJ, 521, L141

Kelch, W. L., Linsky, J. L., & Worden, S. P., 1979, ApJ, 229, 700

Kippenhahn, R., & Weigert, A., 1990, Stellar Structure and Evolution (Springer, Berlin)

Kuperus, M., 1972, Sol. Phys., 22, 257

Lites, B. W., Rutten, R. J., & Kalkofen, W., 1993, ApJ, 414, 345

Mitra-Kraev, U., & Benz, A. O., 2001, A&A, 373, 318

Muller, R., Roudier, Th., Vigneau, J., & Auffret, H., 1994, A&A, 283, 232

Musielak, Z. E., & Ulmschneider, P., 2001, A&A, 370, 541

Musielak, Z. E., & Ulmschneider, P., 2002a, A&A, 386, 606

Musielak, Z. E., & Ulmschneider, P., 2002b, A&A, 386, 615

Musielak, Z. E., Rosner, R., & Ulmschneider, P., 1989, ApJ, 337, 470

Musielak, Z. E., Rosner, R., Stein, R. F., & Ulmschneider, P., 1994, ApJ, 423, 474

Musielak, Z. E., Rosner, R., Gail, H. P., & Ulmschneider P., 1995, ApJ, 448, 865

Musielak, Z. E., Rosner, R., & Ulmschneider, P., 2000, ApJ, 541, 410

Musielak, Z. E., Rosner, R., & Ulmschneider, P., 2002, ApJ, 573, 418

Narain, U., & Ulmschneider, P., 1990, Space Sci. Rev., 54, 377

Narain, U., & Ulmschneider, P., 1996, Space Sci. Rev., 75, 453

Parker, E. N., 1981, in Solar phenomena in stars and stellar systems, Proc. Adv. Study Inst. Bonas, eds. R. M. Bonnet, & A. K. Dupree, (D. Reidel, Dordrecht) 33

Parker, E. N., 1991 in Mechanisms of Chromospheric and Coronal Heating, eds. P. Ulmschneider, E. R. Priest, & R. Rosner, (Springer, Berlin) 615
Priest, E. R., 1991, in Mechanisms of Chromospheric and Coronal Heating, eds. P. Ulmschneider, E. R. Priest, & R. Rosner, (Springer, Berlin) 520
Rammacher, W., & Ulmschneider, P., 2003, ApJ, 589, June 1
Rutten, R. G. M., 1986, A&A, 159, 291
Rutten, R. G. M., Schrijver, C. J., Lemmens, A. F. P., & Zwaan, C., 1991, A&A, 252, 203
Schrijver, C. J., & Zwaan, C., 2000, Solar and Stellar Magnetic Activity, (Cambridge University Press, Cambridge)
Schrijver, C. J., Coté, J., Zwaan, C., & Saar, S. H., 1989, ApJ, 337, 964
Skartlien, R., Carlsson, M., & Stein, R. F., 1994, in Chromospheric Dynamics, Proc. Mini-Workshop, ed. M. Carlsson, Inst. Theor. Astroph., Oslo, p. 79
Solanki, S. K., 1993, Space Sci. Rev., 63, 1
Stenflo, J. O., 1994, Solar Magnetic Fields (Kluwer, Dordrecht)
Strauss, H. R., 1991, Geophys. Res. Let., 18, 77
Theurer, J., 1998, Ph. D. Thesis, Univ. Heidelberg, Germany
Theurer, J., Ulmschneider, P., & Cuntz, M., 1997, A&A, 324, 587
Theurer, J., Ulmschneider, P., & Kalkofen, W., 1997, A&A, 324, 717
Tu, C.-Y., & Marsch, E., 1997, Sol. Phys., 171, 363
Tu, C.-Y., & Marsch, E., 2001a, A&A, 368, 1071
Tu, C.-Y., & Marsch, E., 2001b, JGR, 106, 8233
Ulmschneider, P., 1996 in Cool Stars, Stellar Systems and the Sun, eds. R. Pallavicini, & A. K. Dupree, Astr. Soc. Pacific Conf. Ser. 109, 71
Ulmschneider, P., & Musielak, Z. E., 1998, A&A, 338, 311
Ulmschneider, P., Theurer, J., & Musielak, Z. E., 1996, A&A, 315, 212
Ulmschneider, P., Theurer, J., Musielak, Z. E., & Kurucz, R., 1999, A&A, 347, 243
Ulmschneider, P., Musielak, Z. E., & Fawzy, D. E., 2001, A&A, 374, 662
Ulmschneider, P., Fawzy, D., Musielak, Z. E., & Stępień, K., 2001, ApJ, 559, L167
Vernazza, J. E., Avrett, E. H., & Loeser, R., 1981, ApJS, 45, 635
Wood, B. E., Linsky, J. L., & Ayres, T. R., 1997, ApJ, 478, 745
Wunnenberg, M., Kneer, F., & Hirzberger, J., 2002, A&A, 395, L51
Zhugzhda, Y., Bromm, V., & Ulmschneider, P., 1995, A&A, 300, 302
Ziegler, U., & Ulmschneider, P., 1997, A&A, 327, 854

The Solar Corona

B.N. Dwivedi

Department of Applied Physics, Institute of Technology, Banaras Hindu University, Varanasi 221005, India
email: dwivedi@banaras.ernet.in, bholadwivedi@yahoo.com

Abstract. The solar corona is an extremely hot (10^6 K or about 0.1 keV), almost fully ionised plasma which extends from a few thousand km above the Sun's visible surface or photosphere (6000 K) to where it freely expands into the solar system as the solar wind. The exact reasons for its high temperature are still being debated despite more than 50 years of research, but magnetic fields are believed to be responsible for this heating. This article reviews some recent progress in our understanding, using data from spacecraft (Yohkoh, SOHO and TRACE) and from ground-based eclipse observations.

1 Introduction

To most people, the Sun might seem like a uniform ball of gas, the essence of simplicity. Although it does not have any well defined surface, it can be divided into several regions (see Fig. 1). Its radiation, upon which all life on the Earth ultimately depends, derives from nuclear reactions deep in the Sun's core. The energy leaks out (by radiation and then by convection) very gradually throughout the Sun's interior towards the visible surface where it escapes to space and a small part of it reaches the Earth. There is a steady decrease of temperature from 15 million K at the centre to 6000 K at the surface, which is also known as the photosphere. Above the surface is a tenuous atmosphere. The lower part, visible as a bright red crescent during total eclipses, is the chromosphere. Beyond the chromosphere is the pearly white corona, extending out for millions of kilometres (Fig. 2).

In keeping with physical expectations, the Sun's temperature drops steadily from its core to the photosphere. But then a surprising thing happens. The chromosphere's temperature steadily rises again to 10 000 K (e.g., Ulmschneider, this volume). Even more startling, the temperature in the corona jumps to 1 million K. Parts of the corona associated with sunspots are even hotter. This fact was first recognised in the 1940s, when unfamiliar spectral lines that had been observed since the nineteenth century were identified with those emitted by iron atoms that have lost several of their normal retinue of 26 electrons, a situation that could only exist at temperatures of 1 million K or more. The temperature of the corona is in fact so high that it emits copious amounts of X-rays and extreme ultraviolet radiation which can only be observed from above the Earth's atmosphere with rockets and satellites, as was eventually found in pioneering studies soon after World War II. In view of the fact that the energy must originate beneath the photosphere, how can this be? One would not expect at first sight

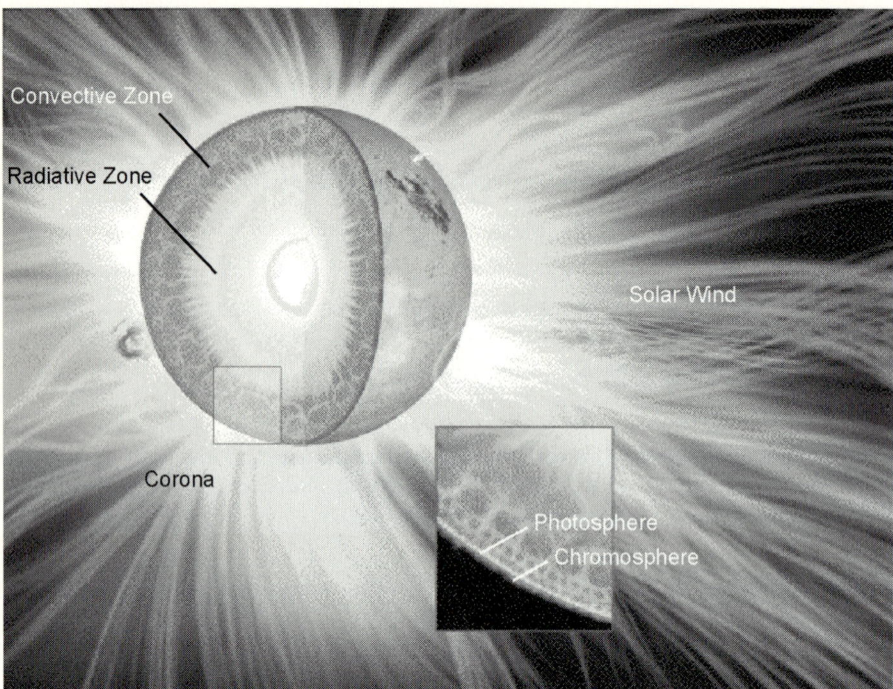

Fig. 1. Regions of the Sun's core and atmosphere (adapted from Dwivedi & Phillips 2001). Solar energy derives from nuclear reactions in the core of the Sun at about 15 million K, the pressure being 300 billion times the atmospheric pressure on the Earth. This energy (in the form of X-rays and gamma-rays) is transmitted through the solar interior, first a region known as the radiation zone and then the convection zone (outer 200 000 km, the solar radius is 700 000 km). Because of the high densities in the solar interior, this energy is continually absorbed and re-emitted. As a result the photons take about 10 million years to reach the surface or photosphere of the Sun, a narrow region of the solar atmosphere where the temperature is about 6000 K from which most of the Sun's visible radiation escapes to the outer space. A slightly hotter region called the chromosphere lies immediately above the photosphere, but the bulk of the solar atmosphere consists of the corona where the temperature exceeds one million K (much hotter locally). The rotation and convection in the solar interior combine to produce a dynamo action, where magnetic fields are periodically generated in a 22-year cycle. The exact mechanism behind this solar cycle is poorly understood. The magnetic field leads to the formation of active regions, extending from the photosphere (where sunspots appear) to the corona (regions of enhanced temperature and density), and are the sites of flares, sudden releases of energy resulting in extremely high temperature, ionised gas (or plasma) and emission of particles, and mass motions (flares are frequently associated with coronal mass ejections in the form of giant bubbles of plasma expanding into interplanetary space)

Fig. 2. The white-light corona: Computer-processed image of the total solar eclipse in India on October 24, 1995 (during a period of minimum solar activity) (courtesy of E. Hiei)

to get warmer when walking away from an energy source like a coal fire. This extraordinary mystery of the Sun's hot corona has intrigued astronomers for the past half-century. It is particularly strange that a puzzle like this should exist for the only star we can study at close hand, and for which we might expect to have a complete and detailed knowledge. But this mystery is not confined to the Sun: many stars with properties like the Sun's appear to have X-ray-emitting atmospheres where the temperature is at least as high as that in the solar corona (Dwivedi & Phillips 2001).

Over the years there has been a steady improvement of our understanding about the heating of the solar corona. It is known that magnetic fields observed and measured in the photosphere are implicated, since where the fields are stronger the corona is also hotter. Two main possibilities are emerging from both observations and theory: either the field converts its energy into heat by many small-scale reconnections (the same process that is involved in major explosive energy releases called solar flares) or by damping of magnetic waves of various sorts. Large and sophisticated spacecraft like the Solar and Heliospheric Observatory (SOHO) (Domingo, Fleck & Poland 1995; Dwivedi & Mohan 1997) have been launched in recent years to look for clues, particularly those associated with tiny flare-like phenomena (now dubbed as nanoflares). But ground-based observations like those made during eclipses also have a role to play, since one is not generally restricted with the amounts of collected data as with spacecraft for which there is a limit to the amount of data that can be telemetered down to a ground station. Ground based observations enabled video-rate electronic imaging of the corona to be done, for example, which is not possible from currently operating spacecraft. Many such recent observations have helped us to gain a clearer picture of the processes going on. There have also been advances in our theoretical understanding of coronal heating, including how it is possible to heat a gas like the solar corona by electrical means even though its conductivity is very high. These advances have implications not only for solar physics but also for atomic and nuclear spectroscopy, cosmic magnetometry, neutrino astrophysics, asteroseismology and space weather.

2 History of Coronal Studies

For a long time solar eclipses offered the only opportunity of studying the corona, and astronomers mounted expeditions to remote areas of the Earth to observe them, risking not only life and limb but also cloudy skies at the appropriate time – totality never lasts more than about seven minutes, usually much less. Nowadays, it is possible to observe at least the inner corona routinely with special telescopes called coronagraphs (e.g., Bhatnagar, this volume), in which the light from the intensely bright photosphere is masked artificially by means of an opaque disk. The corona's white-light emission is simply sunlight from the photosphere which has been scattered off fast-moving free electrons in the corona into our line of sight. The effect is similar to the scattering by tiny dust particles in a sunbeam which renders them visible to us. The density of the corona is extremely small so that almost all the sunlight escapes without being scattered. The corona, like the air in a dusty room, is almost perfectly transparent. Nevertheless, about one out of every million photons leaving the Sun strikes an electron in the corona and is scattered. Both free electrons and dust grains in the corona do the scattering, and so give rise to two main components of the coronal emission.

The electron-scattered component, known as the K-corona (standing for the German word Kontinuum), dominates from near the photosphere out to about two solar radii from the Sun's centre, or about 700 000 km above the photosphere. Its spectrum is a featureless continuum like that of the photosphere but without the Fraunhofer lines. The dust-scattered component has a spectrum that resembles the photospheric spectrum with the Fraunhofer lines; it is known as the F (for Fraunhofer) corona. Both the K and F components decrease in intensity with increasing distance from the Sun, but beyond 2.5 solar radii from the Sun's centre the F component is more intense than the K component. Figure 3 shows how the surface brightness (i.e., brightness per unit area of the sky) of the F and K components varies with distance from the centre of the Sun. From the brightness of the corona we may calculate that the electron density in the low corona is about $10^8\,\mathrm{cm}^{-3}$, falling off with increasing distance from the photosphere. Such densities are many trillions times smaller than that of the gas composing the Earth's atmosphere; in fact, coronal densities are low enough to be considered an almost perfect vacuum in laboratories.

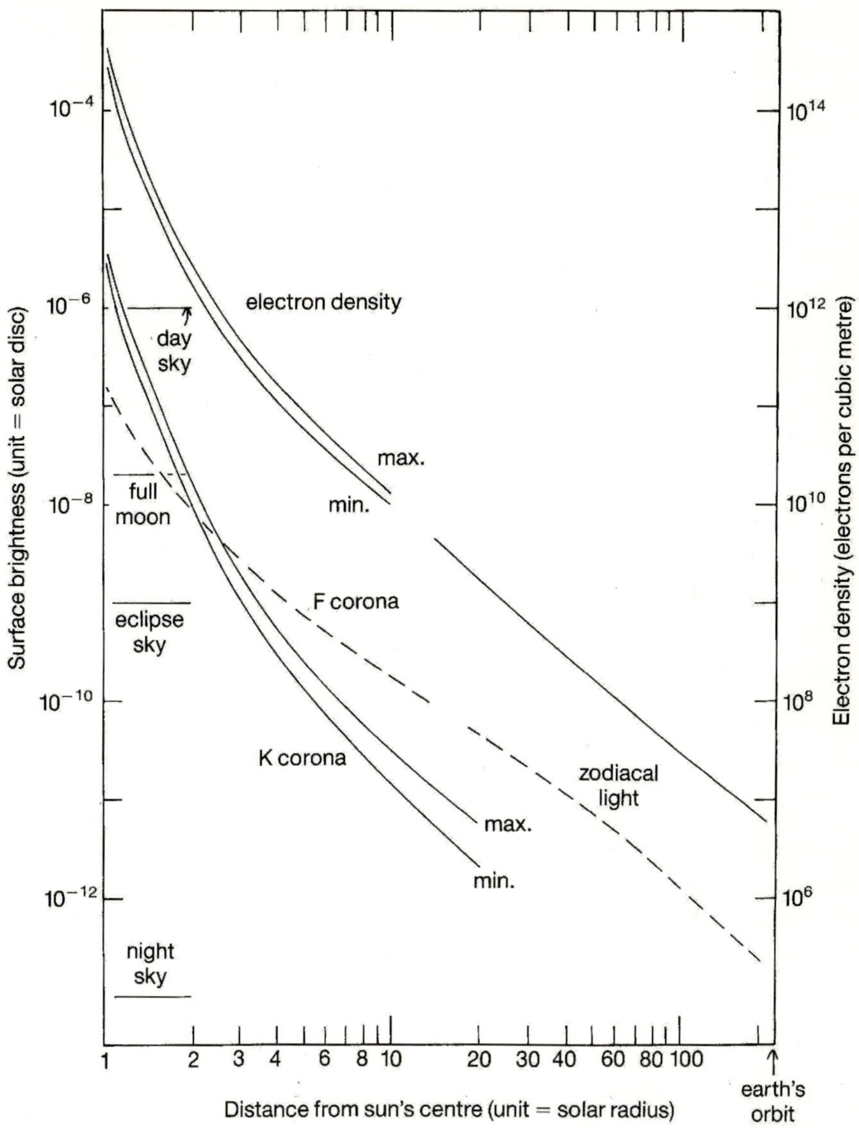

Fig. 3. Variation of the coronal surface brightness and electron density with distance from the Sun's centre (in units of the solar radius of 696 000 km). The surface brightness (in units of the mean solar disc) is shown on the left vertical scale, while the electron density (in electrons per cubic meter) on the right scale. The K-corona surface brightness is shown for the solar maximum and minimum (equatorial values for the latter). The F-corona values (*dashed* line) connect with the zodiacal light that is also produced by scattering by dust. For comparison, the surface brightness values of the sky for day and night and during a total solar eclipse are indicated, as well as for a night with a full moon. Electron densities are plotted separately for solar maximum and minimum for small distances (from Phillips 1995)

The first clues that the corona might be an unusually hot environment were revealed during total eclipses in the nineteenth century. C. A. Young and W. Harkness studied the corona spectroscopically during a total eclipse in 1869 and found a bright emission line at 530.3 nm (now known as the 'green' line since this is the region of the visible spectrum where it is located) which could not then be identified with spectral lines of known elements. Several more unidentified lines became evident in spectra obtained during subsequent eclipses, and a new element named 'coronium' was suspected to be the reason for these spectral lines. As the years passed, however, it became clear that coronium could not be easily fitted into the periodic table of elements and its existence as an element became discredited. The lines of coronium were eventually reproduced when extremely hot spark sources produced in the laboratory were spectroscopically observed. The clinching argument was the discovery by B. Edlén that the 'green' coronal line was in fact due to iron with 13 of its electrons stripped off. Such a situation is only possible if the temperature is about 1 million K. The green and red coronal lines are so-called forbidden lines in the spectra of Fe XIV (Fe^{13+}) and Fe X (Fe^{9+}) respectively. 'Forbidden' means that the electron transitions involved are highly improbable by certain quantum-mechanical selection rules.

This very high temperature of coronal gas has an important consequence. Like the rest of the Sun, by far the most abundant element in the corona is hydrogen with a small amount of helium and much smaller amounts of heavier elements. At a temperature of a million K or more, much of the hydrogen gas exists not in a neutral form, but in a fully ionised state, with protons and electrons moving independently of each other. The same is true of helium which also exists in a fully ionised state in the corona. The corona is in fact what is called a plasma, consisting of charged particles such as protons, electrons, helium nuclei, and small numbers of partially ionised atoms like iron. This is different from most gases we are familiar with on the Earth where the constituent particles are neutral atoms or molecules. A plasma has a wide range of phenomena associated with it which are not observed in neutral gases, particularly associated with magnetic fields, a point which we shall come back to later (also see Phillips & Dwivedi 2003).

3 X-rays and Ultraviolet Emission from the Solar Atmosphere

The ultraviolet emission (Fig. 4) that the solar corona is hot enough to produce was first detected with instruments built by R. Tousey and his colleagues at the U. S. Naval Research Laboratory in the immediate post-war period using captured German military (V2) rockets. Solar X-ray emission was first detected by T. R. Burnight in 1949 using a pinhole camera on board a rocket. Thereafter there was a rapid increase in our knowledge of the Sun's atmosphere from data collected by U. S. and Soviet spacecrafts in the 1960s and 1970s dedicated to solar observations, particularly the manned NASA Skylab mission of 1973–1974. Ultraviolet and X-ray telescopes on board Skylab gave the first high-

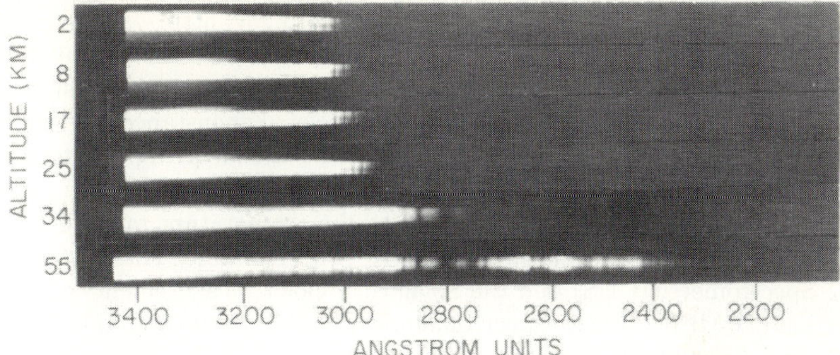

Fig. 4. The first photograph of the solar ultraviolet spectrum. It was obtained with an instrument built by the U. S. Naval Research Laboratory on a V-2 rocket launched in October 1946, and shows how, as the rocket altitude increases, the spectrum extends to progressively shorter wavelengths because of the decreasing atmospheric absorption. The broad dark lines at wavelength 280.3 and 279.6 nm (2803 and 2796 Å) are due to the magnesium h and k lines. (courtesy U. S. Naval Research Laboratory, Washington, DC)

resolution images of the corona as well as the chromosphere (a highly structured part of the atmosphere between the photosphere and corona where the temperature is about 10 000 K) and an intermediate transition region (thought to be a thin layer separating the chromosphere and hot corona). Images of active regions (the photospheric counterparts of which are the sunspot groups) showed a complex of loops which varied greatly over their several-day lifetimes, while ultraviolet images of the 'quiet' Sun (i.e., away from active regions) showed that the transition region and chromosphere resembled a 'network' appearance previously known from images in the light of a strong visible-wavelength spectral line due to once-ionised calcium atoms, Ca II in the chromosphere. The X-ray images showed that the quiet-Sun corona was characterised by diffuse large-scale arches, stretching across several million kilometres.

The spatial resolution of the spacecraft instruments has steadily improved ever since to extremely impressive levels, not far short of that which can be achieved with ground-based solar telescopes. The Japanese Yohkoh spacecraft, launched in 1991, had on board a soft X-ray telescope made jointly by U. S. and Japanese scientists. It provided images of the Sun and in particular, flares at wavelengths of 0.2 to 2 nm with an angular resolution of about 2 arc-seconds (equivalent to 1450 km on the Sun: the mean solar diameter is 32 arc-minutes, or just over half a degree) (e.g., http://www.lmsal.com/SXT/). These X-ray images combined with those from the radio part of the spectrum, particularly those from the Nobeyama radio telescope array in Japan (working at wavelengths of a few centimetres), show the close correspondence of emitting regions.

The ESA/NASA satellite SOHO, launched in December 1995 into an orbit about the inner Lagrangian point situated some 1.5 million km from the Earth on the Sunward side, has on board twelve instruments which get an uninterrupted view of the Sun (e.g., http://sohowww.nascom.nasa.gov, http://sohowww.estec.esa.nl, http://umbra.gsfc.nasa.gov/eit), unlike the instruments on Yohkoh which was in a low-Earth orbit. There are several imaging instruments, sensitive from visible light wavelengths to the extreme-ultraviolet. For instance, the Extreme-ultraviolet Imaging Telescope (EIT) uses normal incidence optics to get full Sun images several times a day in the wavelengths of lines emitted by coronal ions Fe IX, Fe X, Fe XII, Fe XV (emitted in the temperature range 6×10^5 to 2.5×10^6 K), as well as the chromospheric He II 30.4 nm line. The Coronal Diagnostic Spectrometer (CDS) and the Solar Ultraviolet Measurements of Emitted Radiation (SUMER) are two spectrometers operating in the extreme ultraviolet region, capable of getting temperatures, densities and other information from spectral line diagnostics. The Ultraviolet Coronagraph Spectrometer (UVCS) has been making spectroscopic observations of the extended corona from 1.25 to 10 solar radii from the Sun's centre, determining empirical values for densities, velocity distributions and flow velocities of hydrogen, electron, and several minor ions. The striking difference in the width of line profiles seen on the disc and in a polar coronal hole from UVCS and SUMER instruments, is a new observational fact. This led to the discovery of the large velocity anisotropy observed in coronal holes and its interpretation as solar wind acceleration by ion-cyclotron resonance (Tu et al. 1998).

The extremely broad O VI line yields velocities up to 500 km s^{-1}, which corresponds to a kinetic temperature of 200 million K (Kohl et al. 1998; Wilhelm et al. 1998). The Large Angle and Spectroscopic Coronagraph (LASCO) instrument observes the white-light corona with high resolution out to distances of more than 20 million km. Movies from LASCO show the large-scale (magnetic) structures in the corona as they rotate with the rest of the Sun (with a (synodic) period of about 27 days as seen from the Earth), as well as the large ejections of coronal mass in the form of huge bubbles, moving out with velocities of up to 1000 km s^{-1} that, on colliding with the Earth in particular, give the well-known magnetic storms and associated phenomena that have become a matter of widespread concern for telecommunications in recent years. Another recent spacecraft which has given us spectacular images of the corona is the Transition Region and Coronal Explorer (TRACE), operated by the Stanford-Lockheed Institute for Space Research (Handy et al. 1999), launched in 1998 (e.g., http://vestige.lmsal.com/TRACE/POD/TRACEpodoverview.html; or http://www.lmsal.com/solarsites.html). It is able to resolve coronal structures in the ultraviolet down to about 1 arc second (725 km). Images from TRACE have revealed that active-region loops are often thread-like features no more than a few hundred km wide. There is a clear relation of these loops and the larger arches of the general corona to the magnetic field measured in the photospheric layer. The crucial role of this magnetic field has only been realised in the past decade. The fields dictate the transport of energy between the surface of the Sun and the corona. The loops, arches and holes appear to trace out the Sun's magnetic field (see Figs. 5 and 6).

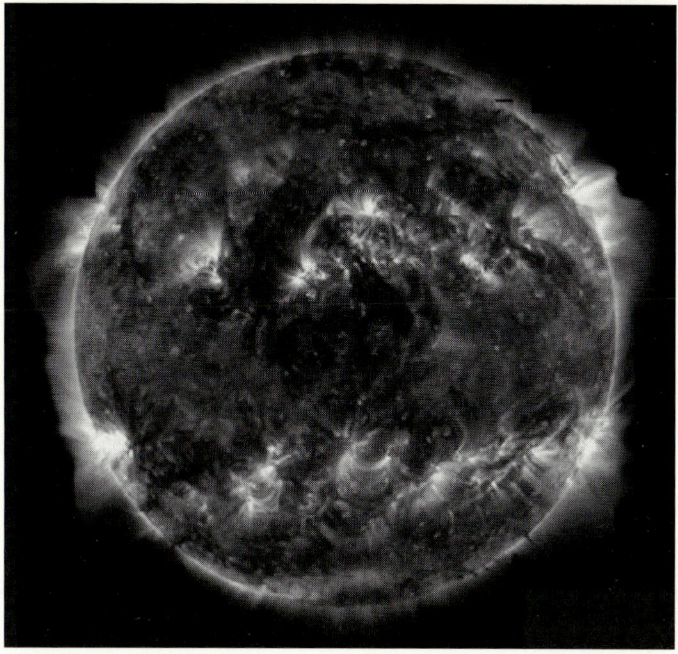

Fig. 5. Million degree solar corona: Image taken on October 10, 1998 by the TRACE spacecraft in the EUV light Fe IX 171 Å. It shows bright structures associated with active regions and smaller-scale structures, where the magnetic field concentrations are more localised. On the Sun's edge (or limb), the bright structures are more clearly seen in the form of loops and follow magnetic fields that emerge in one magnetic polarity, arch around and connect again to the photosphere in the opposite polarity. Sunspots may occur at one or both locations where the field meets the photosphere (courtesy of the TRACE Consortium)

4 Coronal Heating

Magnetic fields are thought to originate in the solar interior by a dynamo process associated with convection currents and the differential rotation of the Sun (low latitudes have a slightly shorter rotation period than higher latitudes) (e.g. Venkatakrishnan, this volume). The fields are buoyed up to the photosphere in the form of rope-like structures which pierce the photosphere at sites such as sunspot groups and extend outward into the solar atmosphere. The entire corona (as revealed in X-ray and EUV images) is pervaded by magnetic field and in fact, the various forms of the corona are determined by the geometry of the local magnetic field loops, giant arches, coronal holes (funnel-shaped regions within which the field opens out into interplanetary space and along which fast solar wind streams flow). The reason for this is that the charged particles making up the corona move in helical paths up and down field lines and so are closely tied to them. For the general solar corona, where the magnetic field is about 1

Fig. 6. Coronal loops, seen in the ultraviolet light (Fe IX 171 Å) by the TRACE spacecraft, extending 120 000 km off the Sun's surface (courtesy of the TRACE consortium)

milli-Tesla (10 G), the radius of gyration is only a few centimetres, far less than the scale of coronal structures.

One vital piece of information that we are still unable to measure is the corona's magnetic field strength. We can measure, with considerable accuracy, the photospheric magnetic field, using instruments called magnetographs on solar telescopes that work on the principle of Zeeman effect (magnetic splitting of the spectral lines) (e.g., Bhatnagar, this volume). This can be done for small regions so that a complete magnetic field map of the Sun's visible surface (a magnetogram) can be constructed. These are routinely available in, for example, the Solar-Geophysical Data Bulletin issued by the US National Oceanic and Atmospheric Administration (NOAA). Vector magnetographs can deduce all three components of the photospheric magnetic field. Although, eventually, infrared measurements may give important information, in practice, the only way at present in which the coronal field can be deduced is through extrapolations of the photospheric field through the assumption, for example, of a potential ($\nabla \times \boldsymbol{B} = 0$, i.e., current-free) or force-free ($\boldsymbol{J} \times \boldsymbol{B} = 0$) field. It is clear, however, from photospheric magnetograms that the field in active regions is more complex than that in quiet regions. It is also known that the active region corona is appreciably hotter (typically 4×10^6 K, depending on the nature of the active region) than in quiet regions (2×10^6 K, less in the coronal holes at the poles). There does seem, then, to be a qualitative relation between field strength and heating.

Recent high resolution observations of MHD wave activity in the corona reported by Aschwanden et al. (1999), Nakariakov et al. (1999) and Schrijver

& Brown (2000) allows us to develop a new method for the measurement of the coronal magnetic field by using MHD *coronal seismology*. Nakariakov & Ofman (2001) have recently developed a new method for the determination of the absolute value of the magnetic field strength in coronal closed magnetic structures, based on the analysis of flare-generated oscillations of coronal loops. Interpretation of the oscillations observed in terms of global standing kink waves allows to connect the period of the oscillations and the loops length with the magnetic field strength in the loops. For loop oscillations observed with TRACE on July 14, 1998 and July 4, 1999, they estimated the magnetic field strength as 4–30 G. Using TRACE 171 Å and 195 Å images of the loop, taken on July 4, 1999 to determine the plasma density, they estimated the magnetic field in the loop as 13 ± 9 G. It is, however, noted that improved diagnostic of the loop length, the oscillation period, and the plasma density in the loop will significantly improve the method's precision.

A considerable theoretical problem with magnetic field heating is the fact that it requires the diffusion of the magnetic field, which implies a resistive plasma. However, the coronal plasma is, on the contrary, highly conducting. Using Spitzer's (1962) classical expression for plasma resistivity (=1/conductivity σ) at temperature T (K),

$$1/\sigma = 10^3 \times T^{-1.5} \quad \text{Ohm m} , \tag{1}$$

we find for $T = 2 \times 10^6$ K that $1/\sigma = 4 \times 10^{-7}$ Ohm m is only a factor 20 or so higher than a highly conducting solid like copper at room temperature. To illustrate the effect of this high conductivity, we use the induction equation of magnetohydrodynamics (MHD),

$$\frac{\partial \boldsymbol{B}}{\partial t} = \nabla \times (\boldsymbol{v} \times \boldsymbol{B}) + \eta \nabla^2 \boldsymbol{B} , \tag{2}$$

where $\eta = 1/(\mu\sigma)$ is the magnetic diffusivity. If the first term on the RHS can be neglected (i.e., if v, the plasma velocity, is very small), then the diffusion time τ for a magnetic field is given by

$$\tau = \frac{L^2}{\eta} . \tag{3}$$

With $\mu \approx \mu_0 = 1.26 \times 10^{-6}$ H m^{-1} (the permeability of free space) and L of the order of the dimensions of the visible structures in the corona, we find that the time τ is extremely long (many million years). Only if the characteristic distance L over which diffusion occurs is as short as a few meters, does the diffusion time become as small as a few seconds. Expressed another way, the first (advection) term in the induction equation (2) is generally much larger than the second (diffusion) term. The magnetic Reynolds number R_m defined by

$$R_m = \frac{LV}{\eta} = \frac{VB/L}{\eta B/L^2} \sim \frac{|\nabla \times (\boldsymbol{v} \times \boldsymbol{B})|}{|\eta \nabla^2 \boldsymbol{B}|} , \tag{4}$$

measures how tied the magnetic field is to the plasma. Here L is the typical length scale and V the typical velocity. One normally has $R_m = 10^6$–10^{12} for the corona unless the length scales are very small. Only when one has small scales can magnetic reconnection be achieved.

It is known that very small length scales do occur in the region of neutral points or current sheets, where there are steep magnetic field gradients which give rise to large currents. It is thought, then, that such geometries are important for coronal heating if this is by very small energy releases, known as nanoflares (Parker 1988). Some 10^{16} J would be released in a nanoflare, i.e., 10^{-9} of a large solar flare, and many energy releases such as this occurring all over the corona, quiet regions as well as active regions, could account for the heating of the corona. Most likely this mechanism would not apply to coronal hole regions where the field lines are open to the interplanetary space.

The above reasoning applies equally to the competing wave heating, in which magnetohydrodynamic waves generated by photospheric motions (e.g., granular or supergranular convective motions) are damped in the corona. In this case, we need conditions such that the magnetic field changes occur in a shorter time than, say, the Alfvén wave transit time (i.e., the typical time necessary for an Alfvén wave to cross a coronal structure) across a closed structure like an active region or quiet Sun loop (assuming magnetic field of $B = 100$ G, a plasma density n of 10^9 cm^{-3} one has an Alfvén speed of $v_A = B/\sqrt{4\pi n m_H/2} = 10^9$ cm s^{-1} where the Alfvén transit time for a loop length of 10^{10} cm would be 10 s). It has, however, recently been suggested that small-scale reconnection occurring in the chromospheric network creates high-frequency Alfvén waves, and that these waves may represent the main energy source for the heating of the solar corona and generation of the solar wind (Axford & McKenzie 1997). However, if these waves exist, they will be absorbed preferentially by the minor heavy ions with low gyro-frequencies, and thus it is unclear whether there is actually enough wave energy left over for the heating and the acceleration of the major solar wind ions, namely protons and alpha particles, in the extended corona after the absorption by heavy ions (Cranmer 2000). Tu & Marsch (2001) have recently studied this problem with the multi-fluid model, which includes the self-consistent treatment of the damping of the waves as well as the associated acceleration and heating of the ions.

4.1 Coronal Heating by Nanoflares

Movies made by concatenating TRACE images of active regions reveal a vast wealth of detail, with coronal loops showing continual brightenings and motions. The rapid variability of coronal structures uncovered recently and indeed since Skylab is an important clue to the corona's nature and the origin of its high temperature. Some earlier observations with rocket-borne instruments by Brueckner & Bartoe (1983) showed the presence of localised dynamic events in the transition region in which material is accelerated with velocities of up to 400 km per second. The energy contained in some of the more important of these so-called turbulent events and jets amounts to a millionth of that of a strong flare, or about 10^{19} J. It is possible that shock waves generated by jets could contribute to the heating of the corona. Enough energy and mass are contained in the jets, assumed to occur over the whole Sun, to satisfy the requirements of not only the corona but also its dynamic extension, the solar wind – a stream of protons, electrons and other charged particles which moves outward into the deepest reaches of the solar system at speeds of between 400 and 800 km per second (e.g., Manoharan, this volume). Ultraviolet jets (e.g., Innes et al. 1997) are only one of many sorts of dynamic phenomena occurring all over the Sun, in quiet regions (i.e., far from sunspot groups) as well as active regions. Microflares, discovered from a balloon-borne X-ray detector by Lin et al. (1984), are another example – in this case impulsive bursts of very short-wavelength X-ray emission which are like miniature versions of full fledged flares. Their frequency and energies would suggest that they too could participate in the heating of the corona. Shimizu (1995), has studied the Yohkoh SXT data for active regions, to find numerous small brightenings in active region loop structures having energies of the order 10^{20} J, i.e., comparable to microflares (also see, Shimizu & Tsuneta 1997).

Eugene Parker, at the University of Chicago and famous for his theoretical prediction of the solar wind in the 1950s, has theorised that numerous even smaller events – nanoflares – below the detectability levels of spacecraft instruments, could explain coronal heating very effectively (Parker 1988). There is now observational evidence from recent data which support this idea, since tiny flare-like events with energies of down to 10^{17} J, or about ten times the energy of a nanoflare, have now been observed. Could their combined energy over the whole solar corona be sufficient to explain the corona? The quiet-Sun corona's total radiative output is about 3×10^{18} W, with about the same amount of power being transferred to the photosphere by thermal conduction. Present estimates of the total energy of various sorts of dynamic phenomena actually detected by spacecraft like SOHO and TRACE, although very rough still, are 20 percent of this amount. This is encouraging since recent SOHO and TRACE observations have revealed that magnetic reconnection is ubiquitous not only in the corona and the chromosphere but also in the photosphere (e.g., Schrijver et al. 1997). The remainder could be accounted for by flare-like pulses, below the detectability thresholds of present instrumentation.

The nanoflare heating, however, provides the ground for debate. For example, the occurrence frequency of microflares and nanoflares have been found to be $dN/dW \sim W^{-\alpha}$ (where N is the number of events (micro/nanoflares) and W is the energy per event), with the power-law index $\alpha \approx 1.5$–$1.6 < 2$ (Shimizu 1995; Shimojo & Shibata 1999; Aschwanden & Parnell 2002), suggesting that the total energy released by nanoflares is not enough for heating the corona. Krucker & Benz (1998) found $\alpha > 2$ from SOHO/EIT data (Benz & Krucker 2002), and Parnell & Jupp (2000) also found $\alpha > 2$ using TRACE data. Yashiro & Shibata (2001) analysed Yohkoh/SXT data to find the relation between the average gas pressure p and the magnetic field strength B of active regions, and obtained $p \sim B^{0.78}$. This supports the Alfvén wave heating model rather than the nanoflare model if a simple situation is assumed. Shibata & Moriyasu (2002) note that magnetic reconnection can generate Alfvén waves (Yokoyama & Shibata 1996; Yokoyama 1998; Takeuchi & Shibata 2001a, b) and hence the nanoflare model may be unified with the Alfvén wave model. Exploring the generation of spicules and coronal heating by Alfvén waves (Kudoh & Shibata 1999; Saito, Kudoh & Shibata 2001), Shibata & Moriyasu (2002) find that if Alfvén waves with an amplitude of more than 1 km s^{-1} are generated at the photospheric level, spicule generation, nonthermal line width in the transition region and corona, and coronal heating are all explained consistently. They also examine the heating of a loop by Alfvén waves, using the self-consistent MHD simulations of nonlinear Alfvén wave propagation in a loop with heat conduction and radiative cooling (Moriyasu et al. 2002), and apply the result to the generation of a coronal loop in emerging flux regions. The results show that the heating is due to both slow and fast mode MHD shocks which are generated by nonlinear mode coupling with Alfvén waves, and also that the time scale of appearance of a hot coronal loop in emerging flux is roughly consistent with TRACE observations of emerging flux regions.

4.2 Coronal Heating by Waves

The nanoflare hypothesis for the heating of the corona thus looks a very plausible one. However, many theorists have concentrated on the idea that heating by waves is dominant. An early theory that the corona is heated by sound waves or sonic shock fronts was discarded in the late 1970s when it was established that shock waves would all be dissipated in the chromosphere, leaving no energy for the corona itself. Waves associated with the Sun's coronal magnetic field are much more plausibly involved in heating processes. Such waves could take the form of Alfvén waves, which are like the waves which travel along elastic bands when stretched. But more generally it is thought that the waves important for heating processes are magnetohydrodynamic (MHD), that is they share characteristics of sound and Alfvén waves. Plasma physicists recognise two sorts of MHD waves, fast-mode and slow-mode. MHD waves generated by photospheric motions (e.g., granular or supergranular convective motions) are damped in the corona. In this case, we need conditions such that the magnetic field changes occur in a shorter time than, say, the Alfvén wave transit time across a closed structure like an active region or quiet Sun loop.

The literature for wave heating of the corona is considerable, but we may briefly summarise it by stating that the waves, generated by turbulent motions in the solar convection zone or at the photosphere, may be surface waves in a loop geometry, or body waves which are guided along the loops and are trapped. Recent developments were already discussed in the last section. The work of Porter, Klimchuk & Sturrock (1994) shows that short-period fast-mode and slow-mode waves (periods less than 10 s) could be responsible for heating since only for them are the damping rates high enough. Theoretical predictions indicate that waves with very short periods, perhaps only a few second long, are the most effective in heating. The other suggestion, already discussed, of course, has been that small-scale reconnection occurring in the chromospheric network creates high-frequency Alfvén waves, and that these waves may represent the main energy source for the heating of the corona and generation of the solar wind (Axford & McKenzie 1997). Axford et al. (1999) and McKenzie, Axford & Banaszkiewicz (1997) also proposed that the picoflare in the chromosphere can be the source of Alfvén waves which eventually accelerate high speed solar wind.

Either way, a considerable problem with magnetic field heating is the fact that it requires the diffusion, and therefore reconnection, of magnetic field which in turn implies that the coronal plasma has a certain (small) amount of electrical resistivity. However, the corona is, to the contrary, highly conducting. The corona's electrical conductivity (defined to be the inverse of resistivity), as discussed earlier, is not too unlike that of room-temperature solid copper. It can be calculated that the diffusion time for a magnetic field is a few seconds when the length scale is as short as a few meters. Very small length scales do occur in the corona where there are steep magnetic gradients which give rise to large currents. This problem applies equally to the two theories which are currently thought to explain the corona's high temperature, nanoflares and damping of magnetic waves.

The very high conductivity of the corona is predicted by 'classical' plasma physics calculations in which it is generally assumed that the resistivity of a plasma is in the form of so-called Coulomb collisions between charged particles such as electrons and protons – the flow of electrons composing an electric current is inhibited by the continual deflections that the electrons suffer by the electrostatic forces due to charged particles in their path. However, if the plasma is turbulent, the resistivity could be much larger as electrons might suffer collisions not with other particles but with plasma waves of various sorts. This is strongly indicated to be the case with laboratory plasmas such as those in fusion devices. There is now evidence that this is also true for the solar corona. A key observation was made in 1999 when a fine loop was seen by the TRACE spacecraft to perform damped oscillations as a result of a nearby powerful flare. The period of the oscillations indicated that indeed the conductivity is not nearly as high as would be calculated on the usual classical assumptions, and so the conductivity of the solar plasma may not be the problem it was once thought to be (e.g., Nakariakov et al. 1999).

4.3 Fieldwork

Although the nanoflare hypothesis of coronal heating may be observationally plausible, MHD waves may well contribute significantly. It is, for example, unlikely that nanoflares could heat the corona in the regions of open field lines such as those occurring in coronal holes at each of the solar poles (since a reconnection would merely accelerate plasma rather than heat it), yet it appears that the corona is still hot in these open field regions. It is therefore important to look for signatures of wave motions, particularly short-period MHD waves as these are probably the most important in heating processes. To look for these wave motions we need imaging at frequencies higher than the expected frequencies of wave motions. For this purpose, ground-based instruments operating during total solar eclipses can out-perform spacecraft instruments in terms of fast imaging since spacecraft imaging is necessarily rather slow. Some non-periodic variations in coronal brightness have been reported from the SOHO LASCO coronagraph over periods of about 30 minutes. However, theoretical results indicate that MHD waves having very short periods, of a few seconds, are the only ones significant for coronal heating. If such short-period waves are important, there must still be considerable interest in observing the visible-light corona during total solar eclipses from the ground, since one can use high-speed electronic cameras to obtain rapid imaging of particular coronal structures. J. Pasachoff, a pioneer in this work, has performed experiments at various eclipses around the world since the 1980s. Analysis of his best results indicate the presence of a slight peak in Fourier spectra at frequencies of 0.5–1 Hz (Pasachoff & Ladd 1987). This has been seen in more recent eclipses, including the 1998 eclipse in the Caribbean. Other measurements using ground-based white-light coronagraphs have been taken, notably by S. Koutchmy, and searches were made for periodic modulations in both intensity and velocity of the green line, with evidence of periods equal to 43, 80, and 300 s (Koutchmy, Zhugzhda & Locans 1983; Koutchmy et al. 1997).

Phillips et al. (2000) have developed an instrument consisting of a pair of charge coupled device (CCD) cameras which images the white-light corona at subsecond speeds. The instrument, called the Solar Eclipse Coronal Imaging System (SECIS), uses an adapted PC to grab the digital data from the cameras at image rates up to 44 frames a second. Preliminary trials were done using the 40 cm coronagraph at Sacramento Peak in 1998, and the instrument was fully tested during the eclipse of August 11, 1999 (Phillips et al. 2000) on the coast of the Black Sea, a small town called Shabla in Bulgaria, almost exactly on the centre line of the eclipse path. Unlike most of the other parts of Europe, there were clear skies and the instrument worked extremely well. Over the 2 minutes and 23 seconds of totality, some 12 728 images of the corona were obtained, half of them in white-light, and the other half in the light of the green line. Analysis of the vast quantities of data, which has involved some instrumental corrections, has uncovered very subtle signs of oscillatory behaviour which could be linked to MHD wave motions. Fourier analyses using two different methods show the presence of periodic modulations in a few locations, generally the tops of loop structures which may therefore indicate a standing-wave phenomenon. The periods are between 2 and 10 seconds. The experiment has been successfully repeated during the much more favourable eclipse of June 21, 2001, from a location near Lusaka, Zambia (Phillips 2001, private communication).

5 Conclusions

In conclusion, the high temperature of the Sun's corona is almost certainly due to either wave heating or heating by nanoflares. Although the evidence now favours nanoflares for the bulk of coronal heating, waves may also play an important role.

At present this can only be investigated using ground-based instruments since the periods of MHD waves effective for coronal heating are likely to be very small (a few seconds). Spacecraft imaging is too slow to search for such periodicities.

References

Aschwanden, M. J. & Parnell, C. E. 2002, ApJ, 572, 1048

Aschwanden, M. J., Fletcher, L., Schrijver, C. J., & Alexander D. 1999, ApJ, 520, 880

Axford, W. I., & McKenzie, J. F. 1997, in Cosmic Winds, and the Heliosphere, eds. M. S. Matthews, A. S. Ruskin & M. L. Guerrieri (University of Arizona Press, Tucson) 31

Axford, W. I., McKenzie, J. F., Sukhorukova, G. V. Banaszkiewicz, M., Czechowski, A., & Ratkiewicz, R. 1999, Space Sci. Rev. 87, 25

Benz, A. O., & Krucker, S. 2002, ApJ, 568, 413

Brueckner, G. E., & Bartoe, J.-D. F. 1983, ApJ, 272, 329

Cranmer, S. R., 2000, ApJ, 532, 1197

Domingo, V., Fleck, B., & Poland, A. I. 1995, Sol. Phys., 162, 1

Dwivedi, B. N., & Mohan, A. 1997, Current Science, 72, 437

Dwivedi, B. N., & Phillips, K. J. H. 2001 Scientific American, 284, 40 (June issue)

Handy, B. N., Acton, L. W., Kankelborg, C. C., et al. 1999, Sol. Phys., 187, 229

Innes, D. E., Inhester, B., Axford, W. I., & Wilhelm, K. 1997, Nature, 386, 811

Kohl, J. L., Noci, G., Antonucci, E., et al. 1998, ApJ, 501, L127

Koutchmy, S., Zhugzhda, Y. D., & Locans, V. 1983, A&A, 120, 185

Koutchmy, S., Hara, H., Suematsu, Y., & Reardon, K. 1997, A&A, 320, L33

Krucker, S., & Benz, A. O. 1998, ApJ, 501, L213

Kudoh, T., & Shibata, K. 1999, ApJ, 514, 493

Lin, R. P., Schwartz, R. A., Kane, S. R., Pelling, R. M., & Hurley, K. C. 1984, ApJ, 283, 421

McKenzie, J. F. Axford, W. I., & Banaszkiewicz, M. 1997, Geophys. Res. Lett., 24, 2877

Moriyasu, S., Kudoh, T., Yokoyama, T., & Shibata, K. 2002, preprint

Nakariakov, V. M., & Ofman, L. 2001, A&A, 372, L53

Nakariakov, V. M., Ofman, L., DeLuca, E. E., Roberts, B., & Davila, J. M. 1999, Science, 285, 862

Parker, E. N. 1988, ApJ, 330, 474

Parnell, C. E., & Jupp, P. E. 2000, ApJ, 529, 554

Pasachoff, J. M., & Ladd, E. F. 1987, Sol. Phys., 109, 365

Phillips, K. J. H. 1995, Guide to the Sun, Chapter 5 (Cambridge University Press)

Phillips, K. J. H., Read, P. D., Gallagher, P. T., et al. 2000, Sol. Phys., 193, 259

Phillips, K. J. H., & Dwivedi, B. N. 2003, in Dynamic Sun, Chapter 17, ed. B. N. Dwivedi (Cambridge University Press), 335

Porter, L. J., Klimchuk, J. A., & Sturrock, P. A. 1994, ApJ, 435, 482

Saito, T., Kudoh, T., & Shibata, K., 2001, ApJ, 554, 1151

Schrijver, C. J., & Brown, D. S. 2000, ApJ, 537, L69

Schrijver, C. J., Title, A. M., van Ballegooijen, A. A., Hagenaar, H. J., & Shine, R. A. 1997, ApJ, 487, 424

Shibata, K., & Moriyasu, S. 2002 preprint

Shimizu, T. 1995, PASJ, 47, 251

Shimizu, T., & Tsuneta, S. 1997, ApJ, 486, 1045

Shimojo, M., & Shibata, K. 1999, ApJ, 516, 934
Spitzer, L. 1962, Physics of Fully Ionized Gases, Interscience Publ. (John Wiley, New York)
Takeuchi, A., & Shibata, K. 2001a, ApJ, 546, L73
Takeuchi, A., & Shibata, K. 2001b, Earth, Planets, and Space, 33, 605
Tu, C.-Y., Marsch, E., Wilhelm, K., & Curdt, W. 1998, ApJ, 503, 475
Tu, C.-Y., & Marsch, E. 2001, A&A, 368, 1071
Wilhelm, K., Marsch, E., Dwivedi, B. N., Hassler, D. M., Lemaire, P., Gabriel, A. H., & Huber, M. C. E. 1998, ApJ, 500, 1023
Yashiro, S., & Shibata, K. 2001, ApJ, 550, L113
Yokoyama, T. 1998, in Solar Jets and Coronal Plumes, ESA SP-421, 215
Yokoyama, T., & Shibata, K. 1996, PASJ, 48, 353

The Solar Wind

P.K. Manoharan

Radio Astronomy Centre, Tata Institute of Fundamental Research,
P. O. Box 8, Udhagamandalam (Ooty) 643001, India
email: mano@racooty.ernet.in

Abstract. This chapter is intended to be an overview of the aspects of solar wind, with particular emphasis on the properties of the solar wind within about 1 AU of the Sun. The topic is split into two parts on the basis of two types of solar wind flows, (i) quasi-stationary winds and (ii) transient flows. The first part covers the basic ideas of the coronal heating and acceleration of the quasi-stationary wind. It also reviews the large-scale properties and long-term changes of quasi-stationary structures of the solar wind. The second part describes the characteristics of transients in the solar wind and highlights results on the study of radial evolution of transients generated by coronal mass ejections in the inner heliosphere. A brief discussion of the spectral characteristics of micro-turbulence in the solar wind is also included.

1 Introduction

The solar wind is an ionised, magnetised gas which continuously flows radially outward from the Sun in all directions and fills the interplanetary space. It is composed of mainly protons and electrons with trace quantities of heavier ions. It begins at zero velocity at the corona, which is at a kinetic temperature of about 2×10^6 K. The major acceleration of the solar wind to supersonic speed occurs within a heliocentric distance of about 15 solar radii ($R_\odot$) and near the orbit of the Earth the speed of the solar wind is typically 400 km s^{-1} with a density of about 5–10 protons cm^{-3}. However, the speed and density can have large variations, which are determined by the complex structure of the coronal magnetic field. The coronal field evolves during the 11-year cycle of solar activity. At low heliographic latitudes, the solar wind is dominated by low- and high-speed flows from quasi-stationary field structures and is also often perturbed by interaction between low- and high-speed streams and transient flows produced by coronal mass ejections (CMEs) from the Sun. This chapter gives simple physical descriptions of (i) the hot corona, (ii) the acceleration of the solar wind, (iii) the interplanetary magnetic field, (iv) the temporal and spatial variations of the quasi-stationary wind, and (v) the structural evolution of transients propagating in the solar wind.

2 The Hot Solar Corona

The visible surface of the Sun, called the *photosphere* has an equivalent black-body temperature of ~ 5700 K. Above the photosphere lies the *chromosphere*, where the temperature rises rapidly with increasing height from about 5000 to 25 000 K. The thickness of the chromosphere is about 2–5×10^3 km. The *corona* is the outer region of the solar atmosphere at altitudes $> 10^4$ km, with a temperature in the range 1–2×10^6 K. The sharp increase in temperature, from the photosphere through the chromosphere to the corona, is the most important physical characteristic of the corona, which could not possibly be heated by radiation from the relatively cool photosphere. The direct corollary of the hot corona is the extreme state of ionisation. For example, most of the coronal lines are due to atoms of iron, nickel, calcium and even argon ionised 10 to 15 times, and such high degree of ionisation require temperatures of 1–2×10^6 K (Dwivedi, this volume). Several characteristics observed in the corona, including its white-light appearance, its emission spectrum, and its magnetically determined structure, are a direct consequence of its ionic property. The microwave radio emission, X-rays and extreme ultraviolet (EUV) radiation of the corona also confirm temperatures upwards of million K present throughout the corona.

2.1 Coronal Magnetic Fields

The morphology of the corona is essentially determined by the complex structure of solar magnetic fields, which extend through the solar atmosphere to the corona and out into the heliosphere. The corona, thus structured, is very tenuous and its density varies in the range 10^8–10^{11} particles cm^{-3} at the coronal base (e.g., Newkirk 1967; Koutchmy 1994). The corona on the whole occupies an enormous volume, but radiates very little visible light, i.e., about one millionth as bright as the solar disk brightness (Koutchmy & Lamy 1985). It can be therefore seen only during a total solar eclipse (or when the solar disk is artificially occulted in a coronagraph). The brightness of the corona is composed of three main components: (1) The Thomson-scattered photospheric radiation from coronal electrons is a continuum and called white-light or K-corona (K from the German word, *kontinuum*). (2) The E-corona is by the emission from the coronal ions, especially, in highly ionised states. (3) The Fraunhofer- or F-corona is from interplanetary dust particles and shows an absorption spectrum. The Fraunhofer lines are due to the scattering of sunlight by dust particles along the line of sight.

The appearance of the K-corona is closely related to the Sun's magnetism and varies greatly between the maximum and minimum of the solar cycle. At sunspot maximum, i.e., at times when sunspots are most numerous, the coronal brightness is rather uniformly distributed around the photosphere. Figure 1 shows typical white-light images observed with the Large Angle and Spectrometric Coronagraph (LASCO) on board the SOHO spacecraft (Brueckner et al. 1995) during maximum and minimum of solar activities. During the solar maximum, the corona has a nearly spherically symmetric appearance and a closer look reveals more structures with condensation, enhancements, helmets and streamers distributed rather evenly over all latitudes. In contrast, at sunspot minimum, the corona is elongated along the solar equatorial direction and shows 'polar plumes' (thin hair-like structures), which resembles the lines of force of a bar magnet. The coronal intensity declines steeply with increasing distance from the

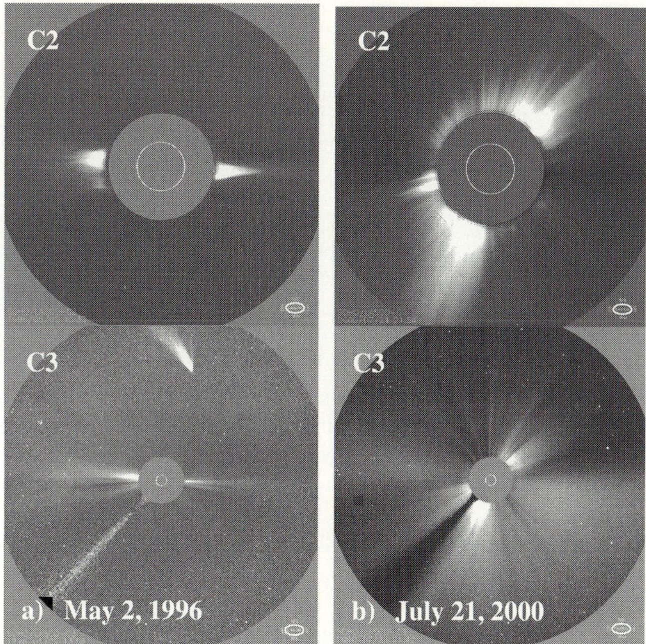

Fig. 1. White-light images of the corona observed by the LASCO/SOHO C2 and C3 coronagraphs for representative dates: (a) at solar minimum and (b) at solar maximum. The size of the photospheric disk is shown as circle at the middle of the occulting disk of C2 and C3 coronagraphs, which cover fields of view of about 6 and 30 $R_\odot$, respectively. A comet with its tail pointing away from the Sun is seen in the C3 field of view

Sun indicating a sharp decrease in plasma density within distances of about $10R_\odot$.

Since the cool photosphere produces very little intensity at X-ray wavelength, an X-ray image also clearly reveals the coronal structures on the solar disk. Figure 2 shows representative soft X-ray images observed with the Yohkoh satellite (Tsuneta et al. 1991) during minimum and maximum periods of the solar cycle. These images have been taken on the same days as the white-light images shown in Fig. 1. It is evident that during the maximum, the corona is rather bright and structured in the form of arcades, loops, and helmets. These structures outline the lines of force of the coronal magnetic field and can be basically characterised by two kinds of configurations, *closed* and *open* field lines. A closed field line is anchored in the photosphere at two opposite magnetic polarity regions, extending into the corona as a loop or arch, whereas open field lines are rooted at single points in the photosphere and reach out into the interplanetary space. In X-rays, the brightness difference seen over the image indicates a difference in density (i.e., the emissivity is proportional to square of the density). In other words, denser material is confined to closed field regions and less dense material is found in the open field regions. Therefore, the dark regions seen on the X-ray image (e.g., near the polar regions of the Sun during minimum of activity)

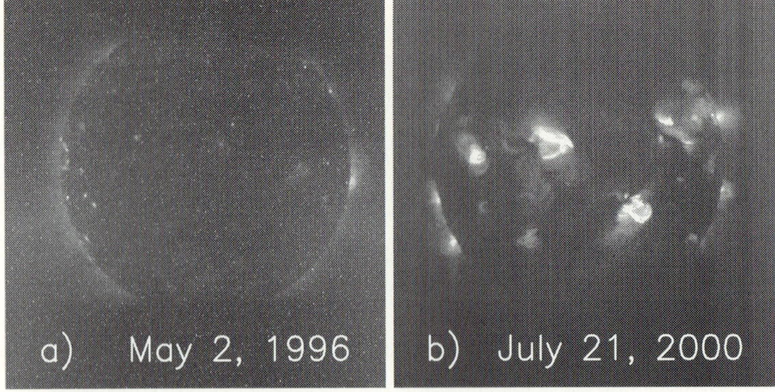

Fig. 2. X-ray images of the Sun taken by the soft X-ray telescope on Yohkoh spacecraft: (a) at solar minimum and (b) at solar maximum. They have also been observed on the same days as the white-light images displayed in Fig. 1

indicate the open corona and these regions are known as 'coronal holes' (e.g., Hundhausen 1977).

The coronal magnetic field can not be measured directly. However, photospheric magnetograms, obtained from Zeeman splitting measurements, are being made on a routine basis and they can be extrapolated in the frame of physical approximations, such as force-free or potential magnetohydrostatic fields, to infer the coronal field conditions (e.g., Hoeksema, Wilcox & Scherrer 1983; Hakamada 1995; see Fig. 1 in Venkatkrishnan's article in this volume). The Faraday rotation signal from spacecrafts or natural radio sources while passing through the coronal plasma can also give an estimation of the coronal field strength at heliocentric distances of less than $10R_\odot$ (Bird & Edenhofer 1991).

2.2 Coronal Heating

One of the great puzzles of solar physics is how the corona is heated and maintained at temperatures greater than a million K. This process requires energy of the order of 10^{22} W. It is now generally agreed that the magnetic field of the Sun is responsible for the coronal heating (Dwivedi, Ulmschneider, this volume). Such fields can transport energy in a form other than heat, thereby overcoming the usual thermodynamic restrictions arising due to the relatively low-temperature photosphere. A tenuous, almost fully-ionised gas (i.e., plasma) which constitutes the corona is an excellent electrical conductor, so that the magnetic field lines extending from the Sun out into the corona are frozen-in and any motion of coronal plasma carries the field line with it – or conversely, any disturbance given to the magnetic field relocates the coronal plasma. This relates the mechanical energy per unit volume of the corona, $\frac{1}{2}nm_HV^2$ to the magnetic energy, $B^2/8\pi$, where n is the number density, m_H is the mass of the proton, V is the bulk speed of the plasma, and B is the field intensity. In other words, the coronal material can distort the field lines if more mechanical energy is available (in the form of gross kinetic energy) than magnetic energy (resulting from the distortion of field lines). It is to be noted that the coronal gas can move along but not across magnetic field lines. The photospheric motion can produce sufficient distortions of the magnetic field threading into the corona to cause major disturbances in the coronal material. For example, within the tenuous coronal plasma, the magnetic pressure exceeds the thermal pressure by a factor of at least 100 and the coronal temperature is higher where the fields are stronger. But, the way in which energy is dissipated in the corona is still strongly debated (e.g., Einaudi & Velli 1994).

Associating coronal heating to magnetic fields, however, requires conversion of magnetic energy to thermal energy. To achieve this conversion the magnetic field lines should be able to diffuse through the plasma and the scale-size/rate of diffusion is inversely proportional to resistivity. The heating by magnetic field therefore demands the corona to have a considerable amount of electrical resistivity (i.e., the corona should not be a perfect conductor). Binary Coulomb interactions (i.e., collisions) can lead to resistivity, and the thermal conductivity due to such Coulomb collisions implies that the conducted heat should dominate the energy flux in the low corona. However, interplanetary observations have shown that the heat flux is negligible compared to the convective energy flux (Schwenn 1990). On the other hand, theory suggests that small-scale turbulence and fluctuations produced by micro-instabilities can lead to a much larger energy transport through turbulent resistivity.

The motion of magnetic field footpoints caused by the solar convection cells (i.e., granules) produces magnetohydrodynamic (MHD) waves – Alfvén waves – that move along the magnetic field lines into the corona and dissipate to produce the heating. For example, in a model computation (Moore et al. 1992), Alfvén wave heating (i.e., the waves getting trapped in) the corona and escaping into interplanetary medium have been demonstrated, respectively, for conditions below 10^6 K and higher temperatures. Although, the amount of heat energy required is still controversial, Alfvén waves appear to be important in the heating of the coronal plasma. Another heating mechanism is a large number of nano/micro flares, each releasing energy in the range 10^{24}–10^{26} erg, occurring on the solar surface can be an additional source of energy to heat and maintain the temperature of the corona (Klimchuk & Porter 1995; Cargill & Klimchuk 1997). Moreover, it is possible that heating mechanisms in open and closed magnetic field regions may be different. A number of ideas have been proposed (Ulmschneider, this volume), it is certain that the ultimate source of heat is the kinetic energy of turbulent motion of convective cells and without the convection, there would be no coronal heating or corona.

3 Coronal Expansion/ Solar Wind

The rarefied corona is a good thermal conductor, so that the temperature can be maintained at a very high level up to several solar radii from the Sun (i.e., the temperature gradient is very small). However, the decreasing density with increasing heliocentric distance produced by the gravity of the Sun causes a steep gradient in the gas pressure. As a result, hydrostatic equilibrium can not be maintained in the corona and the pressure gradient aids to push the coronal material outward until a continuous flow occurs and the expanding gas known as *solar wind* flows radially outward in all directions and fills the interplanetary space.

The existence of the solar wind was first proposed by Biermann (1951) based on observation of comet tails roughly pointing into the direction opposite to the Sun (e.g., see Fig. 1). Chapman (1957) worked out a model for the solar wind flow, assuming that the corona, having been heated by unknown process, maintained the static equilibrium by conducting the heat outwards. Chapman's model led to an excessive gas pressure even at infinity. Since the pressure of the interstellar gas was not adequate to maintain the corona in a state of hydrostatic equilibrium, Parker (1964) postulated that the entire atmosphere of the Sun should be in a state of continual expansion. In a steady state, the equation of motion and the equation of continuity for the spherically symmetric flow of solar wind are given by,

$$nm_H V \tfrac{\mathrm{d}V}{\mathrm{d}r} + \tfrac{\mathrm{d}}{\mathrm{d}r}(2nkT) + \tfrac{nm_H GM_\odot}{r^2} = 0 \,, \tag{1}$$

$$nVr^2 = n_0 V_0 r_0^2 \,, \tag{2}$$

where n is the number of proton-electron pairs per unit volume, m_H is the mass of the proton (the electron mass is neglected), V is velocity, r is distance from the centre of the Sun, k the Boltzmann constant, T is temperature, $M_\odot$ mass of the Sun, and G the gravitational constant. The factor 2 in the pressure gradient is to account for the pressure exerted by electrons and protons. In addition, one has an energy equation and the energy per unit mass of the coronal gas is the sum of the thermal, kinetic and gravitational energies,

$$E = \frac{3kT}{m_H} + \frac{1}{2}V^2 - \frac{GM_\odot}{r} \,. \tag{3}$$

Parker (1964) also investigated a generalised function for the variation of temperature with distance from the Sun,

$$T(r) = T_0 r^{-b} \,. \tag{4}$$

At the base of the corona, the sum of energies, E, is negative and the system remains stable. With increasing distance, the gravitational potential decreases as $1/r$ and the thermal energy, governed by the temperature gradient, $T(r)$, declines rather gradually with distance for power-law indices, $b < 1$. For a realistic value of $b \approx 0.3$, the energy, E, becomes positive at distances beyond $r \approx 10R_\odot$, and solar wind flows with the supersonic speed. One can show that the gravitational field of the Sun acts as a nozzle (like in a rocket engine) for the flow. For many years, the study of the solar wind acceleration has been actively carried out and various observational efforts (based on both ground-based and space-borne instruments) and theoretical models have been produced. These have improved the assumptions concerning the energy balance in the corona and succeeded to explain the measurements near the Sun, close to the Earth's orbit and at the distant heliosphere (Coles et al. 1991a).

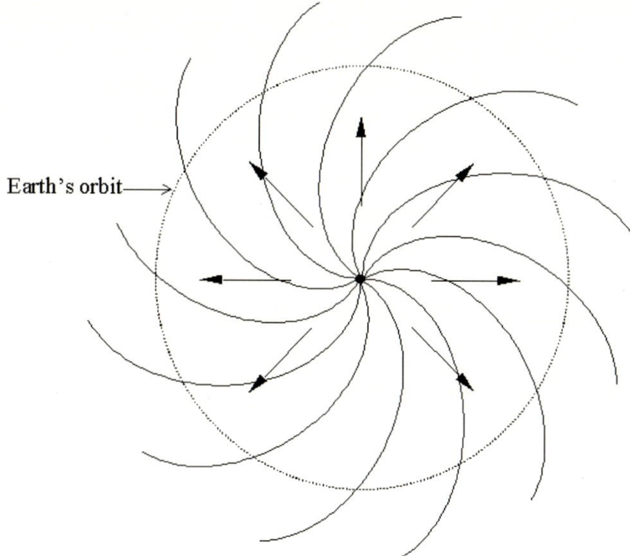

Fig. 3. The spiral pattern of the interplanetary magnetic field in the ecliptic plane for a constant speed of the solar wind of about 425 km s^{-1}. The orbit of the Earth is shown

4 Interplanetary Magnetic Fields

The coronal magnetic field and the properties of the solar wind are intimately related. In regions where the coronal field is strong and transverse to the general flow direction, it can obstruct this outward flow of coronal plasma. In contrast, in regions where the field tends to be more radial and open, the solar wind can flow and because of its very high electrical conductivity, it can drag the coronal magnetic field into interplanetary space and form the interplanetary magnetic field. The radial flow of the wind and the rotation of the Sun (~ 27 days near the solar equator as seen from the Earth) make the field lines form an Archimedean spiral pattern. For a typical speed of about 425 km s^{-1}, the solar wind takes about 4 days to reach the orbit of the Earth and during this time, the Sun rotates by about 55° westwards. Figure 3 shows the spiral shape of the interplanetary field and the relative location of the Earth. At Earth's distance ($1\,\mathrm{AU} \approx 215 R_\odot$), the interplanetary field strength is about 5×10^{-5} Gauss, which is about 10 000 times weaker than the geomagnetic field. However, the Earth's magnetic field drops off steeply with height (above the Earth's surface) and becomes nearly equal to the interplanetary field at ≈ 12 Earth radii. At this height, the interaction between the solar wind and the magnetosphere of the Earth occurs and the solar wind plasma gets pumped into the magnetosphere where it travels along the field lines and reaches the polar regions of the Earth, causing geomagnetic storms and aurorae.

Accurate interplanetary magnetic field measurements are obtained from *in situ* data and routine equatorial field measurements are made by Earth orbiting satellites. For the first time, the Ulysses spacecraft mapped the magnetic field in the polar cap regions of the heliosphere (http://ulysses.jpl.nasa.gov/science/results.html). The polar fields observed by Ulysses can be extrapolated back to the Sun. These measurements indicate that at high northern and southern heliographic latitudes, the polarity of the interplanetary magnetic field lines coincides with the polarity of the photospheric field. The heliospheric current sheet (HCS) divides the regions of north and south polarity in the heliosphere (Neugebauer 1995). The interplanetary magnetic field becomes weaker with distance from the Sun but shows local maxima and minima related to the solar activity as well as to the interaction of fast- and slow-speed solar wind streams.

5 Solar Wind Measuring Techniques

The solar wind properties (e.g., speed, temperature and density) at distances closer to the Earth and beyond are obtained primarily from in situ measurements, which are taken along the one-dimensional scan during the fortuitous encounter with spacecraft. The spacecraft observations have been made only down to a heliocentric distance of 0.3 AU ($\approx 64R_{\odot}$), which is the perihelion distance of the Helios satellite. Moreover, spacecraft measurements are confined to the ecliptic plane and Ulysses is the first and only space mission to probe the high latitude heliosphere (Bame et al. 1992; Phillips et al. 1994; http://swoops.lanl.gov/). However, in the space between the Sun and Earth, the available information on the solar wind flow has been obtained from various remote sensing methods. The interplanetary scintillation (IPS) technique (Hewish, Scott & Wills 1964) is one among them. It is important to note that this technique provides the three-dimensional structure of the solar wind at various distances in the inner heliosphere (≤ 1 AU) and in the later part of this article, solar wind properties within about 1 AU of the Sun obtained from the scintillation measurements are discussed. Therefore, a brief description of the scintillation technique is given in the following subsection.

5.1 Interplanetary Scintillation

The interplanetary scintillation (IPS) technique exploits the scattering of radiation from distant point-like radio sources (e.g., quasars or radio galaxies of angular diameter ≤ 0.4 arcsec) by the electron-density irregularities in the solar wind. That is, the plane wavefront gets phase modulated by the refractive index variations caused by the density fluctuations in the solar wind (Tatarski 1961) and the resultant diffraction pattern caused by the scattered radio waves, which drifts past the observer with the velocity of the solar wind, produces temporal intensity fluctuations on the ground as shown schematically in Fig. 4. The temporal spectrum of intensity fluctuations, $P(f)$, i.e., the power spectrum of intensity measured at the input of a radio telescope, is the Fourier transform of the autocorrelation function of intensity.

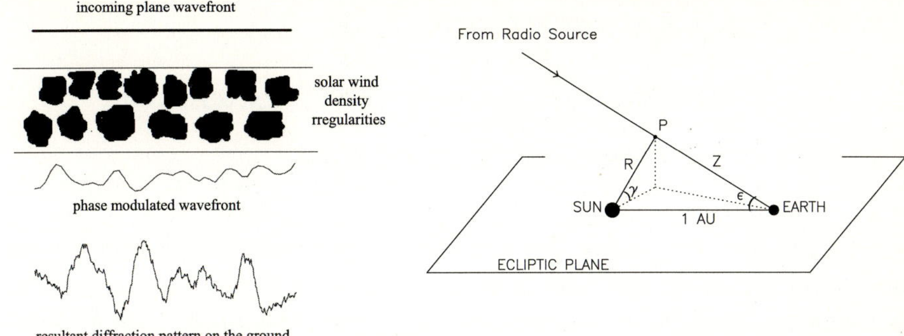

Fig. 4. Schematic diagram illustrating the formation of interplanetary scintillation by random diffraction screen and blurring of scintillation. The left panel illustrates the formation of scintillation and the right panel shows the geometry. The angle between the Sun, Earth and the radio source is the solar elongation, ϵ (heliocentric distance to the radio path $= \sin\epsilon$ AU). The angle γ gives the heliographic latitude

The radio-wave scattering in the solar wind can be classified into two categories: (1) strong scattering when the rms phase fluctuations, $\phi \geq 1$ radian and (2) weak scattering for $\phi < 1$ radian (Manoharan 1993). The scattering strength, which is strong at the near-Sun region, decreases with distance from the Sun. For example, in the case of Ooty IPS measurements at 327 MHz, the 'strong-to-weak' transition occurs at a distance of $\sim 40R_\odot$ from the Sun and the transition region moves further away from the Sun for decreasing frequency of observation. In the weak-scattering region, the temporal spectrum, $P(f)$, is linearly related to the electron-density fluctuations spectrum, $\Phi_{N_e}(\kappa)$ and the relation can be inverted either by model fitting (e.g., Manoharan & Ananthakrishnan 1990) or by an inverse Abel transform (e.g., Coles & Harmon 1978).

In the weak-scintillation regime, the Born approximation is applicable. Thus the extended medium between the radio source and the observer can be considered to consist of thin layers perpendicular to the line of sight, the observed intensity fluctuations being the sum of contributions from all layers. The contribution from each layer is weighted by the local level of turbulence, $C^2_{N_e}$, which decreases rapidly with distance from the Sun, $C^2_{N_e} \sim R^{-4}$ (Manoharan 1993). Such a steep gradient means that most of the scattering power occurs at the point of closest approach of the line of sight to the Sun, and IPS measurements are heavily weighted to the solar wind in the region of closest solar approach.

When the diffraction pattern caused by the scattering moves past the observer with a velocity, $V_p(z)$, the wavenumber κ_x appears as the temporal frequency, $f = V_p(z)\kappa_x/2\pi$, where $V_p(z)$ is the projected velocity of the solar wind along the x-axis of a scattering layer at a distance z from the ground. The observed temporal spectrum of intensity variations is due to the integration of the scattering strength and density spectrum over the x–y plane as well as the summation along the propagation path, z,

$$P(f) = (2\pi r_e \lambda)^2 \int_0^\infty \frac{dz}{|V_p(z)|} \int_\infty^\infty d\kappa_y C_{N_e}^2 \Phi_{N_e}(\kappa_x, \kappa_y, z) \times F_{\text{diff}}(\kappa_x, \kappa_y, z) F_{\text{source}}(\kappa_x, \kappa_y, z) \;, \tag{5}$$

where r_e is the classical electron radius and λ the wavelength of observation. $F_{\text{diff}}(\kappa_x, \kappa_y, z) = 4\sin^2(\kappa^2 z\lambda/4\pi)$ is the Fresnel propagation filter, which attenuates wavenumbers smaller than $\kappa_f \approx (2\pi/\lambda z)^{1/2}$ and it does not alter the shape of temporal spectrum at large wavenumbers (Manoharan, Kojima & Misawa 1994). The term F_{source} is the squared modulus of the radio source visibility function, given by $\exp(-\kappa^2 z^2 \theta_0^2)$ for a symmetrical Gaussian brightness distribution of half maximum diameter, $\Theta_s = 2.35\theta_0$ and it cuts off the spectrum at wavenumbers above $\kappa_s = 1/(z\theta_0)$. The spatial spectrum of density fluctuations is of power-law nature and contains inner-scale term, $\Phi_{N_e}(\kappa) = \kappa^{-\alpha}\exp(-\kappa^2/\kappa_i^2)$, where $S_i \approx 3/\kappa_i$ is the inner-scale or cut-off scale of the turbulence (Manoharan, Kojima & Misawa 1994; Manoharan et al. 2000; also refer to Sect. 6.4 and Fig. 12).

The rms of intensity variations due to IPS is the integral of the power spectrum and for a radio source of unit flux density, the scintillation index, m, is estimated by,

$$m^2 = \int_0^{f_c} P(f)\, df \;, \tag{6}$$

where f_c is the cut-off frequency of the temporal spectrum where the scintillation equals the noise level. In the weak-scattering case, the scintillation index, m, is linearly related to the rms of electron-density fluctuations, δN_e (Manoharan 1993). The temporal power spectrum, $P(f)$, can also be suitably calibrated (i.e., using (5)) to estimate the speed of the solar wind and shape of the density spectrum (Manoharan & Ananthakrishnan 1990; Manoharan, Kojima & Misawa 1994).

Depending on the observing wavelength, the IPS method can provide solar wind conditions in the three-dimensional heliosphere of radius a few solar radii to about 1 AU. Routine IPS measurements are being carried out at the Radio Astronomy Centre, Tata Institute of Fundamental Research, Ooty, India and Solar–Terrestrial Environment Laboratory (STEL), Nagoya University, Japan (e.g., Manoharan et al. 2000) and they provide: (1) the scintillation index, m, a measure of density-turbulence level of the solar wind (Manoharan 1993), (2) the turbulence spectrum in the spatial-scale range 10–500 km (Manoharan, Kojima & Misawa 1994), and (3) the speed of the solar wind. As mentioned above, the scintillation of a single-antenna system, having good signal to noise ratio, can provide the speed and density turbulence of the solar wind (e.g., Manoharan & Ananthakrishnan 1990; Manoharan 1993). In the case of a multi-antenna system, the speed of the solar wind is estimated by cross correlating IPS signals from a pair of antennas (e.g., Coles & Kaufman 1978; Kojima & Kakinuma 1990; Rickett & Coles 1991). Figure 5 shows examples of Ooty measurements, (a)

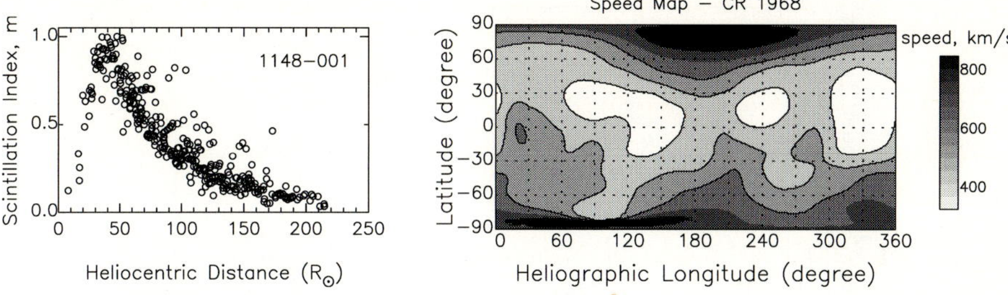

Fig. 5. Variation of scintillation index as a function of distance from the Sun for the quasar 1148-001 at 327 MHz (left) and the two-dimensional map of the speed of the solar wind for the Carrington rotation 1968 (right). A large number of solar wind speed estimates obtained from the Ooty data have been projected onto a sphere of radius $\sim 100R_{\odot}$ to produce the speed map

the scintillation index variation with distance from the Sun for a given radio source, 1148-001, which has a diameter of about 15 milli arcsec and (b) a two-dimensional map of the speed of the solar wind projected onto the heliosphere of radius $\sim 100R_{\odot}$. It may be noted that in the weak-to-strong transition region, the scintillation index peaks (i.e., around $40R_{\odot}$) and as the Sun is approached, it decreases due to the source size smearing. The large variations in the scintillation index from the mean value indicate the high level of solar wind turbulence due to transients at respective heliocentric distance.

Although IPS measurements are integrated along the line of sight (i.e., along the radio path), observations on a grid of large number of radio sources (i.e., towards various lines of sight) can readily detect and provide the image of the large-scale structure of the quasi-stationary solar wind as well as propagating interplanetary disturbances (Manoharan 1998; Manoharan et al. 2001). In the case of a propagating disturbance (e.g., a coronal mass ejection or a co-rotating interaction region), the IPS technique detects the compression region (or sheath) between the shock ahead and the pushing driver gas. Thus, the increase in scintillation with respect to the background solar wind flow can identify the presence of disturbance in the interplanetary space at various distances from the Sun (Manoharan et al. 2001). At Ooty, normally scintillations towards 300 to 400 radio sources are recorded each day and the distribution of their normalised scintillation indices (generally denoted by g = (observed scintillation)/(expected scintillation)) give the image of the density turbulence in the interplanetary medium (refer to Figs. 15 and 16).

The scintillation method has also been effectively used to determine the solar wind properties in the acceleration region, $\leq 20R_{\odot}$ (as mentioned above, high-frequency scintillation measurements can probe the near-Sun solar plasma; e.g., Armstrong et al. 1990; Armstrong & Woo 1981; Ekers & Little 1971; Scott, Coles & Bourgois 1983; Tyler et al. 1981; Yamauchi et al. 1998) as well as up to a heliocentric distance of ~ 1 AU and their latitudinal variation at various phases of the solar cycle (Kojima & Kakinuma 1990; Rickett & Coles 1991; Manoharan 1998).

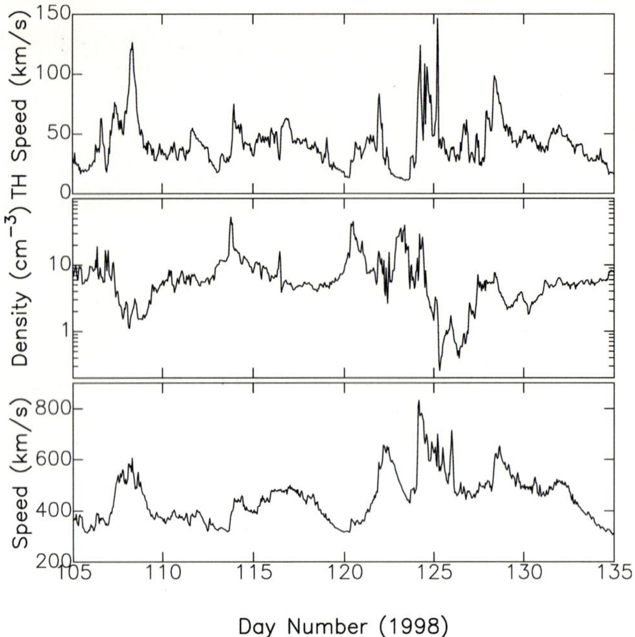

Fig. 6. Hourly averages of solar wind speed, density and thermal speed measured at ~ 1 AU by the *Wind*/SWE spacecraft (courtesy of http://web.mit.edu/space/www)

6 Solar Wind in the Inner Heliosphere

As mentioned earlier, the corona exhibits quasi-steady structures in a variety of forms: streamers extending radially outward, loops that begin and end on the Sun, and voids or holes with less denser material. The characteristics of these structures imprinted on the solar wind are being carried by the steady flow, which forms the *quasi-stationary* solar wind. However, in addition to the quasi-steady flow, there are coronal transients of many kinds. These are usually associated with explosive activities (e.g., violent flares, coronal mass ejections) and they add excess mass and magnetic field into the interplanetary medium, causing strong *transients* in the solar wind. Specifically, coronal mass ejections (CMEs), involving energies (kinetic plus magnetic) in the range $\sim 10^{30}$–10^{32} erg, occur several times a day during solar activity maximum and have a strong solar cycle variation in phase with the sunspot number (Howard et al. 1986; St. Cyr et al. 1999). Fast CMEs travelling in the solar wind, with a speed in excess of the propagation speed of any MHD wave, produce shock transients. In the following sections, the basic properties of the quasi-stationary and transient solar wind are described.

Table 1. Property of solar wind

Property (at 1 AU)	Low-speed wind	High-speed wind
Association	?	coronal holes
Speed	≤ 400 km s^{-1}	600–800 km s^{-1}
Density	5–10 cm^{-3}	~ 3 cm^{-3}
Structure	filamentary	uniform
Temperature	$T_p \sim 4 \times 10^4$ K $T_e \sim 1.3 \times 10^5$ K	$T_p \sim 2 \times 10^5$ K $T_e \sim 10^5$ K
Solar minimum	$\pm 35°$ latitude region	polar region
Solar maximum	all latitudes	close to poles; low-latitude coronal holes

6.1 Quasi-stationary Solar Wind

The quasi-stationary flow generally describes the large-scale behaviour of the solar wind over periods of many days to several solar rotations. Its properties are characterised in terms of the flow speed of the plasma. Figure 6 displays the solar wind speed, density and thermal speed of ions observed by the *Wind*/SWE spacecraft (Ogilvie et al. 1995; http://web.mit.edu/space/www) at near-Earth space over a period of about a month (i.e., between the days 105 and 135 of 1998). The thermal speed of the ion is proportional to the square root of the temperature, $v_{\rm th} = \sqrt{3kT/m_H}$. As seen in the figure, the low-speed flow tends to be cooler and denser than that of the high-speed wind. Table 1 presents the typical average properties of low- and high-speed solar winds observed close to the Earth's orbit.

The steady flows of high-speed streams originate above the large coronal hole regions of low-density, single polarity, weak magnetic field, where the lines of force open out into the interplanetary space. The source of low-speed winds is not well determined. It is likely that it may be flowing from the edge of the coronal hole as well as from the other structures near active regions. A detailed investigation on the origin of the low-speed wind by Kojima et al. (1999) has shown that the low-speed flow originates from the vicinity of a single polarity side of the active region and not from the traditional helmet streamer cusp. Figure 7 (M. Kojima, private communication 2001) shows the latitude-longitude plot of potential-field lines extrapolated from the photosphere (bottom) to the source surface (top). The source surface defines a height at which the field is assumed to be radial while computing the coronal magnetic field from photospheric field observations with a potential field model (e.g., Hoeksema, Wilcox & Scherrer 1983). The grey-scale plot of the solar wind speed is given on the source surface and the black area indicates wind speed < 370 km s^{-1}. The magnetic topology thus plays an essential role in determining the flow pattern of the solar wind.

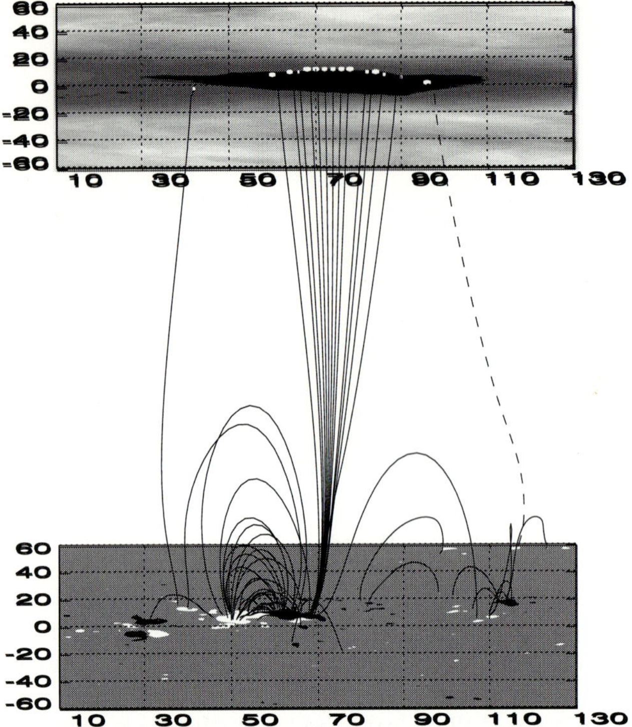

Fig. 7. Magnetic field lines extrapolated between the photosphere (bottom) and the source surface (top) (e.g., Hakamada 1995). The vertical and horizontal axes in degrees represent heliographic latitude and longitude, respectively. The map at the top shows the solar wind speed plot at the source surface. The speed estimates have been obtained from the interplanetary scintillation measurements (Kojima et al. 1999; also refer to Sect. 5.1). In this plot, the high-speed ($\sim$ 800 km s^{-1}) to low-speed winds are shown by the white to black grey scale. In the bottom image, black and white represent magnetic polarities. It may be noted that the low-speed wind originates from the single polarity side of the active region (M. Kojima, private communication)

6.2 Radial Evolution of the Quasi-stationary Wind

In order to understand the acceleration behaviour of the solar wind, many workers have attempted to measure the speed as a function of heliocentric distance (see review by Bird & Edenhofer 1991; Grall et al. 1996). Based on IPS observations, Coles and his co-workers (Coles et al. 1991a; Coles & Esser 1992) measured the speed-distance profiles for a number of streams in the distance range of 10 to 90 $R_\odot$ and compared them with the solar wind acceleration model (e.g., Parker 1964), which included the addition of momentum by ponderomotive forces from Alfvén waves. They found that more than half of the profiles agree with the above MHD model and about one fourth of the profiles showed delay in the acceleration, which require further improved models. A typical plot of the radial evolution of speed of the solar wind in the acceleration region is shown in

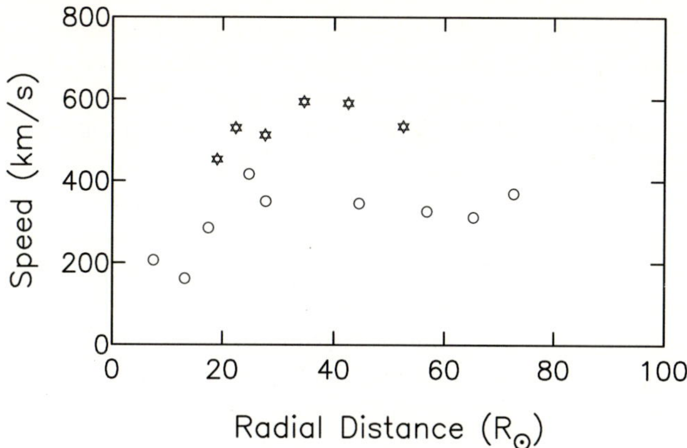

Fig. 8. Acceleration of the solar wind: (a) low-speed flow from the equatorial region (open circles) and (b) high-speed wind from the high-latitude coronal hole region (star symbols). The speed estimates have been obtained from the IPS measurements at 2 and 8 GHz (Y. Yamauchi, private communication)

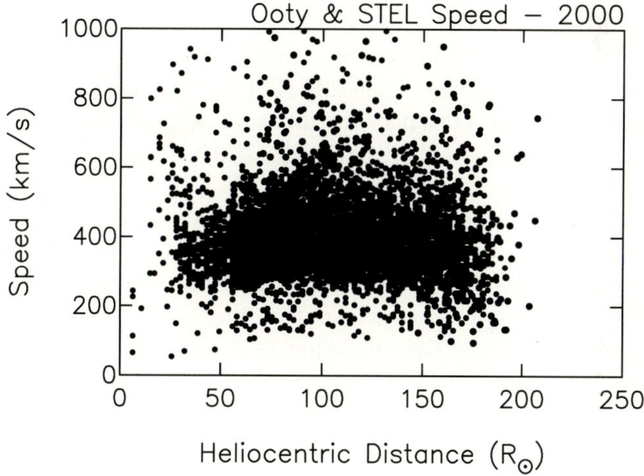

Fig. 9. Solar wind speeds observed at the maximum of the solar activity cycle. A 'speed-distance' plot obtained from IPS measurements taken over the year 2000 at Ooty and STEL. The large variations in the estimates indicate the typical condition of the activity at the solar maximum

Fig. 8. The measurements shown in the figure have been obtained from the IPS of 2 and 8 GHz during 1994 (Y. Yamauchi, private communication). The speed estimates for the high-speed and low-speed streams, respectively, are from the polar coronal hole regions and the equatorial regions of the Sun. The speed of the low-speed flow gradually increases and attains $\sim$ 400 km s^{-1} at a distance of about $20R_\odot$ from the Sun and the increase in speed of the high-speed plasma is steeper than that of the low-speed plasma.

In the case of quasi-stationary flows, along a given streamer, the speed of the solar wind shows no significant acceleration or deceleration at heliocentric distances greater than $\sim 40R_\odot$ (e.g., Coles et al. 1991a). In the equatorial region, although the average speed of the solar wind is $\sim$ 400 km s^{-1} in the distance range of 50–200 $R_\odot$, the routine monitoring of the speed over a long period shows a large variation as illustrated in Fig. 9. This 'speed-distance' plot based on speed estimates obtained using IPS data from Ooty and STEL during the year 2000, i.e, at the maximum of the current solar cycle (The Ooty measurements have been made by the author and the STEL data are from M. Kojima, private communication). The large variations of speed are likely to be due to low- and high-speed streamers and transients present in the solar wind and they characterise the typical conditions of the heliosphere during the maximum of the solar activity.

6.3 Latitudinal Variations

Observations of IPS are also effectively used to study the latitudinal evolution of speed at various phases of the solar cycle. Figure 10 gives 10° latitude averages of the solar wind speed at distances greater than $50R_\odot$ obtained between 1986 and 2000, covering a period of about one and half solar cycles (i.e., cycle 22 in full and the first half of current cycle 23). This plot provides evidence that the latitudinal variations of speed reflect the large-scale structure of the solar corona and its changes with phase of the solar cycle. During solar minimum, the large-scale magnetic field is predominately dipolar. At high solar latitudes, the open magnetic field configuration of the coronal holes allows the solar wind plasma to expand rapidly into interplanetary space. Therefore, in the polar coronal hole regions, flow speeds of $\sim$ 800 km s^{-1} are observed (e.g., see the plots of 1988 and 1994 in Fig. 10). In contrast, the middle and low latitudes regions are dominated by a high-density 'streamer-belt', where the closed field largely confines the coronal plasma and the low-speed wind originates near the streamer belt. Moreover, the streamer belt, which marks the equatorial region of the north and south poles (i.e., the dipole equator), evolves into an interplanetary (or heliospheric) current sheet (HCS) that separates the flows originating in the two hemispheres.

The above described simple dipole structure seems to exist for two or three years around the solar minimum and as the level of activity increases, polar coronal holes move towards the equator, and the quadrupole component of the solar magnetic field increases, producing a tilted dipolar component of the field and a substantial latitudinal warp in the near-equatorial streamer belt and consequently in the heliospheric current sheet. The 'speed-latitude' profile tends to become flat. During two or three years around the maximum of the solar activity, the dipolar component of the solar magnetic field becomes very weak, coronal holes and their associated high-speed solar wind disappear, and low-speed flows dominate the solar wind at all latitudes (e.g., see the plots around the beginning of 1990 and 2000). Following the solar maximum, the dominant dipolar component returns. However, the field is now reversed in direction and the dipole axis

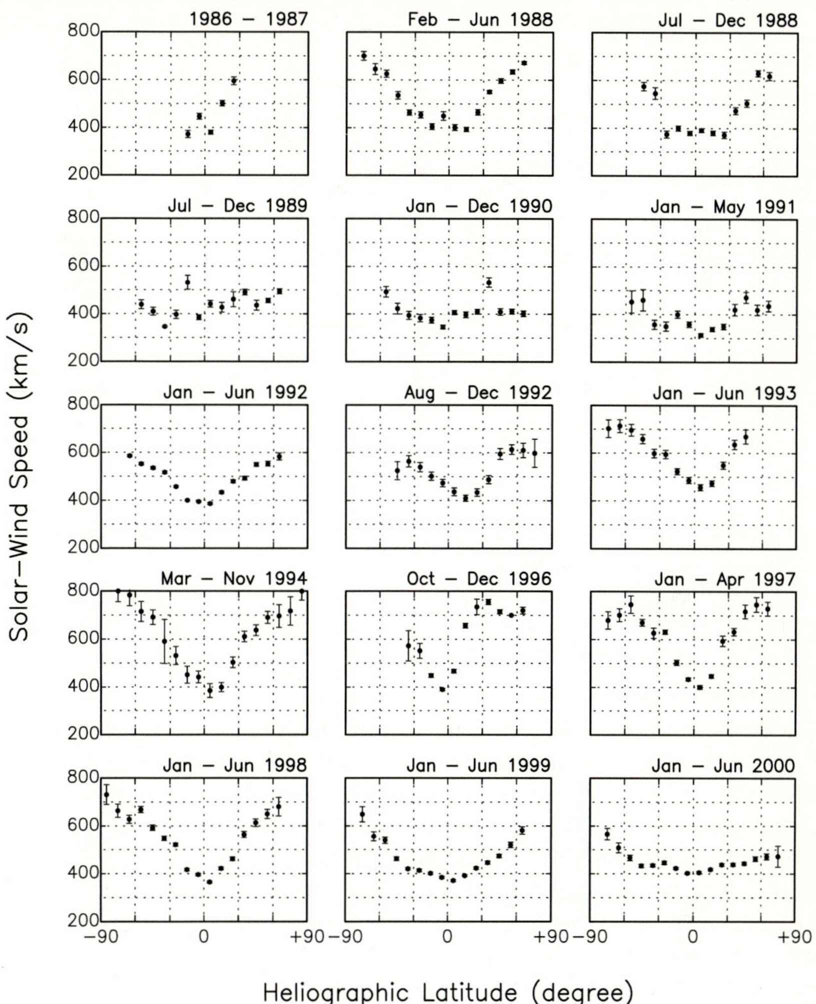

Fig. 10. Latitudinal variations of the solar wind speed observed over a period of about one and half solar activity cycles. These speed estimates have been obtained from the IPS observations from Ooty (e.g., Manoharan 1998)

is tilted by as much as 30° from the rotation axis of the Sun. The tilt persists for the next two or three years until the solar minimum is again reached.

6.4 The Density Turbulence Spectrum

Various scattering experiments have indicated that in the near-Sun region, $\leq 10R_{\odot}$, the density microstructures of scales less than 50 km are field aligned and become increasingly anisotropic the closer the Sun is approached. An axial ratio of as high as 10 is observed within $6R_{\odot}$ (Armstrong et al. 1990; Grall et al. 1997). Moreover, the microstructures at scales of about 10 km are more anisotropic,

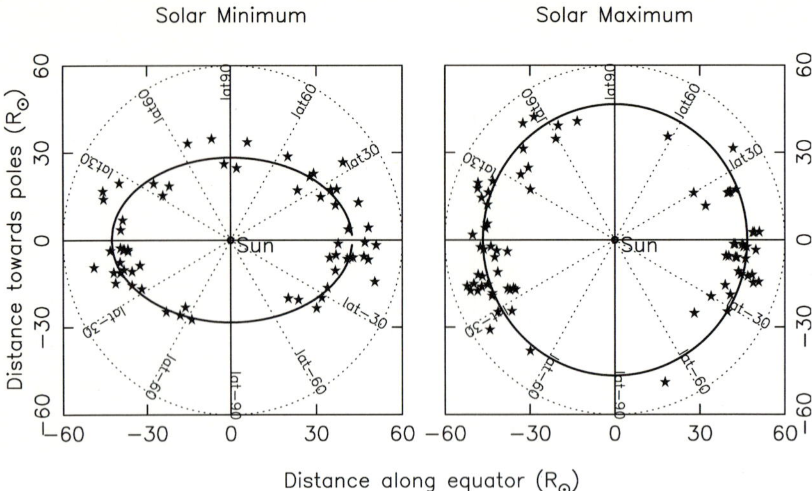

Fig. 11. Contours of constant density fluctuations (δn_e) observed during (a) solar minimum and (b) solar maximum of activities. At a given distance, during solar activity minimum, the δn_e value decreases from equatorial region to the pole by factor of about 2.5, whereas the δn_e distribution is nearly spherically symmetric near the maximum of activity (e.g., Manoharan 1993)

whereas large-scale structures ($\sim$ 100–1000 km) tend to be nearly isotropic (Grall et al. 1997). At distances close to the Sun, $< 50R_\odot$, large variations in the speed as well as in density are observed and these variations are more pronounced in low-speed streams than in high-speed flows originating from the coronal hole regions (see review by Bird & Edenhofer 1991).

The density fluctuations (δn_e) in the solar wind also show strong latitudinal variation. The δn_e measurements close to the Sun, $< 10R_\odot$, by Coles et al. (1995) as well as further out to distances, $\sim 45R_\odot$, by Manoharan (1993) have shown similar changes with heliographic latitude. Figure 11 shows the contours of constant δn_e around the Sun during solar minimum and maximum periods. This figure has also been obtained from many years of IPS measurements (Manoharan 1993). It illustrates that when the solar activity is minimum, a given value of the δn_e appears closer to the Sun at the poles than at the equatorial regions and the contour of the constant δn_e appears like an ellipse around the Sun. During the maximum of solar activity, however, at a given distance from the Sun, on an average a uniform distribution of δn_e is seen at all heliographic latitudes and its constant value contour appears like a circle. During the solar cycle minimum, at $\sim 40R_\odot$ from the Sun, the value of δn_e is about 2.5 times larger in the equatorial region than that in the polar region. The coronagraph observations have also suggested a similar trend that at a given heliocentric distance, the average density in the solar corona is smaller in the polar region than in the equatorial region (Coles et al. 1995).

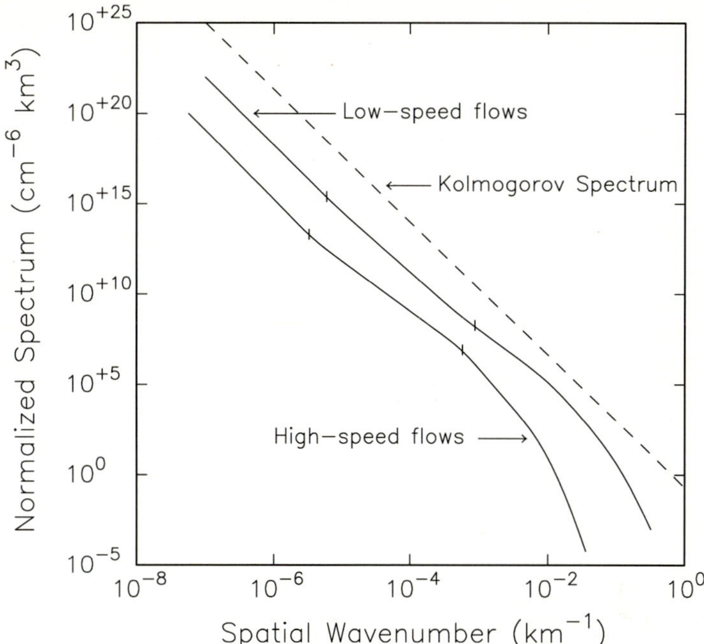

Fig. 12. Composite model spectra of density turbulence derived from the IPS and spacecraft measurements for the low-speed and high-speed flows (Manoharan, Kojima & Misawa 1994). The wavenumber of the spectral break is indicated by a tick mark

The spectral characteristics of density fluctuations in the solar wind differ significantly between low- and high-speed streams. In general, the shape of the spectrum approaches that of fully developed turbulence, that is, a Kolmogorov spectrum, $\Phi_{n_e} \sim \kappa^{11/3}$, where κ is the spatial wavenumber. Spacecraft and IPS data, in the spatial scale range of $1/\kappa \approx 10$–10^6 km, have revealed that the prevailing spectral shape of turbulence differs in three distinct scale-size ranges. In the case of high-speed plasma, as shown in Fig. 12, nearly a Kolmogorov spectrum of $\alpha \approx 11/3$ (where α is the power law index) is observed at spatial scale $> 10^5$ km, followed by a flattening, $\alpha \approx 3$, in the range of 10^3–10^5 km, and a steeper spectrum, $\alpha \geq 3.8$, at scales less than 1000 km. The steepening at small-spatial scale of the spectrum occurs on scales of the order of ion (or proton) inertial length, $r_p = V_a/\omega_p$, where V_a is the Alfvén speed and ω_p is the proton cyclotron frequency. The spectral flattening at mid scales suggests an enhancement associated with one of the possible plasma instabilities, (e.g., ion cyclotron or magnetoacoustic) at high wavenumbers (Coles et al. 1991b; Manoharan, Kojima & Misawa 1994; Yamauchi et al. 1998). The steepening, at scales smaller than 1000 km, may be attributed to an increase in the Alfvén speed and the low-level of the density in the solar wind from the coronal hole resulting in large inertial- or inner-scale size. That is, the size of the inner scale goes as $n_p^{-1/2}$. In the case of the low-speed solar wind, the spectral characteristics differ at the mid-range and small scales, respectively, having slopes of $\alpha \approx 3.5$ and 2.8.

7 Solar Wind Transients

There are two known sources causing transients in the solar wind: (1) co-rotating interaction regions formed at the interface of fast and slow streams and (2) coronal mass ejections. Studies of the three-dimensional structure of these transients are essential for the interpretation of energetic particles in the interplanetary space and in particular, for the understanding of the role played by interplanetary disturbances in stimulating the recurrent and non-recurrent geomagnetic storms.

7.1 Co-rotating Interaction Regions

When a coronal hole is located near the equator, as the Sun rotates, the fast and low-speed flows will alternate and with increasing heliocentric distance along the spiral path, the fast wind will overtake and compress the low-speed plasma ahead of it. Since the magnetic field frozen in the solar wind prevents the interpenetration of streams, compressive co-rotating interaction regions (CIRs) are formed as shown in Fig. 13. A CIR thus can cause a forward shock propagating at the leading edge and a reverse shock at its trailing edge (refer to inset of Fig. 13). The steepening of these shock waves takes place, typically, at distances greater than about 1 AU. The low-speed wind ahead of the interaction region is pushed against the forward shock and the reverse shock tends to decelerate the high-speed wind causing compression. The basic internal structure of a CIR therefore primarily depends on the properties of the low-speed wind being swept and compressed (Siscoe & Intriligator 1993). Understanding the structural evolution of CIRs as functions of heliocentric distance and longitude has come from in situ measurements available over two decades (see review by Gosling 1996). The recent high-latitude Ulysses mission has provided results on the latitudinal distribution of CIRs, (i) the formation of CIRs being prominent in the latitude range 0–20 degrees and (ii) a decline in the formation of CIRs with increasing heliographic latitude (Gosling 1996; Phillips et al. 1994).

The CIRs play an important role in causing recurrent storm activities in the Earth's magnetosphere. The intensity of the geomagnetic storms, particularly, depends on the southward-directed component (B_z) of the interplanetary magnetic field, which can merge with the field lines on the day side of the Earth's magnetosphere (e.g., Tsurutani et al. 1992). However, various studies on the effectiveness of geomagnetic storms show that large storms are associated with solar wind effects from coronal mass ejections, CMEs (e.g., Gosling 1997; Webb 1995). The CMEs generally occur from regions of closed magnetic field configurations (e.g., Hundhausen 1998) and they cause transient disturbances of variable strength in the solar wind depending on their speed, mass and geometry of the magnetic field carried with them. There are excellent reviews available on the

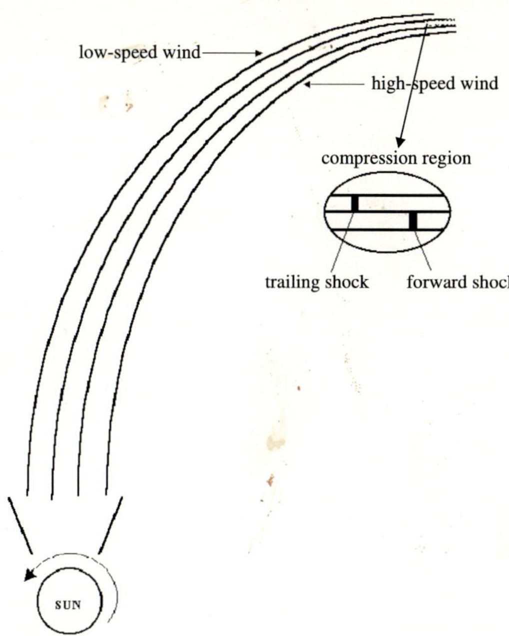

Fig. 13. Interaction stream structure in the equatorial region formed between low- and high-speed streams

various aspects of CMEs (Crooker, Joselyn & Feynman 1997; Dryer 1994; Webb 1995).

7.2 Coronal Mass Ejections

A typical coronal mass ejection (CME) carries off about 10^{15}–10^{16} g of solar mass and it has a spatial-scale size of $\sim 1R_\odot$ in the low corona. The ejection speeds, in the near-Sun region (i.e., $< 10R_\odot$), range from a few tens of km s^{-1} to as high as 2000 km s^{-1} with an overall average speed of about 450 km s^{-1} (Hundhausen 1997). CMEs occur much more frequently near the equatorial region of the Sun than at high latitudes and the rate of occurrence of CMEs also shows a dependence on the solar cycle. That is, there may be ∼4–5 events per day during solar maximum and only about one ejection per day near the minimum of the solar cycle (Howard et al. 1986; St. Cyr et al. 1999). When a CME is observed, one or more of the following solar events may be observed in association with it, such as, eruptive prominences, long duration X-ray emissions often followed by type II and IV metric radio bursts, intense microwave emissions, solar energetic particle events in the interplanetary space, and quasi-stationary post-flare loops at the mass ejection site (Kahler, Sheeley & Liggett 1989; Sheeley et al. 1983; Chertok, Gnezdilov & Zaborova 1992; Reames 1995).

There are a number of signatures which identify CMEs in the solar wind: (1) magnetic clouds, (2) strong fields, but low field variance, (3) low plasma beta,

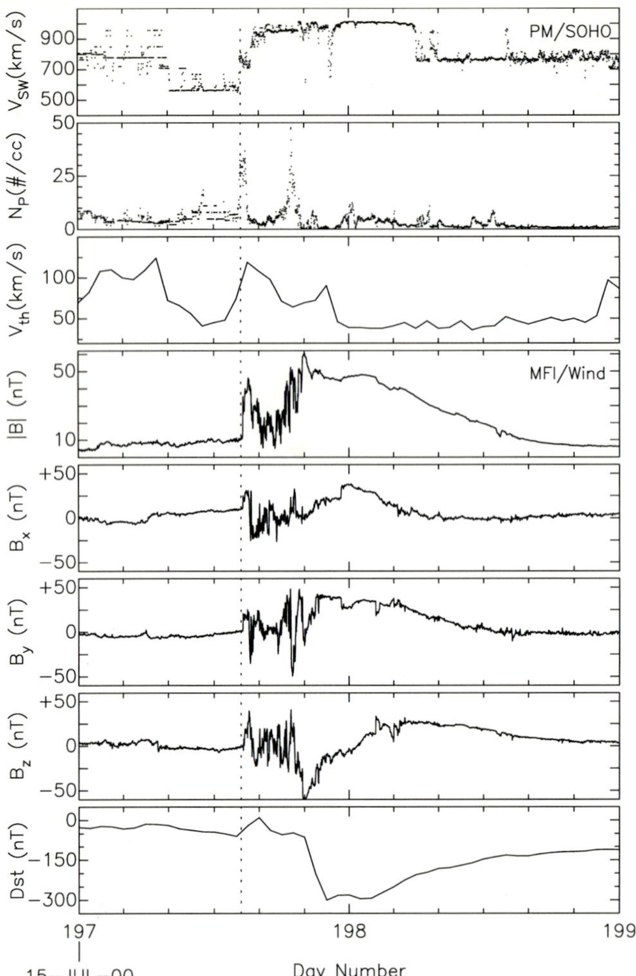

Fig. 14. Solar wind parameters associated with the passage of a shock disturbance generated by the halo CME on July 14, 2000. From top to bottom, solar wind parameters from the space mission *SOHO*/PM (solar wind proton speed, $V_{\rm sw}$, density, N_p, and thermal speed, $V_{\rm th}$), magnetic field data obtained from *Wind*/MFI spacecraft. The bottom panel of figure shows the plot of the geomagnetic index, Dst. The vertical line indicates the arrival time of the shock at the spacecraft

(4) counterstreaming of electrons and protons, (5) excessive helium abundances, and (6) unusual ionisation states. One or more of the above signatures may appear jointly at any particular event. A relatively high-speed CME with respect to the ambient speed of the solar wind, however, also produces compressions, which cause excessive turbulence at the leading edge of the propagation and may eventually steepen into a shock wave with increasing distance from the Sun. Figure 14 shows the plasma parameters, associated with the shock produced by

an Earth-directed CME, observed by the *Wind* spacecraft at near-Earth space. The figure also includes the plot of the geomagnetic index, Dst, to illustrate the severity of the storm that occurred in the Earth's magnetosphere.

There are ample possibilities for a considerable evolution of the physical properties (speed, density, mass and magnetic field) of CMEs on their way from the Sun to Earth. Hence, the details on the propagation of CMEs in the region between the manifestations near the Sun and in situ measurements near the Earth are crucial for making progress about understanding of interplanetary transients and their geoeffectiveness. The direct measurements of transients are however not possible in the inner heliosphere and the study of the evolution of CME transients with radial distance depends solely on remote sensing techniques.

The remote sensing scintillation methods are extremely sensitive to density turbulence in the solar wind and they can be used as powerful probes to detect the increase of the turbulence caused by compression in the propagating transients. For example, in the near-Sun region, Doppler scintillation techniques have been employed to study transients (Huddleston, Woo & Neugebauer 1995, Woo & Schwenn 1991, Woo 1993) and at distances beyond 0.2 AU, the IPS method has been used to detect and track CIR features and CME associated transients (Tappin, Hewish & Gapper 1984; Watanabe & Schwenn 1989; Tokumaru et al. 2000; Janardhan et al. 1996; Manoharan et al. 1995, 2000, 2001). Although, scintillation measurements are limited by the line of sight integration, an important advantage of this method, as compared to in-situ measurements using spacecraft, is that it is useful to study the three-dimensional structure of transients at a range of heliocentric distances. Further, a systematic monitoring of scintillations, over the sky by sampling a grid of radio sources on a day to day basis, can provide the map of transient plasma as shown in Fig. 15, which includes scintillation images from Ooty measurements and while-light images taken with LASCO/SOHO coronagraphs for the July 14, 2000 halo CME. Another example of propagation of a partial halo CME on June 25, 1992 moving to the west of the Sun is shown in Fig. 16. These scintillation images have been made by the author using the data from the Ooty Radio Telescope. It is evident from the above figures that the size of the CME transients in the solar wind increases linearly with distance from the Sun. The evolution of size with distance, $a_{\rm CME} \sim {\rm R}^{1.0}$, suggests that pressure balance is maintained between the CME driver gas and the ambient solar wind (Manoharan et al. 2000, 2001). It may also be noted that the spectrum of turbulence of the compressed plasma ahead of the driver gas shows a much flatter spectrum than that of quasi-stationary low-speed wind (see Fig. 12).

The coordinated study using near-Sun radio imaging techniques, scintillation observations and space mission data, has proved to be essential in understanding the propagation of CME associated transients in the inner heliosphere (e.g., Manoharan et al. 2001). For example, 'speed-distance' profiles obtained from such coordinated studies are shown in Fig. 17 for solar wind transients associated with two different CMEs. Their speeds demonstrate that the deceleration of the CMEs does not follow a simple radial law over the entire distance range in the

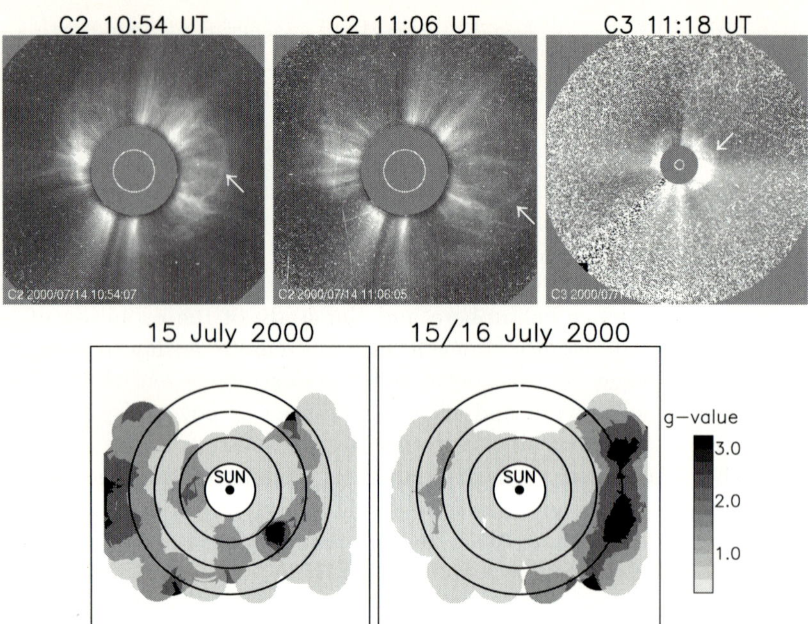

Fig. 15. White-light (top) and scintillation (bottom) images of the halo CME of July 14, 2000. The white-light images are from LASCO/SOHO C2 and C3 coronagraphs, respectively, with fields of views of $\sim 6R_\odot$ and $\sim 30R_\odot$. The inner circle in the LASCO images indicates the position and size of the photosphere. In the scintillation images, the Sun is at the centre and the concentric circles are at radii 50, 100, 150, and 200 $R_\odot$. In the IPS measurements (bottom images), time increases from the right (i.e., measurements at the west side of the Sun) to the left (i.e., east side of the Sun). That is, the extreme west region of the image is observed in the early hours of the day and with time, region observed moves closer to the Sun around mid day and and extreme eastern region is observed late in the evening. The halo CME seen at about $50R_\odot$ in the west expanded and moved to a greater heliocentric distance at a later time as seen in the eastern side observations. It may also be noted that the size of the disturbance increases with distance from the Sun (Manoharan et al. 2001). For the definition of *g-value* see Sect. 5.1

inner heliosphere, but, indicates a two-level deceleration: (i) a low decline in speed at distances within or about 100 solar radii and (ii) a rapid decrease at larger distances from the Sun. The observations at Ooty, as well as STEL, have shown similar two-level decelerations for a large number of transients. Although, the slopes of the speed in these transients have similar trends, the differences in power-law index between different transients (at different segments of their propagation path) can be attributed to the input energy associated with the CME eruption as well as the dynamics of the ambient solar wind. In other words, the initial speed of the CMEs and the physical properties of the solar wind plasma along their propagating path play an essential role in determining the characteristics of the speed-distance profile of the transients. The average

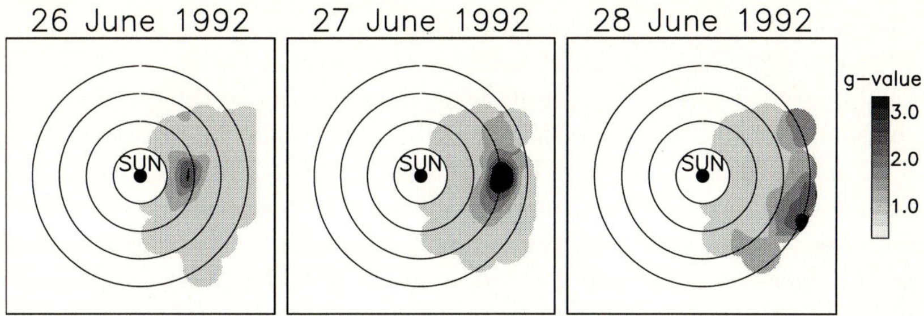

Fig. 16. Ooty scintillation images of a partial halo CME on June 25, 1992. The format of the figure is same as Fig. 15. The speed of the CME in the distance range of ~80–200 $R_\odot$ decreased from $V \sim 1400$ to 500 km s^{-1} and it also showed a power-law proportionality with distance from the Sun, $V \sim R^{-0.8}$ (Manoharan et al. 2000). It is likely that the initial speed of the CME in the near-Sun region should have been greater than 1400 km s^{-1} (also see Fig. 17)

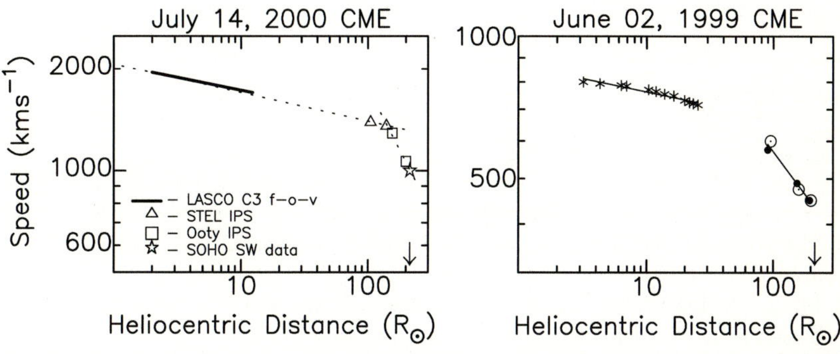

Fig. 17. Radial evolution of speeds of disturbances associated with the halo CME on July 14, 2000 and the partial halo CME on June 2, 1999. In these log-log plots, arrow marks on the x-axis indicate the heliocentric distance of the Earth orbit, i.e., 1 AU $\approx$ 215 $R_\odot$. Initial speeds of these CMEs are, respectively, ~2000 km s^{-1} and ~1000 km s^{-1}. In the right panel, stars represent LASCO white-light measurements, while filled and open circles are Ooty IPS measurements

speed of the ambient solar wind naturally divides transients into fast and slow categories, where fast ones get decelerated and slow CMEs are accelerated or carried away by the ambient solar wind flow (e.g., Gopalswamy et al. 2000).

Further, as illustrated in the above figure, the deceleration in the first phase ($< 100R_\odot$), for a given typical mass of the CME, would amount to an energy loss of $\geq 10^{32}$ erg. Such an amount of energy dissipation has not caused the CME to slow down considerably, suggesting that the energy spent along the propagation path is partly compensated by the energy stored within the CME. Manoharan et al. (2001) have suggested that the magnetic energy of the flux rope associated with the CME would have been utilised in assisting the propagation of the CME into the interplanetary medium. At large distances from the Sun (e.g., $> 100R_\odot$), when the stored magnetic energy is significantly reduced, the transient is expected to go through a rapid decline in speed, which is in fact shown by the rapid drop of the observed speed-distance curve.

In summary, we have given an overview of properties of the quasi-stationary solar wind and of the interplanetary disturbances. In particular, the radial evolution of speed and size of the disturbances represents an important step in quantifying their propagation properties in the inner heliosphere. The combined use of remote sensing IPS methods, radio observations, together with modelling, and in situ measurements using satellites such as STEREO will certainly enhance the understanding of the three-dimensional structure of the heliosphere.

Acknowledgements

The white-light images used in this chapter are by the 'Courtesy of the SOHO/LASCO consortium'. SOHO is a project of international cooperation between ESA and NASA.

References

Armstrong, J. W., & Woo, R. 1981, A&A, 103, 415

Armstrong, J. W., Coles, W. A., Kojima, M., & Rickett, B. J. 1990, ApJ, 358, 685

Bame, S. J., McComas, D. J., Barraclough, B. L., Phillips, J. L., Sofaly, K. J., Chavez, J. C., Goldstein, B. E., & Sakurai, R. K. 1992, A&AS, 92, 237

Biermann, L. 1951, Z. Astrophys., 29, 274

Bird, M. K., & Edenhofer. P. 1991, in Physics of the inner heliosphere - I, eds. R. Schwenn & E. Marsch, (Springer-Verlag, Berlin), 13

Brueckner, G. E., Howard, R. A., Koomen, M. J., et al. 1995, Sol. Phys., 162, 357

Cargill, P. J., & Klimchuk, J. A. 1997, ApJ, 478, 799

Chapman, S. 1957, Smithsonian Contr. Astrophys., 2, 1

Chertok, I. M., Gnezdilov, A. A., & Zaborova, E. P. 1992, in Proc. SOLTIP I Symp., Liblice, eds. S. Fischer & M. Vandas, Czechoslovak Acad. Sci., Vol. 2, 39

Coles, W. A., & Esser, R. 1992, J. Geophys. Res., 97, 19139

Coles, W. A., & Harmon, J. K., 1978, J. Geophys. Res., 83, 1413

Coles, W. A., & Kaufman, J. J. 1978, Radio Science, 13, 591

Coles, W. A., Esser, R., Lovhaug, U.-P., & Markkanen, J. 1991a, J. Geophys. Res., 96, 13849

Coles, W. A., Liu, W., Harmon, J. K., & Martin, C. L. 1991b, J. Geophys. Res., 96, 1745

Coles, W. A., Grall, R. R., Klinglesmith, M. T., & Bourgois, G. 1995, J. Geophys. Res., 100, 17069

Crooker, N., Joselyn, J. A., & Feynman, J. (eds.) 1997, Coronal Mass Ejections, American Geophysical Union monograph 99

Dryer, M. 1994, Space Sci. Rev., 67, 363

Einaudi, G., & Velli, M. 1994, in Advances in Solar Physics, Lecture Notes in Physics 432, (Springer, Berlin), 149

Ekers, R. D., & Little, L. T. 1971, A&A, 10, 310

Gopalswamy, N., Lara, A., Lepping, R. P., Kaiser, M. L., Berdichevsky, D., & St. Cyr, O. C. 2000, Geophys. Res. Lett., 27, 145

Gosling, J. T. 1996, ARA&A, 34, 35
Gosling, J. T. 1997, in Coronal Mass Ejections, Eds. N. Crooker, J. A. Joselyn & J. Feynman, Geophysical Monograph 99, (American Geophysical Union, Washington DC), 9
Grall, R. R., Coles, W. A., Klinglesmith, M. T., Breen, A. R., Williams, P. J. S., & Esser, R. 1996, Nature, 379, 429
Grall, R. R., Coles, W. A., Spangler, S. R., Sakurai, T., & Harmon, J. K. 1997, J. Geophys. Res., 102, 263
Hakamada, K. 1995, Sol. Phys., 159, 89
Hewish, A., Scott, P. F., & Wills, D. 1964, Nature, 203, 1214
Hoeksema, J. T., Wilcox, J. M., & Scherrer, P. H. 1983, J. Geophys. Res., 88, 9910
Howard, R. A., Sheeley Jr., N. R., Michels, D. J., & Koomen, M. J. 1986, in The Sun and the heliosphere in three dimensions, ed. R. G. Marsden, (D. Reidel, Norwell), 107
Huddleston, D. E., Woo, R., & Neugebauer, M. 1995, J. Geophys. Res., 100, 19951
Hundhausen, A. J. 1977, in Coronal Holes and High Speed Wind Streams, ed. J. B. Zirker, (Col. Assoc. Univ. Press, Boulder, CO), 225
Hundhausen, A. J. 1997, in Cosmic Winds and Heliosphere, eds. J. R. Jokipii et al., (Univ. Arizona Press, Tucson, AZ), 259
Hundhausen, A. J. 1998, in The many faces of the Sun, eds. K. T. Strong et al., (Springer-Verlag, New York), 141
Janardhan, P., Balasubramanian, V., Ananthakrishnan, S., Dryer, M., Bhatnagar, A., & McIntosh, P. S. 1996, Sol. Phys., 166, 379
Kahler, S. W., Sheeley Jr., N. R., & Liggett, M. 1989, ApJ, 344, 1026
Klimchuk, J. A., & Porter, L. J. 1995, Nature, 377, 131
Kojima, M., & Kakinuma, T. 1990, Space Sci. Rev., 53, 173
Kojima, M., Fujiki, K., Ohmi, T., Tokumaru, M., Yokobe, A., & Hakamada, K. 1999, J. Geophys. Res., 104, 16993
Koutchmy, S. 1994, Adv. Space Res., 14, 29
Koutchmy, S., & Lamy, P. L. 1985, in Properties and interactions of Interplanetary Dust, eds. R. H. Giese, P. Lamy (D. Reidel, Dordrecht), 63
Manoharan, P. K. 1993, Sol. Phys., 148, 153
Manoharan, P. K. 1998, in Proc. of III SOLTIP Symp. (International Academic Publishers, Beijing), 249
Manoharan, P. K., & Ananthakrishnan, S. 1990, MNRAS, 244, 691
Manoharan, P. K., Kojima, M., & Misawa, H. 1994, J. Geophys. Res., 99, 23411
Manoharan, P. K., Ananthakrishnan, S., Dryer, M., Detman, T. R., Leinbach, H., Kojima, M., Watanabe, T., & Kahn, J. 1995, Sol. Phys., 156, 377
Manoharan, P. K., Kojima, M., Gopalswamy, N., Kondo, T., & Smith, Z. 2000, ApJ, 350, 1061
Manoharan, P. K., Tokumaru, M., Pick, M., Subramanian, P., Ipavich, F. M., Schenk, K., Kaiser, M. L., Lepping, R. P., & Vourlidas, A. 2001, ApJ, 559, 1180
Moore, R. L., Hammer, R., Musielak, Z. E., Suess, S. T., & An., C.-H. 1992, ApJ, 397, L55
Neugebauer, M. 1995, Rev. Geophys., 33, 591
Newkirk, G., Jr. 1967, ARA&A, 5, 213
Ogilvie, K. W., Chornay, D. J., Fritzenreiter, R. J., et al. 1995, Space Sci. Rev., 71, 55
Parker, E. N. 1964, ApJ, 139, 72
Phillips, J. L., Balogh, A., Bame, S. J., Goldstein, B. E., Gosling, J. T., Hoeksema, J. T., McComas, D. J., Neugebauer, M., Sheeley, N. R., & Wang, Y.-M. 1994,

Geophys. Res. Lett., 21, 1105
Reames, D. V. 1995, Revs. Geophys. Suppl., 33, 585
Rickett, B. J., & Coles, W. A. 1991, J. Geophys. Res., 96, 1717
Schwenn, R. 1990, in Physics of the inner heliosphere – I, (Springer-Verlag, Berlin), 99
Scott, S. L., Coles, W. A., & Bourgois, G. 1983, A&A, 123, 207
Sheeley, N. R., Jr., Howard, R. A., Koomen, M. J., & Michels, D. J. 1983, ApJ, 272, 349
Siscoe, G., & Intriligator, D. 1993, Geophys. Res. Lett., 20, 2267
St. Cyr, O. C., Burkepile, J. T., Hundhausen, A. J., & Lecinski, A. R. 1999, J. Geophys. Res., 104, 12493
Tappin, S. J., Hewish, A., & Gapper, G. R. 1984, Planet Space Sci., 32, 1273
Tatarski, V. I. 1961, Wave Propagation in a Turbulent Medium (McGraw-Hill, New York)
Tokumaru, M., Kojima, M., Fujiki, K., & Yokobe, A. 2000, J. Geophys. Res., 105, 10435
Tsuneta, S., Acton, L., Bruner, M., Lemen, J., Brown, W., Caravalho, R., Catura, R., Freeland, S., Jurcevich, B., Morrison, M. 1991, Sol. Phys., 136, 37
Tsurutani, B. T., Gonzalez, W. D., Tang, F., & Lee, Y. T. 1992, Geophys. Res. Lett., 19, 73
Tyler, G. L., Vesecky, J. F., Plume, M. A., Howard, H. T., & Barnes, A. 1981, ApJ, 249, 318
Watanabe, T., & Schwenn, R. 1989, Space Sci. Rev., 51, 147
Webb, D. F. 1995, Rev. Geophys. Suppl., 33, 577
Woo, R. 1993, J. Geophys. Res., 98, 18999
Woo, R., & Schwenn, R. 1991, J. Geophys. Res., 96, 21227
Yamauchi, Y., Tokumaru, M., Kojima, M., Manoharan, P. K., & Esser, R. 1998, J. Geophys. Res., 103, 6571

Subject Index

Lecture Notes in Physics

For information about Vols. 1–577
please contact your bookseller or Springer-Verlag
LNP Online archive: http://www.springerlink.com/series/lnp/

Vol. 578: D. Blaschke, N. K. Glendenning, A. Sedrakian (Eds.), Physics of Neutron Star Interiors.

Vol. 579: R. Haug, H. Schoeller (Eds.), Interacting Electrons in Nanostructures.

Vol. 580: K. Baberschke, M. Donath, W. Nolting (Eds.), Band-Ferromagnetism: Ground-State and Finite-Temperature Phenomena.

Vol.581: J. M. Arias, M. Lozano (Eds.), An Advanced Course in Modern Nuclear Physics.

Vol.582: N. J. Balmforth, A. Provenzale (Eds.), Geomorphological Fluid Mechanics.

Vol.583: W. Plessas, L. Mathelitsch (Eds.), Lectures on Quark Matter.

Vol.584: W. Köhler, S. Wiegand (Eds.), Thermal Nonequilibrium Phenomena in Fluid Mixtures.

Vol.585: M. Lässig, A. Valleriani (Eds.), Biological Evolution and Statistical Physics.

Vol.586: Y. Auregan, A. Maurel, V. Pagneux, J.-F. Pinton (Eds.), Sound–Flow Interactions.

Vol.587: D. Heiss (Ed.), Fundamentals of Quantum Information. Quantum Computation, Communication, Decoherence and All That.

Vol.588: Y. Watanabe, S. Heun, G. Salviati, N. Yamamoto (Eds.), Nanoscale Spectroscopy and Its Applications to Semiconductor Research.

Vol.589: A. W. Guthmann, M. Georganopoulos, A. Marcowith, K. Manolakou (Eds.), Relativistic Flows in Astrophysics.

Vol.590: D. Benest, C. Froeschlé (Eds.), Singularities in Gravitational Systems. Applications to Chaotic Transport in the Solar System.

Vol.591: M. Beyer (Ed.), CP Violation in Particle, Nuclear and Astrophysics.

Vol.592: S. Cotsakis, L. Papantonopoulos (Eds.), Cosmological Crossroads. An Advanced Course in Mathematical, Physical and String Cosmology.

Vol.593: D. Shi, B. Aktaş, L. Pust, F. Mikhailov (Eds.), Nanostructured Magnetic Materials and Their Applications.

Vol.594: S. Odenbach (Ed.),Ferrofluids. Magnetical Controllable Fluids and Their Applications.

Vol.595: C. Berthier, L. P. Lévy, G. Martinez (Eds.), High Magnetic Fields. Applications in Condensed Matter Physics and Spectroscopy.

Vol.596: F. Scheck, H. Upmeier, W. Werner (Eds.), Noncommutative Geometry and the Standard Model of Elememtary Particle Physics.

Vol.597: P. Garbaczewski, R. Olkiewicz (Eds.), Dynamics of Dissipation.

Vol.598: K. Weiler (Ed.), Supernovae and Gamma-Ray Bursters.

Vol.599: J.P. Rozelot (Ed.), The Sun's Surface and Subsurface. Investigating Shape and Irradiance.

Vol.601: F. Mezei, C. Pappas, T. Gutberlet (Eds.), Neutron Spin Echo Spectroscopy. Basics, Trends and Applications.

Vol.602: T. Dauxois, S. Ruffo, E. Arimondo (Eds.), Dynamics and Thermodynamics of Systems with Long Range Interactions.

Vol.603: C. Noce, A. Vecchione, M. Cuoco, A. Romano (Eds.), Ruthenate and Rutheno-Cuprate Materials. Superconductivity, Magnetism and Quantum Phase.

Vol.604: J. Frauendiener, H. Friedrich (Eds.), The Conformal Structure of Space-Time: Geometry, Analysis, Numerics.

Vol.605: G. Ciccotti, M. Mareschal, P. Nielaba (Eds.), Bridging Time Scales: Molecular Simulations for the Next Decade.

Vol.606: J.-U. Sommer, G. Reiter (Eds.), Polymer Crystallization. Obervations, Concepts and Interpretations.

Vol.607: R. Guzzi (Ed.), Exploring the Atmosphere by Remote Sensing Techniques.

Vol.608: F. Courbin, D. Minniti (Eds.), Gravitational Lensing:An Astrophysical Tool.

Vol.609: T. Henning (Ed.), Astromineralogy.

Vol.610: M. Ristig, K. Gernoth (Eds.), Particle Scattering, X-Ray Diffraction, and Microstructure of Solids and Liquids.

Vol.611: A. Buchleitner, K. Hornberger (Eds.), Coherent Evolution in Noisy Environments.

Vol.612 L. Klein, (Ed.), Energy Conversion and Particle Acceleration in the Solar Corona.

Vol.613 K. Porsezian, V.C. Kuriakose, (Eds.), Optical Solitons. Theoretical and Experimental Challenges.

Vol.614 E. Falgarone, T. Passot (Eds.), Turbulence and Magnetic Fields in Astrophysics.

Vol.615 J. Büchner, C.T. Dum, M. Scholer (Eds.), Space Plasma Simulation.

Vol.616 J. Trampetic, J. Wess (Eds.), Particle Physics in the New Millenium.

Vol.617 L. Fernández-Jambrina, L. M. González-Romero (Eds.), Current Trends in Relativistic Astrophysics, Theoretical, Numerical, Observational

Vol.618 M. Degli Esposti, S. Graffi (Eds.), The Mathematical Aspects of Quantum Maps

Vol.619 H.M. Antia, A. Bhatnagar, P. Ulmschneider (Eds.), Lectures on Solar Physics

Vol.620 C. Fiolhais, F. Nogueira, M. Marques (Eds.), A Primer in Density Functional Theory

Vol.621 G. Rangarajan, M. Ding (Eds.), Processes with Long-Range Correlations

Druck: Strauss Offsetdruck, Mörlenbach
Verarbeitung: Schäffer, Grünstadt